Sustainable Manufacturing of Advanced Materials

Scrivener Publishing
100 Cummings Center, Suite 541J
Beverly, MA 01915-6106

Advances in Production Engineering

Series Editor: Sandip Kunar (sandip.sandip.kunar@gmail.com)

Advances in Production Engineering" addresses recent developments and research issues related to the production engineering which necessitates the development of new materials and manufacturing processes in order to produce high-quality products at lower prices and in less time. This series aims to provide a scientific platform for researchers, practitioners, professionals, and academicians to discuss the most recent technological developments in metals, polymers, ceramics, composites, biomaterials, nanomaterials, special materials, metals, micro-forming, powder metallurgy, ceramics processing, non-traditional machining, high speed machining, micro and nanomachining, and laser processing. Tribological analysis, friction behavior, modelling, and optimization techniques in materials, machining, and manufacturing are also covered in the series.

Publishers at Scrivener
Martin Scrivener (martin@scrivenerpublishing.com)
Phillip Carmical (pcarmical@scrivenerpublishing.com)

Sustainable Manufacturing of Advanced Materials

Edited by

Gurudas Mandal

Department of Metallurgical Engineering, Kazi Nazrul University, Asansol, West Bengal, India

Swarup Kumar Ghosh

Department of Metallurgy and Materials Engineering, Indian Institute of Engineering Science and Technology, Shibpur, Howrah, West Bengal, India

Snehanshu Pal

Department of Metallurgy and Materials Engineering, Indian Institute of Engineering Science and Technology, Shibpur, Howrah, West Bengal, India

and

Sandip Kunar

Department of Mechanical Engineering, Aditya University, Surampalem, India

Scrivener Publishing

WILEY

This edition first published 2026 by John Wiley & Sons, Inc., 111 River Street, Hoboken, NJ 07030, USA
and Scrivener Publishing LLC, 100 Cummings Center, Suite 541J, Beverly, MA 01915, USA

For more information about Scrivener publications please visit www.scrivenerpublishing.com.

Wiley Global Headquarters
111 River Street, Hoboken, NJ 07030, USA

For details of our global editorial offices, customer services, and more information about Wiley products visit us at www.wiley.com.

The manufacturer's authorized representative according to the EU General Product Safety Regulation is Wiley-VCH GmbH, Boschstr. 12, 69469 Weinheim, Germany, e-mail: Product_Safety@wiley.com.

Library of Congress Cataloging-in-Publication Data

ISBN 978-1-394-44035-1

Cover image: Generated with AI using Adobe Firefly
Cover design by Russell Richardson

Set in size of 11pt and Minion Pro by Manila Typesetting Company, Makati, Philippines

Printed in the USA

10 9 8 7 6 5 4 3 2 1

Contents

Preface

Modern society has benefited from the unparalleled speed of global industrialization and technological advancement. However, there has been a substantial environmental cost associated with these advancements. It is more important than ever to strike a balance between innovation and environmental responsibility as the twenty-first century progresses. In light of this, the production of advanced materials in a sustainable manner becomes an important area of study and business. Emerging innovations in energy, healthcare, transportation, electronics, and other fields are built on advanced materials, such as high-performance composites, nanomaterials, biomaterials, and smart materials. But producing them frequently calls for energy-intensive procedures, dangerous chemicals, and a lot of resources. The advanced materials that hold out hope for a better future could exacerbate climate change and environmental deterioration if they are not purposefully designed for sustainability. Multidisciplinary viewpoints on ways to improve the economic, environmental, and social sustainability of advanced material production are collected in this book. Green synthesis techniques, life cycle evaluations, energy-efficient manufacturing procedures, recycling and circular economy tactics, and the incorporation of digital technologies in the manufacture of sustainable materials are among the subjects discussed.

This book has seventeen chapters.

Chapter 1 offers a comprehensive framework for renovating the existing industries into resilient, sustainable, and future-ready industries. Highlighting the associated drawbacks of the traditional processes, this chapter also explores the viable material processing methods aligning with the greener alternatives which cause minimum ecological harm and reduce undesired environmental burdens. Technological advancements, including smart manufacturing units, renewable energy sources, and additive manufacturing, are also elaborated in brief so that the appropriate technology can be incorporated to replace the traditional ones.

Chapter 2 systematically studies eco-friendly and sustainable methods in the production of advanced materials, providing an overview of environmentally responsible practices across diverse material types and technologies. It presents fundamental principles of green chemistry, sustainable engineering, and environmental assessment metrics, subsequently analysing metals, polymers, ceramics, and hybrid systems and highlighting their industrial significance and environmental implications.

Chapter 3 examines how microstructure influences mechanical properties in three high-carbon hot-rolled steels (HC0, HC1, and HC9) processed as per the industrial processing conditions of wire rods. Moreover, the selection of chemistry and process parameters and their combined effects on steel cleanliness, phase transition, microstructural constituents, and mechanical properties are also thoroughly investigated using optical microscopy, scanning electron microscopy, transmission electron microscopy, Rockwell hardness on the C scale, and uniaxial tensile testing.

Chapter 4 also highlights the role of recyclable, low-carbon steels and green manufacturing practices in enhancing sustainability. Ultimately, high-performance steels remain indispensable for enabling lighter, safer, and more eco-friendly vehicles, underpinning the automotive industry's path toward carbon neutrality and sustainable innovation.

Chapter 5 investigates a sustainable way of manufacturing ferritic stainless steel. Chapter 6 highlights the pivotal role of sustainable steel processing in mitigating the environmental impact of conventional steel production. It provides an in-depth analysis of cutting-edge technologies, including Electric Arc Furnace steelmaking, waste heat recovery-integrated Basic Oxygen Furnace systems, and Direct Reduced Iron processes, designed to minimise environmental impact.

Chapter 7 develops a pneumatic injection technique which can blow Al dross fines into the slag and reduce the Cr_2O_3 content to less than 4wt% in slag. Chapter 8 elaborates on the different mechanisms, materials, fabrication techniques, and sustainability aspects pertaining to bio-inspired coatings designed specifically for steel, highlights pertinent case studies showcasing real-world applications, and addresses future directions, identifying emerging trends and opportunities in this new and exciting area. Chapter 9 covers a review of the historic development of the embedded atom method interatomic potential, and the development methodologies and the recent works being done in this field.

Chapter 10 explores the synthesis, characterization, and routine evaluation of particle composites produced using green processing routes such as solvent-free partying, bio-based binders, low-temperature curing, and additive manufacturing. Chapter 11 discusses about advanced manufacturing methods like additive manufacturing, such as direct ink writing, stereolithography, binder jetting, plasma, and microwave-based fabrication systems.

Chapter 12 analyses circular economy-driven strategies in the sustainable production of advanced materials, with the objective of minimizing resource depletion and environmental impact. It presents an integrated framework that incorporates efficient production systems, sustainable material selection, and sophisticated waste management to facilitate regenerative industry practices. Chapter 13 describes the environmental benefits of Additive manufacturing (AM) technologies and identifies significant application areas that contribute to sustainability across various mechanical and biomedical sectors. Chapter 14 explores the basic concepts, process types (thixocasting and rheocasting), and materials applied in Semi-Solid Metal Casting (SSMC), with an emphasis on the use of aluminum and magnesium alloys. This chapter also brings into focus of sustainability indicators like lifecycle analysis and die life optimization. Applications of SSMC cut across different high-performance industries like automotive, aerospace, and biomedical engineering.

Chapter 15 discusses the application of molecular dynamics (MD) simulations in the design of advanced next-generation metallic systems by microstructural engineering, with a focus on high-entropy alloys (HEAs) and nanocrystalline (NC) metals. Chapter 16 studies how recyclable and renewable composite materials serve as essential components for developing sustainable practices in additive manufacturing (AM). Chapter 17 investigates the connection between eco-friendly bio composites and Additive Manufacturing (AM) as a solution to rising worldwide environmental challenges while demonstrating their sustainable application potential.

This work aims to give researchers, engineers, policymakers, and students a thorough understanding of the innovations, problems, and future directions in the production of sustainable materials. We intend to stimulate the practical solutions that link industrial development with international sustainability objectives by encouraging cooperation across disciplines and sectors.

We express our gratitude to every contributor for their insightful contributions, and we hope that this volume will spur further innovation in the direction of a more resilient, inclusive, and sustainable future.

The Editors
June 2026

Acknowledgment

We would like to express our deepest gratitude to all those who contributed to the successful completion of this book on *Sustainable Manufacturing of Advanced Materials.* First and foremost, we are sincerely thankful to all the authors for their invaluable contributions, continuous support, and insightful feedback throughout this book. Their expertise in sustainable materials and commitment to research excellence were instrumental in shaping the direction of this work.

Our heartfelt thanks go to our peers and colleagues for the many discussions, collaborations, and words of encouragement that enriched our research experience. Furthermore, we wish to express our gratitude to Martin Scrivener and the team at Scrivener Publishing for their efforts in bringing this work to publication.

Lastly, we extend our profound gratitude to our families and friends for their unwavering support, patience, and motivation throughout the preparation of this book.

This book has not only advanced our understanding of the sustainable manufacturing of advanced materials but has also deepened our appreciation for the collaborative nature of scientific research.

Editors

1

Introduction to Sustainability of Manufacturing and Materials Processing Technologies

Sarthak Prasad Sahoo[1]* and Rahul Choudhary[2]

[1]Dept. of Mechanical Engineering, Centurion University of Technology and Management, Odisha, India
[2]School of Mechanical Engineering, KIIT (Deemed to be University), Odisha, India

Abstract

Material processing routes and different manufacturing methods can never be avoided as each and every product, maybe directly or indirectly, associated with these processes. Apart from profit and efficiency, traditional methods are reported to cause unwanted damage to the environment and society, which need to be checked through introducing the concepts of sustainability. By promoting the social and economic responsibilities, sustainable manufacturing aims to achieve reduce environmental impacts, conserve natural resources and hence, to achieve global Sustainable Development Goals (SDGs). Starting with the fundamentals of sustainability, this chapter outlines a multi-dimensional approach to clearly understand the concepts of sustainable manufacturing. Highlighting the associated drawbacks of the traditional processes, this chapter also explores the viable material processing methods aligning with the greener alternatives which cause minimum ecological harm and reduce undesired environmental burdens. Technological advancements including smart manufacturing units, renewable energy sources and additive manufacturing are also elaborated in brief so that the appropriate technology can be incorporated to replace the traditional ones. Again, the effectiveness of the incorporated techniques can be evaluated through different sustainability assessment tools, the outcomes of which can further be used for better decision making and continuous improvements. Towards the end, some real-life success stories of the firms implementing sustainable manufacturing is discussed to motivate the industry experts, policy makers, industrial workforce

**Corresponding author*: sarthakprasad001@gmail.com

Gurudas Mandal, Swarup Kumar Ghosh, Snehanshu Pal and Sandip Kunar (eds.) Sustainable Manufacturing of Advanced Materials, (1–18) © 2026 Scrivener Publishing LLC

and the end users regarding the practical benefits and viability of sustainable practices. Finally, this chapter offers a comprehensive framework for renovating the existing industries into resilient, sustainable and future-ready industries.

***Keywords*:** Sustainable manufacturing, circular economy, Industry 4.0, renewable energy, SDGs

1.1 Introduction

Material processing, a series of operations to be performed on the raw materials to transform them into some intermediate parts or products, is a crucial step in the overall manufacturing process as this process deals with the desired property enhancement in the parts or products as desired by the customers [1, 2]. There are a wide variety of processes or techniques which fall under the umbrella of material processing. Figure 1.1 depicts an overview of different material processing techniques.

According to the growing demand for technological advancement, both the manufacturing industries and material processing techniques are constantly going through evolutions across all the fields starting from the eighteenth century. It can never be hesitated that the adaptation of revolutions at the industry level (starting from the first industrial revolution to the present scenario) resulted in numerous benefits to mankind to upgrade their standard of living and comfort conditions [2, 3]. Though the past revolutions are well known for their remarkable contributions towards the

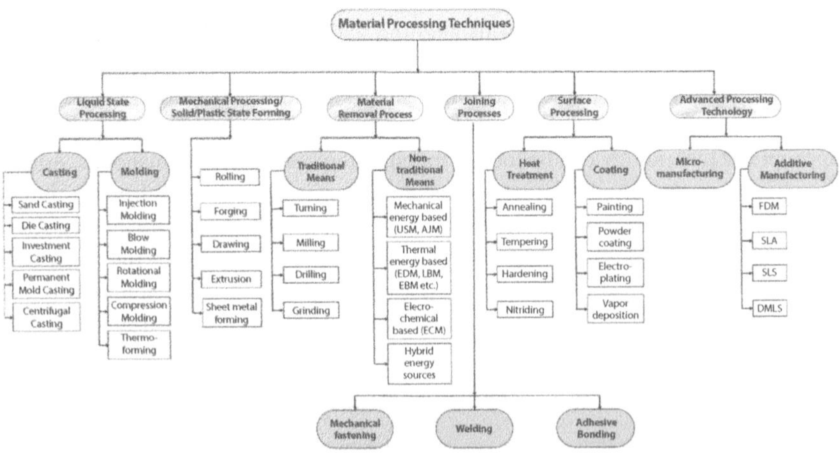

Figure 1.1 A broad overview of various material processing techniques.

upgradation of industrial landscape, Industry 4.0 can be referred as the most significant and transformative avenue for the manufacturing industries by facilitating robust, cost-effective and more efficient manufacturing solutions [3].

1.2 Principles of Sustainable Manufacturing (SM)

With reference to modern manufacturing sectors, sustainability can be stated as the set of manufacturing practices that conserve the naturally available but limited resources, optimize the available energy forms and leaves minimum negative impacts on the environment. Originally, sustainability was termed by the *Brundtland Commission*, which states, "sustainability is the act of meeting the needs of the present without compromising the ability of future generations to meet their own needs" [4]. Sustainability, being the center of attraction in the framework of Industry 4.0, has reformed the traditional industrial processes by extending the scopes for digitalization and integration of advanced technologies. Through this transformations, the manufacturing industries or the business models can optimize the resource and energy consumption to achieve enhanced productivity.

1.2.1 Triple Bottom Lines

The technologies included under sustainability must satisfy the aspects of society, economy and environment, which are readily referred as the "Triple Bottom Lines" [5, 6]. The reduction of adverse ecological footprints associated with the manufacturing sectors can be referred to as environmental sustainability. To ensure this, manufacturers can adopt closed-loop recycling systems, cleaner production processes and energy-efficient technologies which can further result in managing waste efficiency, conserving limited natural resources and minimizing the hazardous greenhouse emissions [7]. Social sustainability aims at facilitating the social welfare for society, communities and the employees as a whole. Raising community engagement, providing fair wages and ensuring harmless working conditions are some of the key attributes of social sustainability. The firms, practicing sustainability, often opt for gender equality, corporate social responsibility and continuous upgradation of existing workforce to promote and improve the social impacts [8]. By adopting strategies like lower operational cost, continuous competitive improvement through sustainable product development and optimization of available resources, economic sustainability always looks forward to long-term economic growth, scalable profitability and innovative and efficient approaches. Also, the expenses associated with waste disposal and energy consumption can be balanced through the economic returns by permitting investments on sustainable manufacturing technologies [8, 9]. As far as the successful implementation of sustainable manufacturing is considered, a perfect balance among these three pillars is, however, essential. Over-focusing on any one attribute at the expense of the rest can destabilize sustainability performance as a whole. Therefore, development of some holistic approaches, which are simultaneously socially comprehensive, environmentally accountable and economically viable, are much crucial for sustainable manufacturing.

1.2.2 Cradle-to-Cradle (C2C) Design and Life Cycle Thinking (LCT)

C2C design successfully eliminates the traditional 'cradle-to-grave' model of linear material consumption, which may often lead to unwanted waste generation imposing adverse impacts on the eco-system. C2C design always encourages to follow the closed-loop systems thereby the materials can be biologically recycled, continuously reused and effectively minimizing the waste generation [10]. LCT involves a comprehensive review of the environmental impacts accompanied with each phase of a product's life (from the phase of initial extraction to the final disposal phase); and hence recommends responsible decision making at each phase. Integration of LCT into traditional product design process enables manufacturers to identify the 'hotspots' of environmental impacts, and hence the improvement strategies can further be implemented [11]. C2C design and LCT can be treated as the baseline concepts in sustainable manufacturing aiming at minimization of environmental defilement across a product's complete life cycle. All together these two strategies support the innovations for sustainable product development by considering the environmental accountability and facilitating a circular economy model in manufacturing.

1.2.3 Circular Economy (CE)

Unlike the conventional linear one-way flow model of material consumption i.e., 'take-make-dispose', CE, a regenerative industrial practice, promotes recycling, reuse, remanufacturing and extension of product life through closed-loop systems, which ensure waste minimization and efficient and optimized use of resources. In sustainable manufacturing, CE principles always follow economic, social, and environmental sustainability by lowering environmental impact, improving resource efficiency and reducing raw material consumption [12]. One such approach could be redesigning the existing products for disassembly and material recovery and enabling reverse logistics to bring the used products back to the production line. This approach redefines the product design by not only conserving natural resources but also generating economic value by integrating innovation in supply chain management and product design [13]. Use of AI, IoT and digital tracking may also promote circularity by assisting in real time resource optimization and improving the product traceability. Apart from this, the principles of CE also align with certain sustainable development goals (SDGs) by promoting accountable production and consumption patterns, creating and supporting green jobs and by minimizing environmental threats [14].

As a note, it can be said that the manufacturing processes target to manufacture the products in an economically sound route by minimizing adverse environmental impacts and optimizing the resource allocation by following the above discussed principles can be readily treated as sustainable manufacturing.

1.3 Environmental Impacts of Traditional Manufacturing

Traditional manufacturing practices at the industry level are always accompanied with global environmental concerns. A huge amount of fossil-extracted energy is consumed by manufacturing units which in turn account for the carbon emissions and polluting the environment. According to a recent study carried out in 2023, China stood at first place, contributing around one-third (about 31%) of total CO_2 emission from energy production sector, due to its major dependency on coal and oils as prime sources of energy [15]. USA and India consecutively shares the second and third position respectively in the same category. Also, conventional machining and manufacturing sectors tremendously augment the level of carbon footprints through fossil-fuel-based electricity consumption [16]. Many of the industrial activities are said to be responsible for generation of waste and pollution. Most of the manufacturing byproducts including sulfur dioxide, nitrogen oxides and VOC (volatile organic compounds) can be treated as major contributors of air pollution [17]. On the other hand, the chemical overflow and waste dumping cause soil degradation while the untreated industrial discharges result in contamination of water bodies [18].

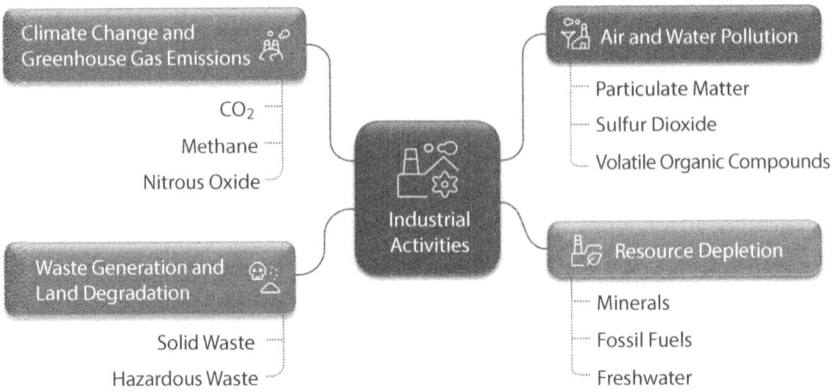

Figure 1.2 Environmental challenges of industrial activities.

The growth of industrial productivity also necessitates the raw materials and fuels to be consumed on a larger lager scale, which are actually limited in quantity. Hence, the limited resources are found to be depleted day by day by the continuous over-exploration of water, fossil fuels and minerals. It has also been reported that a huge portion of natural resources is exclusively consumed by the construction segment which is nearly 40% of the global energy consumption [19]. To check the unwanted natural resource depletion, environmental pollution and adverse change of climatic conditions, energy efficient technologies and sustainable practices must be followed at industry level. No doubt that embedding sustainability with Industry 4.0 can achieve improved operational efficiency and brand reputation by meeting the need of the market, but its main focus is to address the global challenges majorly driven by the industrial activities. Some of these major challenges can be presented through Figure 1.2.

1.4 Sustainable Materials Processing

Substances which impose negligible environmental impact throughout their entire life cycle can be considered as sustainable materials and they also play a vital role in economic growth and social equity by balancing the ecological aspects simultaneously. Though the sustainability of the materials is not inherent, but this can be evaluated at each stage of the material, starting from extraction, processing to final product for customers use and disposal, by considering their total effect on economy, society and environment [20, 21]. Some of the key characteristics of sustainable materials can be listed according to the following [21–26]:

- Abundance or Renewability: Use of materials which are copiously available or can be extracted from renewable sources can readily minimize the dependence on limited available resources.
- Economic Feasibility: The materials need to be feasible for large scale industrial applications with reduced cost and improved performances.
- Low Environmental Impact and Biodegradability: During their extraction, manufacturing and processing stages, materials should consume less energy, less water with reduced carbon emissions. On the other hand, biodegradable materials can be naturally decomposed into water and biomass through microbial actions. Certain materials like starch-based polymers, natural fibre composites and PLA (polylactic acid) can be extensively used in agricultural sectors and packaging firms, as collection and recycling of materials with these sectors are impractical and tedious.
- Recyclability and Reusability: To reduce landfill waste and to enable principles of circular economy, materials must have the ability to be recovered and reused in the material processing cycles without significant degradation of properties. Certain recyclable materials can be re-introduced to the production cycles, dropping the need for virgin resources and consumption of fossil-based energy. Bio-based plastics nowadays show better efficiency after reprocessing with advanced mechanical and chemical recycling.
- Non-Toxicity and Safety: Materials which are safe for human beings and non-toxic in nature can also be categorized under sustainable materials; and thus, they can also ensure a safer working environment.
- Green Chemistry: With a primary objective of minimizing damages to the environment, green chemistry, a science-based approach, aims to design different chemical processes and products by ensuring the use of renewable feedstocks, energy efficiency and waste prevention. Developing coatings and thin films through non-vacuum chemical processes are some of the suitable examples of green chemistry which consume less energy and produce less hazardous by-products.

Therefore, choosing materials wisely is an essential step in sustainable manufacturing practice, which can further result in eco-friendly products meeting different Sustainable Development Goals (SDGs).

1.5 Technological Innovations for Sustainability

To proceed towards the era of sustainable manufacturing, the role of technological innovations is much crucial as they facilitate the pathways for reduced environmental impact, efficient resource management and circular economy principles. Development of eco-friendly materials, focusing on renewable energy sources and adopting smart infrastructures are some of the key areas for sustainable development by reducing hazardous waste and optimizing traditional processes. Let's now briefly understand some vital aspects of recent technological innovations promoting sustainability

- Advanced Manufacturing Technologies (AMTs): These are a set of recently developed innovative technologies which integrate digitalization, data analytics, and automation and material science to enhance resource efficiency, implement circular economy models and to reduce adverse environmental impacts. Some of the popular technologies following sustainability can be listed as [27, 28]:
  i. Additive Manufacturing (AM): A layer-by-layer material deposition technique, popularly known as 3D printing, which allows decentralized production and customized product design with minimal material wastage.
  ii. Green Processing and Automation: To attain enhanced product durability and reduced part replacement, technologies like friction stir welding, precision machining and cryogenic cooling can be adopted as they consume low energy by reducing hazardous emissions to the environment during a production process.
  iii. Smart Manufacturing and IoT: These smart systems work on real time data so that the equipment performances, energy consumption and material flow can be efficiently monitored and optimized to enhance resource allocation and to reduce unwanted down-time and unhealthy emissions.

iv. Advanced Materials and Biotechnology: Recently developed advanced materials possess higher strength, lighter weight, longer durability and bio-based constituents for which life cycle impacts of the products can be minimized. Also, biotechnology empowers bio-manufacturing processes which rely on renewable inputs to produce biodegradable outputs.

With data-driven approach, proper scalability and eco-friendly solutions, AMTs stand at the core of modern sustainable manufacturing practices by facilitating adaptive, low-carbon emitting and resource efficient manufacturing systems.

- Smart Manufacturing and Industry 4.0: Aligning with the principles of Industry 4.0, smart manufacturing technologies aim to transform resource allocation by incorporating Internet of Things (IoTs), big data analytics, artificial intelligence (AI) and cyber-physical systems simultaneously reducing environmental impact. Material usage, energy consumption, and equipment performances can be dynamically monitored by smart manufacturing systems, based upon which the manufacturer can take smart, data-driven and quicker decisions for an optimized resource allocation eliminating unnecessary waste [29]. AI-assisted predictive maintenance also help in reducing unplanned downtime, and hence extending the satisfactory service period of the machinery. Due to the incorporation of automation and sensors in each stage of production, closed loop manufacturing can be performed under Industry 4.0, to minimize the defect rates and maximize the production yield by controlling the material allocation. This is found to be true for additive manufacturing in smart factories [30]. Furthermore, integration of block chain and IoTs into smart supply chain result in higher coordination and transparency among the suppliers facilitating more efficient logistics with a leaner inventory stock [31].
- Integration of Renewable Energy (RE): It is a strategic priority to integrate renewable energy into different manufacturing systems so that the global sustainability goals can be achieved smoothly by enhancing energy efficiency and reducing the hazardous carbon emissions. Recently,

researchers have witnessed significant positive outcomes by combining biomass energy, wind energy, solar energy and geothermal energy with suitable industrial applications and claimed that the traditional energy resilience on fossil fuels can hence be minimized. Several industrial benefits including corporate sustainability, cost saving and higher energy security are achieved by integrating renewable means in expense of higher capital investment and requirement of skilled labor force [32]. Improved energy efficiency and lower emissions to environment are also reported with integrated systems with wind, geothermal or solar energy sources [33]. In the context of battery storage and real time control systems, the CO_2 emissions can be reduced by up to 63% in both residential and industrial settings just by integrating hydropower sectors with energy storage systems [34]. The utmost necessity and feasibility of integrating renewable energy in manufacturing processes must not be ignored. However, successful implementation of this can be possible by facilitating financial supports and suitable policy incentives along with technological advancements.

1.6 Assessment Tools and Metrics

To evaluate the social, environmental and economic impacts across a product's service life, different tools and metrics are to be used to access the sustainability both in manufacturing and material processing. These tools enable manufacturers to evaluate their improvement, to help in decision making and to identify the sectors demanding further improvement. Listed below are some of the major tools and metrics to access sustainability.

- Life Cycle Assessment (LCA): Starting from the extraction of raw materials to the end-life disposal, the environmental impacts associated with each stage can be evaluated through LCA, by quantifying resource consumption, energy use and measure of emissions. However, LCA is often restricted by coarse data, hence improvement is still needed to integrate real-time data for better outcomes [35].
- Sustainability Index (SI) Tools: By combining normalized sustainability metrics across the major four sectors (economy, society, environment and technology), overall

sustainability index is evaluated to compare the processes and examine the policy applicability [36].

- Product and Process Sustainability Index (ProdSI and ProcSI): Evolving around the principles of 6R (reuse, recover, reduce, redesign, recycle and remanufacture), these tools can access sustainability both at product level and process level by combining social, environmental and economic performances, setting a base for closed-loop manufacturing processes [37].
- CAx-Integrated Assessment Tools: To extend the scopes of traditional sustainability indicators, the sustainability metrics are integrated with suitable computer-aided technologies (CAx), which can further evaluate the quality, time requirement, energy used and flexibility [38].
- Exergy-Based Metrics: This provides a thermodynamic-based model to evaluate the material-flow and energy by reducing the impact-shifting between materials and energy. The resource inefficiencies can also be identified and minimized using this approach [39].

An inclusive structure for sustainability evaluation can be made possible by integrating all these tools with a quantified digital model, so that the industrial activities and the sustainable responsibilities can be balanced to proceed towards a greener industrial system.

1.7 Real-World Examples of Sustainable Manufacturing Practices

Not only in theory, but the approaches to sustainable manufacturing can actually bring significant benefits to the firm in real-life as well. To realize this, let's briefly go through some real life examples of sustainable manufacturing practices:

- Sustainability Cone in Car Manufacturing: By combining manufacturing decisions with life cycle thinking (LCT), an automated automotive industry applied sustainability cone framework, by which the company can align with environmental goals with reduced cost using a stage-gate project model [40].

- 6R Approach in Indian Manufacturing Industry: Six R's including reuse, redesign, recover, reduce, recycle and remanufacture are found to be most influential parameters to implement sustainable manufacturing strategy in Indian Manufacturing Industries by enhancing triple bottom line performances [41].
- Data-Driven Sustainability in Tool Design and Quality Assurance: During design of machine tools and some manual operations (like grinding) energy saving, reduced scrap rate and improved process efficiency are noted just by integrating data analysis to the shop floor activities [42].
- Wind Energy Based Dutch Railways: The Dutch Railways (NS), which completely operates on wind power, featured as the world's first train network under the same category which applies concepts of circularity and climate-neutral transformation [43].
- BMW's Recycled Batteries: In its i-series variant, BMW uses batteries which are recycled from the electric vehicles [44].
- Clean Electricity of TATA Power: By installing rooftop solar panels, TATA Power, an India-based power generation company, claimed to generate clean electricity of 421 million watts (in 2021) across 90 cities of the country [45].

1.8 Challenges and Future Directions

Looking at the recent industrial development and technological innovations, it is clear that sustainable manufacturing has the most promising role to balance economic performance, environmental stewardship and social responsibilities. Despite the growing demand and adoption requirement, the full-scale implementation of sustainable manufacturing is still in a dilemma due to several challenges associated with it [46]. Some of the major challenges are discussed below:

- Integration Complexity and Legacy Systems: As many of the traditional manufacturing processes are initially not designed with sustainability principles, redefining and redesigning of these existing systems to integrate different digital tools and green technologies, might be technically complex and costly.

- Fragmented Sustainability Metrics: Due to lack of standardized and universally acknowledged metrics and tools for accessing sustainability, manufacturers face difficulty in benchmarking and decision-making.
- Resource Limitations and Cost Barriers: Adaptation of green technologies, circular processes and renewable systems requires high capital investment, which stands as a strong barrier especially for small and medium-sized enterprises. Many firms find it challenging both from the technical side and from the financial side to incorporate sustainable models.
- Policy and Governing Ambiguity: Lack of suitable incentives from government and improper environmental regulations create more conflicts to rationalize long-term investments towards sustainability models. Well-organized and clear legal frameworks are highly necessary to support industry-wide innovations and compliances.

If the associated challenges are properly addressed, sustainable manufacturing and material processing can be a leading sector under the category of sustainable development. Some of the key areas which are to be explored more by the researchers are listed below:

- Digital and Smart Manufacturing Integration: Integrating recently developed tools like IoT, AI and digital twins into the production systems can enhance waste management, process efficiency and adoptive control for resource allocation.
- Renewable Energy Integration and Electrification: To minimize the impacts of carbon footprints, manufacturing industries must come forward to utilize decentralized and renewable sources of energy such as biomass, wind, geo-thermal and solar.
- Circular Economy and Closed-Loop Systems: Transforming the existing linear models into circular ones is very vital as they include remanufacturing, recyclability, redesigning products for reuse and reverse logistics systems. Choosing biodegradable materials and advanced technologies for recycling can further enhance resource circularity.
- Global Alliance and Information Sharing: The concept of sustainability and its effective outcomes are beyond the

geographical borders. Hence, public-private partnerships, global initiatives and open-source platforms are to be encouraged to harmonize the principles, share best proven practices, and accelerate implementation across industries.

1.9 Conclusion

Through this chapter, it can be clearly understood that a critical shift for the traditional material processing firms and the manufacturing units is highly essential to maximize production efficiency and profit simultaneously emphasizing social responsibility and environmental protection. Long term growth and industrial innovations can be easily achieved by integrating principles of circular economy, three pillars of sustainability and life cycle thinking which can further minimize the ecological footprints. The adverse impacts of traditional material processing can no longer be considered as options, but they need to be properly addressed for the well-being of the whole universe. Resource deletion, hazardous emissions and environmental pollution can be minimized through opening the pathways for green chemistry and sustainable material selection. IoT-assisted operations, renewable energy integration, additive manufacturing and smart factories are some of the healthy options to proceed towards a robust, data-driven and low-carbon production systems. Also, a periodical check for the sustainability measurement is also needed to identify the sectors for improvement and technology requirements. Along with the theoretical approach, some real-life success stories are also highlighted to elucidate the applicability and associated benefits of adopting sustainable practices. At the end, it can never be avoided that the future of manufacturing industries roams around the principles of sustainability, which can be successfully implemented by properly addressing the challenges and transforming them into opportunity by abiding the principles of sustainability and fulfilling the global demands simultaneously.

AI disclosure statement: The author confirms that no artificial intelligence (AI)-generated content or AI-assisted writing tools were used in the preparation of this chapter (including summaries and the abstract). All text, analysis, and conclusions are the result of the author's original work.

References

1. Pandey, P. and Chakrabarti, A., Material Characteristics, Manufacturing Processes and Product Performance: A Review of Some Experimental Studies. *Int. J. Adv. Res. Innovation*, 2015, https://doi.org/10.51976/ijari.311542.
2. Sharma, A.K., Ramacharyulu, D.A., Lakhanpal, S., Shakier, H.G., Reddy, U., Kumari, V., Exploring the Future of Advanced Materials Processing: Innovations and Challenges Ahead: A Review. *E3S Web Conf.*, 505, 01021, 2024, EDP Sciences. https://doi.org/10.1051/e3sconf/202450501021.
3. Ahmmed, M.S., Isanaka, S.P., Liou, F., Promoting synergies to improve manufacturing efficiency in industrial material processing: A systematic review of industry 4.0 and AI. *Machines*, *12*, 10, 681, 2024, https://doi.org/10.3390/machines12100681.
4. Gaughran, W.F., Burke, S., Phelan, P., Intelligent manufacturing and environmental sustainability. *Rob. Comput. Integr. Manuf.*, *23*, 6, 704–711, 2007, https://doi.org/10.1016/J.RCIM.2007.02.016.
5. Javaid, M., Haleem, A., Singh, R.P., Khan, S., Suman, R., Sustainability 4.0 and its applications in the field of manufacturing. *Internet Things Cyber-Phys. Syst.*, *2*, 82–90, 2022, https://doi.org/10.1016/j.iotcps.2022.06.001.
6. Contini, G. and Peruzzini, M., Sustainability and industry 4.0: Definition of a set of key performance indicators for manufacturing companies. *Sustainability*, *14*, 17, 11004, 2022, https://doi.org/10.3390/su141711004.
7. Sembiring, N., Tambunan, M.M., Ginting, E., Sustainable Manufacturing that is Environmentally Friendly. *J. Phys. Conf. Ser.*, 1542, 1, 012004, 2020, May, IOP Publishing. https://doi.org/10.1088/1742-6596/1542/1/012004.
8. Gupta, S., Dangayach, G.S., Singh, A.K., Meena, M.L., Rao, P.N., Implementation of sustainable manufacturing practices in Indian manufacturing companies. *Benchmarking: Int. J.*, *25*, 7, 2441–2459, 2018, https://doi.org/10.1108/BIJ-12-2016-0186.
9. Joung, C.B., Carrell, J., Sarkar, P., Feng, S.C., Categorization of indicators for sustainable manufacturing. *Ecol. Indic.*, *24*, 148–157, 2013, https://doi.org/10.1016/j.ecolind.2012.05.030.
10. Peralta, M.E., Alcalá, N., Soltero, V.M., Weighting with life cycle assessment and cradle to cradle: A methodology for global sustainability design. *Appl. Sci.*, *11*, 19, 9042, 2021, https://doi.org/10.3390/app11199042.
11. Biswas, W. and John, M., The value of life cycle thinking in sustainable engineering education, in: *The Routledge Handbook of Global Sustainability Education and Thinking for the 21st Century*, pp. 372–385, Routledge (Taylor & Francis Group), London, United Kingdom, 2025, https://doi.org/10.4324/9781003171577-29.
12. Ma, H.W., Shih, H.C., Liao, M.I., Circular economy and new research directions in sustainability, in: *Pursuing Sustainability: OR/MS Applications in Sustainable Design, Manufacturing, Logistics, and Resource Management*,

pp. 141–168, Springer International Publishing, Cham, Switzerland, 2021, https://doi.org/10.1007/978-3-030-58023-0_6.
13. Barros, M.V., Salvador, R., do Prado, G.F., de Francisco, A.C., Piekarski, C.M., Circular economy as a driver to sustainable businesses. *Cleaner Environ. Syst.*, *2*, 100006, 2021, https://doi.org/10.1016/j.cesys.2020.100006.
14. Schroeder, P., Anggraeni, K., Weber, U., The relevance of circular economy practices to the sustainable development goals. *J. Ind. Ecol.*, *23*, 1, 77–95, 2019, https://doi.org/10.1111/jiec.12732.
15. https://www.visualcapitalist.com/charted-the-worlds-carbon-emissions-from-energy-production/.
16. Das, S., Roy, K., Das, S., Investigation and Modeling for Energy Consumption During Conventional Machining: A Case Study, in: *Lecture Notes in Mechanical Engineering*, Springer Nature, Singapore, 2021, https://doi.org/10.1007/978-981-16-3497-0_41.
17. Chopperla, S.T., Jupalli, R., Kanraj, D., Bahurudeen, A., Haneefa, M.K., Santhanam, M., Development of an efficient procedure for sustainable low carbon cement manufacturing process. *Appl. Mech. Mater.*, *787*, 142–146, 2015, https://doi.org/10.4028/www.scientific.net/AMM.787.142.
18. Raiyaan, G.I. and Arunachalam, K.D., Pollution Prevention, Sustainability, and Green World, in: *Handbook of Environmental Materials Management*, pp. 1–23, Springer, Cham, 2020, https://doi.org/10.1007/978-3-319-58538-3_208-1.
19. Costa, C., Monteiro, M., Rangel, B., Alves, F.J.L., Industrial and natural waste transformed into raw material. *Proc. Inst. Mech. Eng., Part L: J. Mater.: Des. Appl.*, 231, 1–2, 247–256, 2017, https://doi.org/10.1177/1464420716677087.
20. Ljungberg, L.Y., Materials selection and design for development of sustainable products. *Mater. Des.*, *28*, 2, 466–479, 2007, https://doi.org/10.1016/j.matdes.2005.09.006.
21. Fratzl, P., Introduction: sustainable materials. *Chem. Rev.*, *123*, 5, 1841–1842, 2023, https://doi.org/10.1021/acs.chemrev.3c00091?urlappend=%3Fref%3DPDF&jav=VoR&rel=cite-as.
22. Vimal, K.E.K. and Vinodh, S., Development of checklist for evaluating sustainability characteristics of manufacturing processes. *Int. J. Process Manage. Benchmarking*, *3*, 2, 213–232, 2013, https://doi.org/10.1504/IJPMB.2013.057726.
23. Thomas, A., Francis, M., John, E., Davies, A., Identifying the characteristics for achieving sustainable manufacturing companies. *J. Manuf. Technol. Manage.*, *23*, 4, 426–440, 2012, https://doi.org/10.1108/17410381211230376.
24. Quental, N., Lourenço, J.M., Da Silva, F.N., Sustainability: characteristics and scientific roots. *Environ. Dev. Sustainability*, *13*, 257–276, 2011, https://doi.org/10.1007/S10668-010-9260-X.
25. Vadicherla, T. and Saravanan, D., Sustainable measures taken by brands, retailers, and manufacturers, in: *Roadmap to sustainable textiles and clothing: Regulatory aspects and sustainability standards of textiles and the clothing*

supply chain, pp. 109–135, Springer Nature, Singapore, 2015, https://doi.org/10.1007/978-981-287-164-0_5.

26. Choy, K., Towards sustainable surface processing and products through Green Chemistry and engineering, 2015, https://doi.org/10.4172/2161-0401.C1.007.
27. Chen, B., Wan, J., Shu, L., Li, P., Mukherjee, M., Yin, B., Smart factory of industry 4.0: Key technologies, application case, and challenges. *IEEE Access, 6*, 6505–6519, 2017, https://doi.org/10.1109/ACCESS.2017.2783682.
28. Tanuja, G.G., Srimani, P.S., Rao, A.L.N., Sharma, S., Mashkour, M.S., Singh, N., Innovative Technologies for Sustainable Recycling and Re-manufacturing of Materials and Components. *E3S Web Conf.*, 430, 01130, 2023, EDP Sciences. https://doi.org/10.1051/e3sconf/202343001130.
29. Shankar Raman, A., Haapala, K.R., Raoufi, K., Linke, B.S., Bernstein, W.Z., Morris, K.C., Defining near-term to long-term research opportunities to advance metrics, models, and methods for smart and sustainable manufacturing. *Smart Sustainable Manuf. Syst.*, *4*, 2, 1–24, 2020, https://doi.org/10.1520/ssms20190047.
30. Devito, F., Copani, G., Natalicchio, A., Alami, A.H., Lavecchia, F., Olabi, A.G., Dassisti, M., Business Models and Advanced Additive Manufacturing strategies for better sustainability. *Energy Nexus, 16*, 100337, 2024, https://doi.org/10.1016/j.nexus.2024.100337.
31. Jin, M., Tang, R., Ji, Y., Liu, F., Gao, L., Huisingh, D., Impact of advanced manufacturing on sustainability: An overview of the special volume on advanced manufacturing for sustainability and low fossil carbon emissions. *J. Cleaner Prod.*, *161*, 69–74, 2017, https://doi.org/10.1016/J.JCLEPRO.2017.05.101.
32. Usman, F.O., Ani, E.C., Ebirim, W., Montero, D.J.P., Olu-lawal, K.A., Ninduwezuor-Ehiobu, N., Integrating renewable energy solutions in the manufacturing industry: challenges and opportunities: a review. *Eng. Sci. Technol. J.*, *5*, 3, 674–703, 2024, https://doi.org/10.51594/estj.v5i3.865.
33. Bagherian, M.A. and Mehranzamir, K., A comprehensive review on renewable energy integration for combined heat and power production. *Energy Convers. Manage.*, *224*, 113454, 2020, https://doi.org/10.1016/J.ENCONMAN.2020.113454.
34. Assuad, C.S.A., Deike, L., Liao, Z., Akram, M.A., Integration of Hydropower and Battery Energy Storage Systems into Manufacturing Systems-A Discrete-Event Simulation, in: *IFIP International Conference on Advances in Production Management Systems*, 2023, September, Springer Nature Switzerland, Cham, pp. 549–559, https://doi.org/10.1007/978-3-031-43688-8_38.
35. Eastlick, D.D., Sahakian, M.V., Haapala, K.R., Sustainable manufacturing analysis for titanium components, in: *International Design Engineering Technical Conferences and Computers and Information in Engineering Conference*, 2011, January, vol. 54860, pp. 711–719, https://doi.org/10.1115/DETC2011-48854.

36. Schuch Bork, C.A., de Souza, J.F., de Oliveira Gomes, J., Venancio Pappetti Canhete, V., De Barba, D.J., Methodological tools for assessing the sustainability index (SI) of industrial production processes. *Int. J. Adv. Manuf. Technol.*, *87*, 1313–1325, 2016, https://doi.org/10.1007/s00170-014-6684-8.
37. Badurdeen, F., Jawahir, I.S., Rouch, K.E., A metrics-based evaluation of sustainable manufacturing at product and process levels, 2024, https://doi.org/10.1016/B978-0-12-409548-9.10043-0.
38. Saxena, P., Stavropoulos, P., Kechagias, J., Salonitis, K., Sustainability assessment for manufacturing operations. *Energies*, *13*, 11, 2730, 2020, https://doi.org/10.3390/en13112730.
39. Li, L., Guo, C., Yan, J., Zhao, F., Sutherland, J.W., Exergy-based tool path evaluation method of material and energy flows to support the sustainable-oriented intelligent manufacturing. *Proc. Inst. Mech. Eng., Part C: J. Mech. Eng. Sci.*, *236*, 4, 1960–1972, 2022, https://doi.org/10.1177/0954406220911083.
40. Rödger, J.M., Bey, N., Alting, L., The sustainability cone–A holistic framework to integrate sustainability thinking into manufacturing. *CIRP Ann.*, *65*, 1, 1–4, 2016, https://doi.org/10.1016/J.CIRP.2016.04.033.
41. Shankar, K.M., Kannan, D., Kumar, P.U., Analyzing sustainable manufacturing practices–A case study in Indian context. *J. Cleaner Prod.*, *164*, 1332–1343, 2017, https://doi.org/10.1016/J.JCLEPRO.2017.05.097.
42. Linke, B.S., Garcia, D.R., Kamath, A., Garretson, I.C., Data-driven sustainability in manufacturing: selected examples. *Procedia Manuf.*, *33*, 602–609, 2019, https://doi.org/10.1016/J.PROMFG.2019.04.075.
43. Wijdoogen, W.C., *7 Roles to Create Sustainable Success*, Amsterdam University Press, Amsterdam, Netherlands, 2020, https://doi.org/10.5117/9789082949742.
44. https://www.bmw.bm/en/topics/fascination-bmw/electromobility/sustainability.html.
45. Unlocking a sustainable future: Why digital solutions are the key to sustainable business transformation, Schneider Electric. Accessed: 4/September/2024.
46. Wang, L. and Bi, Z.M., Challenges for better sustainable manufacturing, in: *Advances in Sustainable and Competitive Manufacturing Systems: 23rd International Conference on Flexible Automation & Intelligent Manufacturing*, pp. 1209–1221, Springer International Publishing, Heidelberg, 2013, June, https://doi.org/10.1007/978-3-319-00557-7_99.

2

A Critical Review of Green and Sustainable Approaches in Advanced Materials Manufacturing

Ajithkumar Sitharaj[1]*, Manoj Mohan Prasath V.[2], Sandip Kunar[3], Ganesh N.[4] and Arul Pandiyan[4]

[1]Department of Mechanical Engineering, KPR Institute of Engineering and Technology, Coimbatore, Tamil Nadu, India
[2]Department of Mechanical Engineering, Karpagam Academy of Higher Education, Coimbatore, Tamil Nadu, India
[3]Department of Mechanical Engineering, Aditya University, Surampalem, India
[4]Department of Mechanical Engineering, Sri Shakthi Institute of Engineering and Technology, Coimbatore, Tamil Nadu, India

Abstract

The escalating challenges of climate change, resource depletion, and industrial pollution have rendered sustainability a paramount concern in materials science and engineering. This review systematically examines eco-friendly and sustainable methods in the production of advanced materials, providing an overview of environmentally responsible practices across diverse material types and technologies. It presents fundamental principles of green chemistry, sustainable engineering, and environmental assessment metrics, subsequently analyzing metals, polymers, ceramics, and hybrid systems, highlighting their industrial significance and environmental implications. The efficacy of sustainable processing strategies, including solvent-free methods, renewable feedstocks, energy-efficient techniques, and circular economy models, is assessed. Advanced technologies such as additive manufacturing, green nanotechnology, and dry processing are examined for their environmental and economic advantages. The function of life cycle assessment (LCA) is analyzed as an essential instrument for evaluating environmental impacts, in conjunction with its methodological difficulties. The review emphasizes emerging facilitators such as digitalization, artificial intelligence, and circular economy

**Corresponding author*: ersajithkumarbe@gmail.com

Gurudas Mandal, Swarup Kumar Ghosh, Snehanshu Pal and Sandip Kunar (eds.) Sustainable Manufacturing of Advanced Materials, (19–42) © 2026 Scrivener Publishing LLC

frameworks that can expedite the shift towards sustainable manufacturing. It concludes by delineating research deficiencies and suggesting strategic pathways for forthcoming innovation. This review functions as a critical evaluation and a prospective guide for enhancing a more sustainable and resilient materials economy.

***Keywords*:** Sustainable manufacturing, green chemistry, advanced materials, life cycle assessment, circular economy

2.1 Introduction

Advanced materials manufacturing is becoming a key facet of modern technology and innovation and is used for improving our industries including aerospace, electronics, renewable energy, automotive, construction, and healthcare [1]. These structures - high-strength alloys, lightweight composites, functional polymers and nanostructured ceramics - are all employed to enhance performance, miniaturization and efficiency capabilities in a wide-variety of applications. However, these commodity materials often come at considerable expense to our environment. Gressler *et al.* [2], assessed and compared emerging photovoltaic systems (EPVs) and revealed that although they offer many benefits with regard to lightweight construction, reduced costs in production, reduced energy requirements, and quick payback periods against conventional PV technologies, they also pose environmental problems. They address the use of critical raw materials and hazardous materials related to emerging PV systems which underline the need for better sustainability and circular economy strategies to be implemented within EPVs.

Arvidsson *et al.* [3], observed that the production of advanced materials such as MXenes can have serious environmental effects, especially through increased aquatic ecotoxicity and also inclusion of potentially large amounts of future production. This emphasizes the need to develop new approaches to remedy environmental consequences associated with producing advanced materials. Most traditional manufacturing methods will as well utilize non-renewable resources, dangerous solvents, and significant energy use, proliferation greenhouse gas emissions, toxic waste, and environmental destruction [4]. With climate change worsening, natural resources becoming depleted, and environmental regulations strengthening, it is of utmost importance to reconsider and innovate methods of synthesis and processing for advanced materials. Green and sustainable manufacturing can go further in helping to eliminate environmental damage from industrial activity without hindering technological efficacy or economic feasibility [5]. Green manufacturing focuses on reducing or eliminating toxic materials, increasing energy efficiency, and reducing waste through cleaner production techniques and green sustainable materials. Sustainable manufacturing takes a more holistic approach by taking environmental, social and economic impacts into account throughout the entire product life cycle, from raw material extraction to selected end-of-life recycling or disposal suggestions [6].

In the manufacturing of advanced materials, the reason for, but also the challenge of sustainability is often more complicated due to the complexity chemistries, higher processing temperatures and multi-step processing involved in manufacturing of advanced materials. This article reflects the current direction of green and sustainable practices in manufacturing of advanced materials. New synthesis methods, use of environmentally-friendly solvents, renewable raw materials, energy efficient methods, and enhancements to lessen impact, are reviewed. New processing methods such as additive manufacturing, green nanotechnology, and solvent free coatings that are consistent with eco-design and circular economy principles [7], are evaluated. This chapter covers a diverse range of bulk material types (e.g. metals, polymers, ceramics, and composites) and focuses not just on sustainability issues, but also anticipated applications, technological hurdles, and the positives of using innovative solutions compared to conventional solutions. The article also discusses the importance of life cycle assessment (LCA) as the main tool for evaluating environmental impacts and continue making sustainable production decisions. Jia *et al.* [8], completed a life cycle assessment of the plywood process in China and made eco-design recommendations to support environmental sustainability. While there has been some progress, significant challenges still exist with technical limitations, high initial costs, lack of standardization, and no policies to encourage broad acceptance of sustainability measures. This review wants to provide a strong basis and positive perspective regarding the role of green chemistry and sustainability in the future of advanced materials manufacturing by bringing together existing insights and identifying knowledge gaps.

2.2 Foundations and Drivers of Green and Sustainable Manufacturing

Green manufacturing involves enhanced methods to reduce environmental, economic, and societal impacts of industrial production, while sustainable manufacturing is still about the impacts of industrial production and utilization but involves the consideration of the whole system. Green manufacturing revolves around the improvement of processes to decrease waste, emissions, and energy consumption by using cleaner technologies and safer materials [9]. Green manufacturing is based on the principles of cleaner production, which place importance on source reduction, closed loop systems, and the substitution of hazardous materials with safer materials. Dey *et al.* [10], illustrated that using green technology and automated inspection of a manufacturing–remanufacturing system had decreased carbon emissions, waste, and overall production costs showing that green technology and automated inspection are viable strategies for improving environmental and economic performance. Sustainable manufacturing takes a life cycle and systems viewpoint. It integrates environmental protection with social equity and economic acceptability over the entire lifespan of a product (from raw material extraction, manufacture, use, and end of life). The focus of sustainable manufacturing is to minimize ecological footprint while improving long term envelopment and community well-being. Table 2.1 illustrates the comparison of traditional and green manufacturing approaches in advanced materials.

The application of these frameworks in advanced materials requires a transition from linear and resource heavy production to circular and regenerative systems, which requires rethinking product structure for life, repair/maintenance, and recyclability, while increasing resource efficiency and supply chain transparency through digital technologies and automation technologies. Figure 2.1 outlines a framework for green and sustainable manufacturing systems in advanced materials, and is based on a theoretical architecture that aligns with the Twelve Principles of Green Chemistry and Green Engineering. Green chemistry will encourage innovations in the development of chemical products and processes that will reduce or eliminate the use and generation of hazardous materials [11]. Important principles are waste prevention, safer solvents, energy efficiency, and renewable feedstocks. Green engineering applies the Twelve Principles of Green Chemistry and Green Engineering to the entire design scheme on the system level. They prioritize intrinsic safety, reduce the total amount of resources used, limit the variety and number of materials, and assess the entire life cycle.

In advanced materials production, these concepts are applicable to developments in solvent free polymer formation, low temperature sintering of ceramics, biomimetic materials, and efficient additive manufacturing approaches that minimize waste. Together, these concepts provide a systematic approach for aligning performance aspirations with

Table 2.1 Comparison of conventional vs. green manufacturing approaches in advanced materials [13–15].

Aspect	Conventional manufacturing	Green/sustainable manufacturing
Raw Material Source	Petroleum-based, mined ores, non-renewables	Bio-based, recycled, renewable, or abundantly available resources
Solvents and Reagents	Organic solvents, hazardous chemicals	Water-based, solvent-free, or biodegradable reagents
Energy Consumption	High-temperature, energy-intensive processes	Low-temperature, energy-efficient, sometimes powered by renewables
Waste Generation	Significant solid, liquid, and gaseous waste	Minimized waste through cleaner production, reuse, or circular approaches
Toxic Emissions	VOCs, CO_2, NO_x, heavy metals	Lower emissions, often using closed-loop systems or green chemistry alternatives
Process Efficiency	Often linear and batch-based with losses	Optimized for resource efficiency, continuous flow, or additive manufacturing
Lifecycle Consideration	Focused on performance and cost	Includes environmental and social impacts throughout the product life cycle

environmental and society challenges. When harnessing these concepts, it is important to use strong, quantitative metrics for sustainability adaptation; Life Cycle Assessment (LCA) is established method planning for understanding environmental impact through a continuous lifespan of products, including resources, emissions, and toxicity impacts. These additional performance metrics are employed during all steps of products and processes, for example, the E-factor, atom economy, process mass intensity (PMI), and energy intensity, for increasing the relative environmental convenience and activity of specific processes [12]. Similar analytic tools can be applied to address social and economic concerns, such as Social Life Cycle Assessment (S-LCA), and Life Cycle Costing (LCC), which dissect

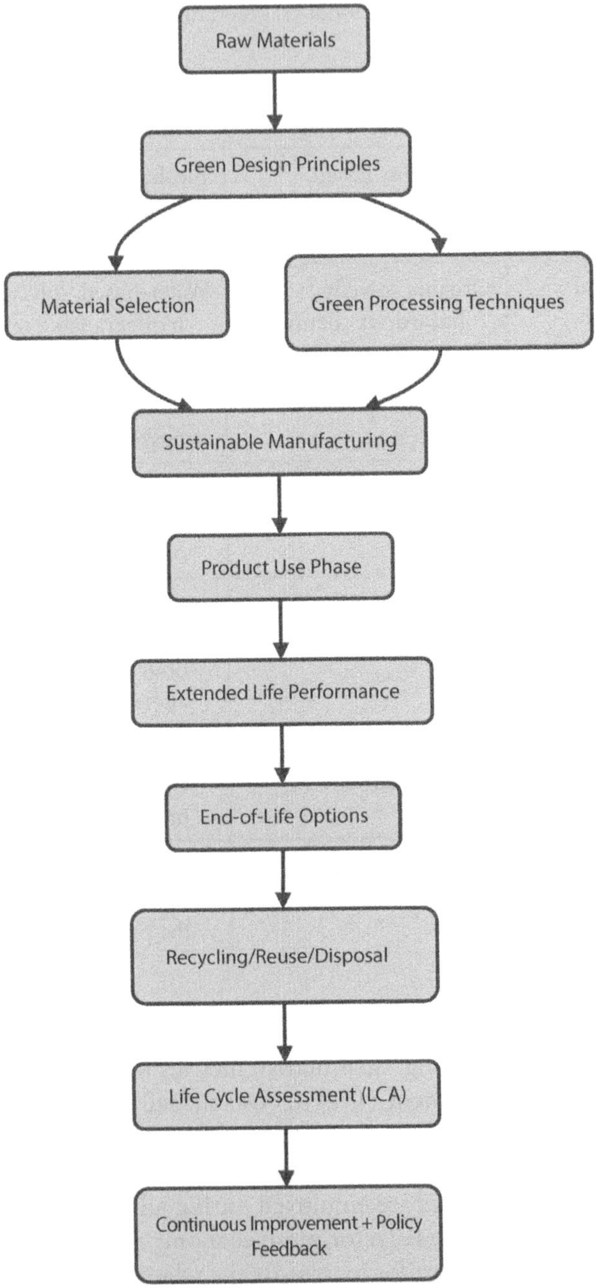

Figure 2.1 Framework for green and sustainable manufacturing in advanced materials.

labor conditions, community impacts, and costs over the life of a product. Emerging and developing integrative metrics exist with sustainable process index (SPI) and Green Chemistry Performance Scorecards, applied for comprehensive multi-dimensional sustainability performance understanding. Continued implementation, development, and standardization of these opportunities is essential to accelerate overall sustainability opportunities with innovative practice, management systems, and policy facilitation to reach its potential to support sustainability in advanced materials manufacturing.

2.3 Overview of Advanced Materials

Advanced materials have many different considerations and tailored materials which are designed to exceed the properties of ordinary materials in demanding applications. They are specifically designed to have better characteristics, such as greater mechanical strength, thermal and chemical stability, better conducting, corrosion resistance, lighter weight, multi-functionality, etc. Because of this, their use in high-performance systems can be very significant in a variety of sectors. Advanced materials can be classified by their type, chemistry, structure, or intended use. The materials include metals, such as high-entropy alloys and shape memory alloys; polymers and polymer composites, such as thermoplastics, thermo-sets, and biopolymers; ceramics and glasses, such as alumina, silicon carbide, and bioactive glasses; and hybrid materials, such as nanocomposites, smart materials, and metamaterials [16, 17].

A very significant category of materials in this area is nanostructured materials. That is, materials that club at the nanoscale that can be developed in a manner such that quantum mechanics and increased surface area/volume ratios are appropriately utilized producing new physical and chemical properties/behavior, especially electrical, optical and catalytic properties. Xu *et al.* [18], evaluated surface engineering tools, such as morphology and defects, that could be used as surface interventions to enhance the catalytic performance of inorganic nanostructures used in energy conversion, energy storage or environmental remediation applications. Advances in bio-based and biomimetic materials have also gained a lot of attention with respect to being a source of stable materials that mimic the high functionality of natural systems [19]. Bio-based and biomimetic materials are typically manufactured with highly designed microstructures or hierarchical architectures to provide subject-performance characteristics. The exponential increase in mark makes it clear that the advancements in the materials design necessarily bring sophistication in contemporary materials to advanced materials and therefore are necessary components of next generation technologies.

2.3.1 Applications and Market Relevance

The effects of advanced materials can be seen in many fields where change and higher performance are required [20]. In aerospace and defense, the use of lightweight composites and high-temperature alloys leads to better fuel consumption, enhancement of payload, and more structural integrity. The electronics and photonics industries use advanced materials like graphene, quantum dots and conductive polymers to create the smallest, fastest and most flexible devices with the best data processing capabilities. In the biomedical industry, environmental-friendly solutions, compatibility with biological systems are leading to advanced materials usage in implants, drug delivery systems, and tissue engineering studies leading to medical care and regenerative medicine advances. Advanced materials are required in the energy sector to improve storage and conversion capabilities such as batteries, supercapacitors, fuel cells, and solar cells are using advanced materials [21].

Demand for these materials is expanding rapidly, driven by global trends such as electrification, renewable energy, smart manufacturing, and sustainable construction. Projections of continued growth of industries of nanomaterials, carbon fibre composites, and advanced ceramics show their strategic importance to modern industry. Integration of advanced materials with digital technologies such as embedded sensors, AI, and the Internet of Thing's innovative applications and business models [22]. The synergies created position materials science as an enabler for Industry 4.0 whereby intelligent interconnected systems rely on advancement in materials to provide superior functionality.

2.3.2 Environmental Challenges in Conventional Manufacturing

Given the potential of advanced materials in the innovative landscape of technologies the way that they are conventionally manufactured has high environmental costs [23]. Many of the processes of manufacturing are energy consuming and either utilizes high temperatures or long reaction times, solvents that are toxic, and often energy-intensive processes in fabricating materials such as sintering, melting, or milling. These also often have quite a large fossil energy component, and create greenhouse gas emissions contributing to climate change. Further, mining essential raw materials like rare earths, lithium, titanium, etc. are environmental and social nightmares for many, as mining destroys habitats, pollutes water systems, and displaces entire communities, particularly in areas with limited regulatory oversight and locality. Nanomaterials are even more problematic due to their nanoscale size and environmental stability and community concerns regarding impacts on human health and environmental impacts [24].

Many of these particles resist most natural degradation processes and their potential behavior in biological and environmental systems is not well understood. The use of volatile organic compounds, heavy metals, or other toxic intermediates involved in the synthesis and manufacturing of advanced composites and coatings present safety issues for handling, storage, and disposal. Waste will also be generated, especially relevant to subtractive manufacturing processes where large quantities of materials board waste must be discarded in order to obtain desired shapes and dimensions of materials, and there is material use inefficiencies.

Addressing these challenges requires a fundamental shift to sustainable manufacturing practices, including adopting circular economy principles that focus on efficiency and reuse, integrating renewable energy sources in production systems, and adopting green chemistry approaches that reduce the use and generation of hazardous materials [25]. Through these efforts, we can take important steps towards reducing the environmental impact of advanced materials and maintaining compatibility with wider sustainability goals.

2.4 Green Synthesis and Processing Techniques

Synthesis and processing techniques using green principles represent a fundamental change in how advanced materials can be conceptualized and produced, favoring sustainability, energy efficiency, and efficient use of resources. Table 2.2 provides examples of green synthesis and processing techniques for different classes of materials. Compared to traditional methods that often apply toxic chemicals, high temperatures, and energy-intensive approaches, environmental-friendly approaches propose cleaner, safer, and sustainable alternatives. Derived from principles of green chemistry and engineering, these methods aim to produce materials with lower or eliminated toxic materials, find renewable resources, and improve material efficiency across the process lifetime [26]. In an era of increased global

Table 2.2 Summary of green synthesis and processing techniques across material classes [17, 29–31].

Material class	Green technique	Sustainability benefit	Example
Metals & Alloys	Electrochemical synthesis, aqueous-based reduction	Avoids high-temperature smelting; reduces emissions	Green synthesis of Ag and Cu nanoparticles
Polymers	Bio-based monomers, enzymatic polymerization	Renewable inputs, biodegradable, less toxic	Polylactic acid (PLA) from corn starch
Ceramics	Sol-gel methods in water, low-temp sintering	Energy savings, safer chemicals	Hydrothermal synthesis of titania ceramics
Composites	Natural fiber reinforcement, solvent-free curing	Lightweight, biodegradable, renewable sources	Hemp fiber-reinforced epoxy composites
Nanomaterials	Plant-extract mediated synthesis, microwave-assisted methods	Reduced use of toxic precursors, efficient energy use	Green-synthesized ZnO nanoparticles
Hybrid/Smart Materials	Layer-by-layer assembly using aqueous or bio-based solutions	Precision control, compatibility with green processing	Biohybrid hydrogels for soft robotics

demand for sustainable manufacturing, green synthesis becomes a foundational aspect for advanced materials in the future.

An important change in this area is the shift from volatile organic solvents to solvent-free or aqueous based methods [27]. Traditional synthesis techniques utilize organic solvents which can be a significant risk to human health and the environment due to their flammable, toxic, and air pollution contributing characteristics. Solvent-free methods, such as mechanochemical synthesis use mechanical energy in the form of grinding or ball milling to promote chemical reactions in the absence of a liquid medium. This approach has been shown to be effective in the synthesis of a number of materials including metal-organic frameworks, ceramics, and polymers, and it provides the opportunity to reduce energy input, shorten reaction times, and produce very few byproducts. Aqueous systems provide an additional environmentally friendly approach. Aqueous-based synthesis method including hydrothermal methods and sol-gel methods conducted in water allows the formation of complex nanostructures, functional oxides and bioinspired materials under relatively mild conditions [28]. Additionally, these methods provide improved safety and allow for better control of reactions; they have good compatibility with applications in the biomedical and pharmaceutical fields. They also provide substantial benefits with scalability and waste disposal, and that they represent a distinct change from the use of harmful solvents in traditional materials chemistry.

The rising dependence on bio-based and renewable feedstocks is also important [32]. Traditionally, materials are derived from finite and environmentally harmful petroleum-based sources, while bio-based materials come from renewable biological derivatives including plants, algae, microbes and agricultural waste. This category includes biopolymers such as polylactic acid, chitosan and cellulose derivatives, bio-based monomers such as glycerol and itaconic acid, and natural reducing or capping agents used in the synthesis of nanoparticles, such as extracts from plants, sugars and amino acids. These renewable feedstocks reduce onerous dependence on fossil fuels and provide alternatives that are inherently more biodegradable and less toxic, consistent with circular economy goals. In many cases, bio-based materials are intended to equal or exceed the performance of their synthetic alternatives such as sustainable carbon materials from lignin for energy storage systems or bio-inspired composites with better mechanical properties for use in construction applications [33].

Another relevant goal in achieving a real minimal energy. Conventional processes such as sintering, melting, and high-pressure processes often require high temperatures leading to high operational costs and carbon emissions. However, low-temperature and low-power methods can be considered as alternatives to these traditional processes. Microwave-assisted synthesis, photochemical reactions and plasma processing, for example, supply energy accurately and directly into the reaction zone with minimal waste. Microwave synthesis, for instance, allows for fast heating at the molecular level, reducing energy and time [34]. Sol-gel methods, which can be performed at room temperature, can produce high-purity, nanostructured materials with little thermal processing or energy. Supercritical fluid methods, especially supercritical carbon dioxide methods, can greatly enhance efficiency because they waste fewer solvents and offer cleaner approaches. Energy efficient methods minimize the environment costs associated with producing materials, they are also easy to mass-manufacture and economical, making them attractive to industry.

Waste minimization and circular design practices are key for facilitating sustainable processing. Manufacturing traditionally follows a linear model of extraction, production and then waste, ultimately wasting significant value and their materials [35]. Green processing takes steps to maximize closed loop systems to achieve reuse, recycling and the recovery of resources [36]. Usage of technology such as catalyst and solvent recycling, and *in situ* recovery of unreacted precursors, and industrial byproducts such as slag or fly ash as value-added functional materials greatly minimize waste. After a product's life cycle, it will last longer if designed with disassembly, modular reuse, and maintenance into its lifecycle. Proper disassembly and modular design components maintain performance at the end of life and support recycling later. Smart programs for tracking through blockchain for supply chains, optimization through simulation software, and real-time waste monitoring through AI applications can facilitate better optimization of resources by providing immediate feedback on efficiencies. The full circular strategy now applied directly to frameworks of synthesis and production can lead materials industry toward much more sustainable levels of resilient and regenerative practices.

2.5 Sustainable Manufacturing Technologies

Sustainable manufacturing technologies are necessary to transition the materials sector from a damaging to an efficient, pollution-free, and circular economy sector. Sustainable manufacturing technologies seek to reduce environmental impacts and to simultaneously enhance manufacturing processes in terms of accuracy, flexibility, and efficiency. Thus, manufacturers can produce high-performance materials while reducing waste and toxic emissions by incorporating novel innovation like digital fabrication, tuning material properties at the nanoscale, solvent-less surface treatment, or writing in an energy-efficient processing environment. This section discusses important and critically emerging technologies that embody sustainable manufacturing.

2.5.1 Additive Manufacturing

Additive manufacturing (AM), otherwise called 3D printing, is a great sustainable technology in today's materials engineering world. Unlike traditional subtractive manufacturing which takes away material from a larger block that creates a lot of waste, additive manufacturing builds an object layer upon layer straight from a digital model while only using the material that the product needs [37]. This process effectively reduces material waste, increases resource efficiency, and allows for complex geometries which are difficult or impossible to achieve with traditional methods. Additionally, the additive manufacturing process reduces the number of steps, shorten supply chains, and allows for production closer to the end user, decreasing emissions associated with transportation.

AM is always used for polymers, metals, ceramics, and composites, and the attention is now on more use of bio-based feedstock and recycled feedstock. There are several methods of AM which are being adapted to more sustainable application, such as, biodegradable polymers, renewable composites, or non-toxic binders, for example, Fused Deposition Modeling (FDM), Selective Laser Sintering (SLS), and Direct Ink Writing (DIW). Energy-efficient processes, such as low temperature sintering and real-time monitoring of each specific process, contribute to more sustainable forms of 3D printing. The on-demand production, mass-customization, and the design for disassembly capability of additive manufacturing supports pursuit of sustainability in the context of the circular economy [38].

2.5.2 Green Nanotechnology

Essentially, green nanotechnology is the overlap of nanoscience and environmental sustainability, allowing for the creation of nanomaterials, and nanoscale processes, to reduce potential harm to humans and the environment. The traditional production of nanomaterials often includes energy-intensive processes and toxic materials. Green nanotechnology uses environmentally sustainable methodologies such as plant-mediated synthesis of nanoparticles, sol-gel processes in aqueous solutions, and bio templating, to develop nanomaterials and processes using natural or safe materials and under less extreme conditions [39].

Green nanotechnology is fundamentally about designing safer nanomaterials by considering their functional aspects as well as the impacts over a lifecycle and environmental considerations. It involves creating biodegradable nanoparticles, surface modified nanomaterials with less ecotoxicity, and scalable processes that consume less solvents and energy.

Applications include applications, such as green electronics. Nanoscale materials and assemble contribute to higher energy efficiencies. Another application oftentimes considered in clean energy devices, including solar or fuel cells, which may experience enhanced performance because of the nano-enabled component [40]. With increasing concerns regarding pollution and toxicity of nanoparticles, green nanotech is a sustainable alternative and ensures ecological integrity as improvements are achieved through nano-innovation.

2.5.3 Eco-Friendly Surface Modification Techniques

Surface modifications are crucial for adding and changing material properties, like corrosion resistance, hydrophobicity, biocompatibility, or adhesion, while keeping bulk properties relatively constant. Traditional surface modification techniques often employ harsh chemicals, elevated temperatures, and create harmful byproducts that negatively impact our planet. In contrast, sustainable surface modification techniques seek to create the desired functional properties using no solvents or energy, and in a non-toxic way [41].

Plasma treatments are a solvent-less dry process that activate material surfaces under controlled conditions and improve properties such as wettability and adhesion without chemical residues remaining [42]. Supercritical CO_2 based coatings are an environmentally friendly alternative to solvent-based coatings that allow continuous and uniform depositions with no volatile organic compounds or VOCs. Other alternative techniques are UV curing, laser-assisted surface texturing, and enzymatically based modifications. Each of these alternative techniques allows for certain functionalities in a way with less environmental impact. All of these techniques are increasingly being used in the biomedical device, packaging, electronics, and aerospace industries, where surface functionality and sustainability is of utmost importance.

2.5.4 Cleanroom and Dry Processing Innovations

Cleanroom technologies are a critical component in the production of materials that are highly pure and accurately controlled, with specific applications in electronics, semiconductors, and biomedical devices. Traditional cleanroom operations require significant resources in energy, deionized water, and chemical solvents, which raises questions around sustainability and a cleaner supply chain. New innovations in cleanroom design and clean or dry processing will help to reduce the environmental and energy footprint of these operations.

Modern and sustainable cleanrooms are equipped with smart ventilation systems, energy recovery units, and LED lighting to reduce energy use. They should be constructed in a modular and adaptive format to operate most efficiently and minimize downtime. Dry processing techniques, such as physical vapor deposition (PVD), atomic layer deposition (ALD), and dry etch, can completely replace wet chemical processing, which reduces the usage of hazardous liquids and removes the source of wastewater [43]. These methods are very suitable for processes on the nanoscale, where precision and cleanliness are essential, but environmental regulations are tightening. Together, these advancements allow high quality production in a controlled environment and with dramatically lower levels of water usage, chemical use, and reduced carbon use [44]. They show how advanced manufacturing facilities can be technologically advanced and environmentally sustainable.

2.6 Case Studies across Material Classes

Real world case studies in metals, polymers, ceramics and hybrid materials show how to implement green and sustainable principles. Advances in sustainability in the metals industry includes electrochemical production of aluminum using inert anodes and minimizing CO_2 emissions compared to the traditional Hall-Héroult process. Bioleaching uses microorganisms that can extract metals such as copper and gold as a more environmentally friendly alternative to chemical methods [45]. Additive manufacturing of titanium alloys used in aerospace and biomedical applications creates less waste, while closed-loop systems maximize recyclable metal powders to contribute to the circular economy goals [46].

In polymers, bioplastics like polylactic acid (PLA) derived from renewable source materials like corn starch and are biodegradable alternatives to petroleum plastics. Natural fiber-reinforced polymers (NFRPs) made from using hemp or flax, combined with bio-based resins are already used in new vehicle parts by carmakers like BMW and Ford to reduce both weight and emissions [30]. Recycled thermoplastic composites allow for high performance re-use of consumer plastics with glass or carbon fibers.

Sustainable approaches in ceramics and glass concentrate on energy and resource efficiencies. Geopolymers, derived from waste feedstock like fly ash, are viable low-carbon and low-emission substitutes for cement which can cure at room temperature, and have verified emissions reductions more than 80% [47]. Cullet — recycled glass — is heavily used in the glass industry, allowing for melting at lower temperatures and fewer requirements for raw materials. Furthermore, low-energy processing, for example, microwave-assisted sintering and aqueous sol-gel approaches, are developing towards more environmentally sustainable ceramic production, including for bioactive coatings for implants.

Hybrid and multifunctional materials have their own sustainability challenges; however, developments do exist. Hydrogels for tissue engineering with cellulose nanocrystals for reinforcement may also integrate biodegradability and performance and are synthesized from water-based and plant-based feedstock [48]. Solvent-free and mechanochemical approaches to creating metal-organic frameworks (MOFs) can minimize reliance on toxic solvents. Agricultural waste-derived activated carbon can be used to fashion sustainable supercapacitors in energy storage. Finally, multifunctional coatings that show self-healing or photocatalytic characteristics will be developed with sustainable approaches using water-based sol-gel or biodegradable materials in a layer-by-layer assembly. This will extend durability and therefore decrease the available material for the lifecycle of a product.

2.7 Life Cycle Assessment (LCA) and Environmental Impact Analysis

As sustainable manufacturing is becoming more popular, Life Cycle Assessment (LCA) has become vital for assessing materials and processes for their environmental impacts over their entire life-cycle from raw material extraction to final disposal [49]. Life Cycle Assessment provides industry the ability to identify environmental hotspots and make better sustainable decisions by measuring the critical indicators of energy use, carbon emissions, water use, and waste generation. Life Cycle Assessment follows methodology as defined by ISO 14040 and 14044, which is composed of four phases: goal and scope definition, inventory analysis, impact assessment, and interpretation. Many tools such as SimaPro, GaBi, OpenLCA, and Umberto utilize global datasets that are made available through databases like Ecoinvent and Agri-footprint. Hybrid, dynamic, and consequential life cycle assessment are more advanced methodologies that can be used to better see emerging materials such as nanomaterials and bio-based composites. Figure 2.2 shows the LCA process in sustainable materials manufacturing.

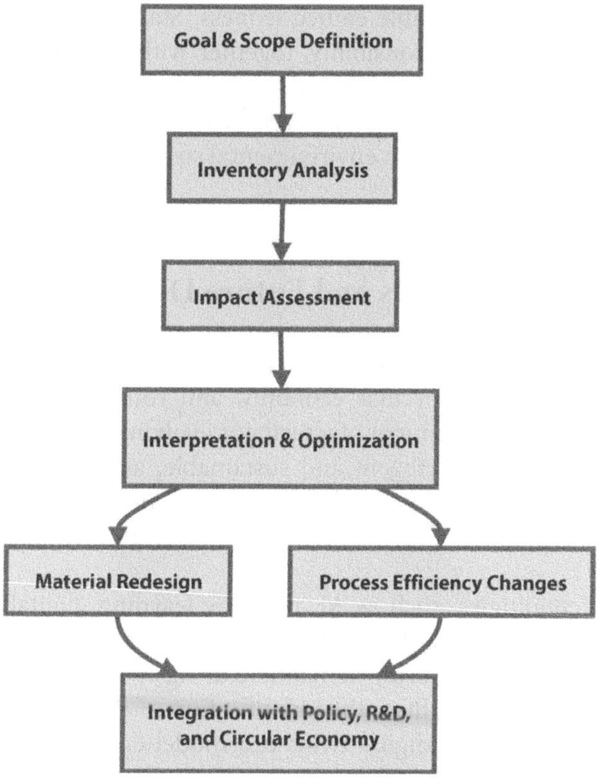

Figure 2.2 Life cycle assessment (LCA) in sustainable materials manufacturing.

Comparative LCA research illustrates the benefits of more sustainable systems of manufacture. The aqueous approach to produce titanium dioxide nanoparticles cut down carbon emissions and toxicity compared to the typical manner of manufacturing. Polylactic acid bioplastics have less global warming potential than petroleum-based plastics, along with several end-of-life options. Additive manufacturing typically achieves less waste and material usage in contrast to conventional machining. Geopolymers emit 40–80% less CO_2 emissions than Portland cement [50]. While there are perceived benefits, there are challenges to the uptake of LCA within material research and applications. There are inadequacies in data with respect to some sustainable materials, complexities in modelling multifunctional products, and social and academic barriers between materials scientists and LCA analysts. Despite the obstructions, the challenges are generating innovation. Initial LCA uptake, emerging Artificial Intelligence (AI) platforms, and trans-disciplinary approaches to sustainability within materials

R&D, are creating sustainable embeddedness. Stakeholder policies, such as extended producer responsibility, together with life cycle sustainability assessment (LCSA), which emphasizes social and economic aspects, strengthens the role of LCA as a necessary tool to contribute to balancing technological innovation with environmental sustainability and continued resilient prospects.

2.8 Emerging Trends and Future Directions

The production of advanced materials is reaching a new stage shaped by digital innovation, the circular economy, and fundamental changes to production models. Green manufacturing is evolving from single, disconnected actions to an intelligent and sustainable, and integrated system. This evolution is defined by two definitions: the coupling of digitalization and sustainability, and the systematic embedding of circular economy strategies.

Digital technologies, such as AI, machine learning, digital twins, and IoT, are able to provide real-time process analytics, predictive maintenance, and environmental monitoring in manufacturing systems [51]. AI-enabled design tools can help facilitate the identification of energy-saving materials and processes, while digital twins can model environmental impacts before production starts. These digital technologies are increasingly coupled with LCA tools and supply chain systems to provide continual up-to-date sustainability assessments of goods and activities. The requisite organizational change required entails new skill sets, data infrastructures, and cross-disciplinary collaboration, which also allows digitalization to act as a critical enabler of intelligent and sustainable factories.

At the same time, the principles of a circular economy change the way material design and business models function. There is an increased focus on recyclability, modularity, and product lifespan. Innovations include: bio-based polymers, recyclable composites and PaaS (Product-as-a-Service) models that allow us to retain material value, across multiple use cycles [52]. Material traceability *via* blockchain or digital passports offers an industry transformation in transparency and compliance. While these developments represent real advances around reducing waste and dependency on resources, further action and research is needed, especially on complex multi-material systems, and creative new recycling technologies as well as greater cooperation across sectors.

In the future, essential gaps need to be addressed. Many green synthesis methods are difficult to scale for industrial applications and information on the long-term consequences of new materials, especially nanomaterials, is often inadequate [53]. The initial discussion of sustainability metrics in material design is still uncommon, with *post-hoc* appraisals remaining the norm. We should be working on active design strategies and forecasting tools that address sustainability, beginning at design. R&D must include important strategic roadmaps that mention interdisciplinary research and development, pilot-scale demonstrations, existing sustainability databases, and enabling policy frameworks. There must also be reform of educational curriculums to build a workforce that has expertise of both technological innovation and sustainability to ensure that future materials manufacturing can be resilient, circular, and environmentally sustainable.

2.9 Conclusion

Shifting toward sustainable production processes in advanced materials manufacture is a critical response to growing environmental, economic, and societal challenges. This review has highlighted the potential to develop sustainable practices through the adoption of green chemistry, renewable feedstocks, energy efficient processing, and life cycle thinking, which can act to reduce ecological footprints and unlock new modes of value creation. Sustainability must be pursued with a holistic view that considers impacts across the entire life cycle of a product (i.e., from raw material extraction, processing, use phase, and post-use). Case studies across classes of materials demonstrate that we can achieve cleaner syntheses, reduced energy consumption, and improved recyclability with equal or improved product performance. A range of modern technologies which include additive manufacturing and green nanotechnology also provide promising platforms through which to pursue this goal.

Life Cycle Assessment has developed into an important tool for the assessment of environmental impacts and sustainable decision making. However, it is still limited, especially in terms of pre-assessment level or new materials. The increased interest in digital technology, artificial intelligence, and a circular economy opens up potential solutions to these challenges. Realistically, changes will happen through collaboration among researchers, industry executives, regulators, and educators. We need to prioritize scalable sustainable technologies, transparent supply chains, and interdisciplinary innovation. Sustainability as a principle of materials development must be adopted to create a better world that is more resilient, equitable, and respectful of the environment.

AI disclosure statement: The author acknowledges the use of AI-based tools for minor language editing or grammatical correction. However, all intellectual contributions, analyses, and conclusions presented in this chapter are entirely the author's own.

References

1. Fernández-Colino, A., Kiessling, F., Slabu, I., De Laporte, L., Akhyari, P., Nagel, S.K., *et al.*, Lifelike Transformative Materials for Biohybrid Implants: Inspired by Nature, Driven by Technology. *Adv. Healthcare Mater.*, 12, 20, 2300991, 2023.
2. Gressler, S., Part, F., Scherhaufer, S., Obersteiner, G., Huber-Humer, M., Advanced materials for emerging photovoltaic systems–Environmental hotspots in the production and end-of-life phase of organic, dye-sensitized, perovskite, and quantum dots solar cells. *Sustainable Mater. Technol.*, 34, e00501, 2022.
3. Arvidsson, R., Peters, G., Hansen, S.F., Baun, A., Prospective environmental risk screening of seven advanced materials based on production volumes and aquatic ecotoxicity. *NanoImpact*, 25, 100393, 2022.
4. Ajithkumar, S. and Arulmurugan, B., Microstructure and mechanical properties of Inconel 686 fabricated by gas metal arc welding-based wire arc directed energy deposition: impact of cryogenic treatments. *J. Mater. Sci.*, 59, 40, 19273–19302, 2024.
5. Gupta, A. and Khanna, A., A holistic approach to sustainable manufacturing: Rework, green technology, and carbon policies. *Expert Syst. Appl.*, 244, 122943, 2024.
6. Nikolakis, N., Catti, P., Chaloulos, A., van de Kamp, W., Coy, M.P., Alexopoulos, K., A methodology to assess circular economy strategies for sustainable manufacturing using process eco-efficiency. *J. Cleaner Prod.*, 445, 141289, 2024.
7. Viles, E., Kalemkerian, F., Garza-Reyes, J.A., Antony, J., Santos, J., Theorizing the Principles of Sustainable Production in the context of Circular Economy and Industry 4.0. *Sustainable Prod. Consumption*, 33, 2022.
8. Jia, L., Chu, J., Ma, L., Qi, X., Kumar, A., Life cycle assessment of plywood manufacturing process in China. *Int. J. Environ. Res. Public Health*, 16, 11, 2037, 2019.
9. Hermundsdottir, F. and Aspelund, A., Competitive sustainable manufacturing - Sustainability strategies, environmental and social innovations, and their effects on firm performance. *J. Cleaner Prod.*, 370, 2022.
10. Dey, B.K., Park, J., Seok, H., Carbon-emission and waste reduction of a manufacturing-remanufacturing system using green technology and autonomated inspection. *RAIRO Oper. Res.*, 56, 4, 2801–2831, 2022.

11. Chen, Y. and Yu, Z., Low-melting mixture solvents: extension of deep eutectic solvents and ionic liquids for broadening green solvents and green chemistry. *Green Chem. Eng.*, 5, 409–417, 2024.
12. Ramadugu, S., Ledella, S.R.K., Gaduturi, J.N.J., Pinninti, R.R., Sriram, V., Saxena, K.K., Environmental life cycle assessment of an automobile component fabricated by additive and conventional manufacturing. *Int. J. Interact. Des. Manuf.*, 18, 2024.
13. Mori, R., Replacing all petroleum-based chemical products with natural biomass-based chemical products: a tutorial review. *RSC Sustainability*, 1, 2023.
14. Bennett, J.A., Campbell, Z.S., Abolhasani, M., Role of continuous flow processes in green manufacturing of pharmaceuticals and specialty chemicals. *Curr. Opin. Chem. Eng.*, 26, 2019.
15. Logesh, B. and Balaji, M., Experimental Investigations to Deploy Green Manufacturing through Reduction of Waste Using Lean Tools in Electrical Components Manufacturing Company. *Int. J. Precis. Eng. Manuf. Green Technol.*, 8, 2021.
16. Diniţă, A., Ripeanu, R.G., Ilincă, C.N., Cursaru, D., Matei, D., Naim, R.I., *et al.*, Advancements in Fiber-Reinforced Polymer Composites: A Comprehensive Analysis. *Polymers (Basel)*, 16, 2024.
17. Baig, N., Kammakakam, I., Falath, W., Kammakakam, I., Nanomaterials: A review of synthesis methods, properties, recent progress, and challenges. *Mater. Adv.*, 2, 2021.
18. Xu, W., Bai, Y., Yin, Y., Surface Engineering of Nanostructured Energy Materials. *Adv. Mater.*, 30, 2018.
19. Zhu, Y., Joralmon, D., Shan, W., Chen, Y., Rong, J., Zhao, H., *et al.*, 3D printing biomimetic materials and structures for biomedical applications. *Bio-Des. Manuf.*, 4, 2021.
20. Ajithkumar, S. and Arulmurugan, B., Investigation of Single and Double Pulse Techniques on Microstructure and Mechanical Anisotropy of Inconel 686 Component Fabricated by Wire Arc Directed Energy Deposition. *Met. Mater. Int.*, 31, 1648–1673, 2024.
21. Jamil, F., Ali, H.M., Janjua, M.M., MXene based advanced materials for thermal energy storage: A recent review. *J. Energy Storage*, 35, 2021.
22. Tran, H., Gurnani, R., Kim, C., Pilania, G., Kwon, H.-K., Lively, R.P., *et al.*, Design of functional and sustainable polymers assisted by artificial intelligence. *Nat. Rev. Mater.*, 9, 866–86, 2024.
23. Ajithkumar, S., Arulmurugan, B., Idhayaraja, S., Robotic Welding and Automation: Cutting-Edge Technology for Industry, in: *Advanced Welding Technologies*, pp. 497–512, Wiley, 2025.
24. Subhan, M., Choudhury, K., Neogi, N., Advances with Molecular Nanomaterials in Industrial Manufacturing Applications. *Nanomanufacturing*, 1, 2021.
25. Omran, B.A. and Baek, K.H., Valorization of agro-industrial biowaste to green nanomaterials for wastewater treatment: Approaching green chemistry and circular economy principles. *J. Environ. Manage.*, 311, 2022.

26. Al-Alimi, S., Yusuf, N.K., Ghaleb, A.M., Lajis, M.A., Shamsudin, S., Zhou, W., *et al.*, Recycling aluminium for sustainable development: A review of different processing technologies in green manufacturing. *Results Eng.*, 23, 102566, 2024.
27. Wu, D., Wang, T., Hu, S., Wu, W., Lu, B., Huang, X., *et al.*, Solvent-free processing of eco-friendly magnetic and superhydrophobic absorbent from all-plant-based materials for efficient oil and organic solvent sorption. *Sci. Total Environ.*, 800, 2021.
28. Koozegar Kaleji, B. and Gorgani, M., Comparison of sol-gel and hydrothermal synthesis methods on the structural, optical and photocatalytic properties of Nb/Ag codoped TiO2 mesoporous nanoparticles. *Int. J. Environ. Anal. Chem.*, 102, 2022.
29. Jiao, K.J., Wang, Z.H., Ma, C., Liu, H.L., Cheng, B., Mei, T.S., The applications of electrochemical synthesis in asymmetric catalysis. *Chem Catal.*, 2, 2022.
30. Ajithkumar, S., Arulmurugan, B., Rajeshkumar, L., Prospects of natural fiber-reinforced polymer composites in engineering and commercial applications, in: *Applications of Composite Materials in Engineering*, pp. 335–64, Elsevier, 2025.
31. Ba-Abbad, M.M., Chai, P.V., Takriff, M.S., Benamor, A., Mohammad, A.W., Optimization of nickel oxide nanoparticle synthesis through the sol-gel method using Box-Behnken design. *Mater. Des.*, 86, 2015.
32. Sitharaj, A. and Balasubramanian, A., Surface modification of root-based fibers and their composites, in: *Surface Modification and Coating of Fibers, Polymers, and Composites*, pp. 157–86, Elsevier, 2025.
33. Dei Rossi, G. and Vergani, L.M., Buccino F. A Novel Triad of Bio-Inspired Design, Digital Fabrication, and Bio-Derived Materials for Personalised Bone Repair. *Materials*, 17, 5305, 2024.
34. Zeng, X., Hu, X., Song, H., Xia, G., Shen, Z.Y., Yu, R., *et al.*, Microwave synthesis of zeolites and their related applications. *Microporous Mesoporous Mater.*, 323, 2021.
35. Sitharaj, A., Arulmurugan, B., Veeraraghavan, V., Idhayaraja, S., Vivekanandh, S.R., Kavin, M., *et al.*, Hybridization in Wire Arc Directed Energy Deposition of Titanium Alloys: A Review, pp. 441–65, 2025.
36. Chai, J., Qian, Z., Wang, F., Zhu, J., Process innovation for green product in a closed loop supply chain with remanufacturing. *Ann. Oper. Res.*, 333, 533–57, 2024.
37. Sitharaj, A., Arulmurugan, B., Dharani Kumar, S., Idhayaraja, S., Manishraj, D., Premkumar, N., Wire-Arc Directed Energy Deposition of Aluminum Alloys: A Review on Properties, Challenges, and Applications, pp. 99–115, 2025.

38. Haines, M.P., Rielli, V.V., Primig, S., Haghdadi, N., Powder bed fusion additive manufacturing of Ni-based superalloys: a review of the main microstructural constituents and characterization techniques. *J. Mater. Sci.*, 57, 14135–14187, 2022.
39. Jha, S.K. and Jha, A., Plant Extract Mediated Synthesis of Metal Nanoparticles, their Characterization and Applications: A Green Approach. *Curr. Green Chem.*, 8, 2021.
40. Ying, S., Guan, Z., Ofoegbu, P.C., Clubb, P., Rico, C., He, F., *et al.*, Green synthesis of nanoparticles: Current developments and limitations. *Environ. Technol. Innov.*, 26, 102336, 2022.
41. Kim, H., Lee, J.E., Jo, S.M., Wooh, S., Universal Large-Scale and Solvent-Free Surface Modification of Inorganic Particles *via* Facile Mechanical Grinding. *ACS Sustainable Chem. Eng.*, 10, 2022.
42. Luque-Agudo, V., Hierro-Oliva, M., Gallardo-Moreno, A.M., González-Martín, M.L., Effect of plasma treatment on the surface properties of polylactic acid films. *Polym. Test.*, 96, 2021.
43. Yuan, B., Wang, Y., Elnaggar, A.Y., El Azab, I.H., Huang, M., Mahmoud, M.H.H., *et al.*, Physical vapor deposition of graphitic carbon nitride (g-C3N4) films on biomass substrate: optoelectronic performance evaluation and life cycle assessment. *Adv. Compos. Hybrid Mater.*, 5, 813–22, 2022.
44. Oke, J.A. and Jen, T.C., Atomic layer deposition and other thin film deposition techniques: From principles to film properties. *J. Mater. Res. Technol.*, 21, 2022.
45. Rendón-Castrillón, L., Ramírez-Carmona, M., Ocampo-López, C., Gómez-Arroyave, L., Bioleaching Techniques for Sustainable Recovery of Metals from Solid Matrices. *Sustainability (Switzerland)*, 15, 2023.
46. Song, T., Chen, Z., Cui, X., Lu, S., Chen, H., Wang, H., *et al.*, Strong and ductile titanium–oxygen–iron alloys by additive manufacturing. *Nature*, 618, 2023.
47. Abbas, R., Abdelzaher, M.A., Shehata, N., Tantawy, M.A., Production, characterization and performance of green geopolymer modified with industrial by-products. *Sci. Rep.*, 14, 2024.
48. Yang, J., Zhao, J.J., Xu, F., Sun, R.C., Revealing strong nanocomposite hydrogels reinforced by cellulose nanocrystals: Insight into morphologies and interactions. *ACS Appl. Mater. Interfaces*, 5, 2013.
49. Rajeshkumar, L., Ajithkumar, S., Arulmurugan, B., Bhuvaneswari, V., Life cycle assessment of mineral fibers and their composites, in: *Synthetic and Mineral Fibers, Their Composites and Applications*, pp. 631–55, Elsevier, 2024.
50. London, M.B., Lewis, G.M., Keoleian, G.A., Life Cycle Greenhouse Gas Implications of Multi Jet Fusion Additive Manufacturing. *ACS Sustainable Chem. Eng.*, 8, 2020.
51. Ajithkumar, S., Arulmurugan, B., Meiyazhagan, R., Welding of nickel alloys interface by finite element modeling and artificial intelligence and envisage by patent landscape. *Int. J. Interact. Des. Manuf. (IJIDeM)*, 2025.

52. Pitchai, U.M.R., Rathinasamy Radhamani, S., Sitharaj, A., Murugan, A., Surface Modification of Lignin by Polyethylene-Block-Poly (Ethylene Glycol), in: *Handbook of Lignin*, pp. 1–16, Springer Nature Singapore, Singapore, 2025.
53. Lu, S., Hou, W., Li, Z., Xia, J., Xie, L., Su, H., Design and optimisation of a green manufacturing-recycling network considering heavy metal pollutants–an electronic assembly case. *Int. J. Prod. Res.*, 60, 2830–2849, 2022.

3

Advanced High-Strength Steel Alloys for Sustainable Applications

Indrajit Dey[1] and Swarup Kumar Ghosh[2]*

[1]*Product Development Group, R & D Department, Jindal Steel, Patratu, Balkudra, Ramgarh, India*
[2]*Department of Metallurgy and Materials Engineering, Indian Institute of Engineering Science and Technology, Shibpur, Howrah, West Bengal, India*

Abstract

This chapter examines how microstructure influences mechanical properties in three high-carbon hot-rolled steels (HC0, HC1 and HC9) processed according to the industrial processing conditions of wire rods. Moreover, the selection of chemistry and process parameters and their combined effects on steel cleanliness, phase transition, microstructural constituents and mechanical properties are also thoroughly investigated using optical, scanning electron microscopy, transmission electron microscopy, Rockwell hardness in C scale and uniaxial tensile testing. Initially, the continuous cooling transformation curves for all three steels are estimated by using JMatPro® software. Concurrently, dilatometry studies are also carried out to determine the actual phase transformation temperatures of these steels. Microstructures of the HC0 and HC1 steels comprise ferrite and pearlite microstructures, whereas, for HC9 steel, it is a multiphase microstructure comprising martensite, retained austenite with some amount of pearlite. Nb microalloying significantly enhances microstructural refinement, affecting austenite grain size, pearlite morphology, and cementite dimensions and thereby resulting in improved mechanical properties of HC1 steel than HC0 steel. Interestingly, the higher Mn and Si in HC9 steel are favorable to lowering the phase transformation temperatures, which resulted in the formation of martensite, and to stabilize retained austenite even at room temperature without allowing undesirable carbide formations. Furthermore, the fine NbC precipitates (15 to 20 nm) in HC9 steel lead to the precipitation strengthening. Finally, the maximum product of strength and

**Corresponding author*: skghosh@metal.iiests.ac.in

Gurudas Mandal, Swarup Kumar Ghosh, Snehanshu Pal and Sandip Kunar (eds.) Sustainable Manufacturing of Advanced Materials, (43–64) © 2026 Scrivener Publishing LLC

elongation is obtained in HC9 steel (≈23,832 MPa%), followed by HC1 (≈13,470 MPa%) and HC0 (≈13,424 MPa%) steel.

Keywords: High carbon steel, Nb microalloying, steel cleanliness, dilatometry, microstructure, mechanical properties

3.1 Introduction

For decades, steel has been used as the dominant construction material in passenger and commercial vehicles. In recent times, the use of low-density materials like Alhttps://www.sciencedirect.com/topics/materials-science/aluminum, Mg, plastics, etc. has confronted the position of steel in either body-in-white as well as chassis for vehicle construction. These requirements have motivated the steel industry for the progress of new high-strength steel grades, which can replace conventional mild steels and compete with such low-density materials in the context of lighter, safer, cost-effective automobile body structures. Indubitably, the fuel consumption and thereby the CO_2 emissions also decrease with a reduction in vehicle weight. Previous research has reported that a reduction in vehicle weight of ≈100 kg results in the savings of ≈0.15-0.5 liters of fuel consumption and thereby a reduction in the CO_2 emissions in the range of ≈4 and 12 g/km [1]. It is noteworthy that apart from reduced fuel consumption, lighter vehicles also facilitate fast acceleration, short braking distances, and good agility. The proportion of high-performance steels has grown significantly in recent years, and the conventional mild steel has decreased even below 30% of the total body weight in the case of advanced vehicles. This leads to a reduction in the vehicle weight of 15-25% [1]. Nowadays, in commercial vehicles, hot-rolled high-strength steels are the choice in frame and chassis parts.

A higher carbon containing high strength steels and multiphase microstructures of bainite, martensite, pearlite and retained austenite have motivated researchers due to their possible applications in high abrasion environments like rail steels, wear-resistant steels, spring steels, wire rope, forging grades, reinforcement tires and high strength bars owing to their remarkable combination of strength and toughness [2–5]. Currently, there is a high demand for high-carbon low-alloy steels for various ultra-high-strength applications where the desired mechanical properties and hardenability can be achieved by allowing more carbon in the steel composition, which in turn replaces the expensive alloying elements like Ni and Cr. The weight reduction can be controlled by the enhanced mechanical properties, which are primarily governed by the refinement of the microstructural features.

A major part of the wire production *via* wire drawing is obtained for structural applications. Eutectoid steels with higher strength are suitable for the commonly used substitutes for this purpose. The synergy of this material from strength and ductility is primarily reliant on its microstructure, where pearlite (P) is the main micro constituent. Pearlite is a nano-composite comprising lamellae of ferrite and cementite. Moreover, during high deformation in wire drawing, this material can experience strong hardening, which is a result of the physical changes of such a type of microstructure [6].

Thermomechanical controlled processing (TMCP) of low and medium-carbon steels is extensively used to manufacture large diameter pipes to transport gas and oil, as well as to enhance the lifecycle and performance of these pipelines for several years. The yield strength and toughness of the pipe materials can be improved by refining the austenite grains and transformation products *via* thermomechanically controlled rolling and various cooling routes, respectively. According to industrial practice, after reheating the slabs at 1100-1200°C followed by descaling, TMCP comprises controlled multistage hot deformation i.e., rolling through roughing stands above the austenite recrystallisation limit temperature (RLT) followed by rolling though finishing stands below the recrystallization stop temperature (RST) and then controlled cooling to room temperature [7–12]. During the initial rolling stages through roughing stands, subsequent hot deformation above the RLT results in the development of fine recrystallized austenite (γ) grains, which further elongate and flatten during the later stage of hot deformation through the finishing stands below the RST, which is also known as pancaking of γ grains. These elongated, flattened and dislocated FCC γ grains are further transformed to fine BCC ferrite (α), BCT martensite, bainite or pearlite depending on the cooling routes [12].

Microalloying with Nb, Ti and V can also be used for the lowering of the prior austenite grain size by solute drag or particle pinning effects at the grain boundaries. Nb depresses the phase transition temperatures and thereby refines the transformation products [6, 7]. In this connection, it needs to be mentioned that although plenty of literature is available on microstructure mechanical property correlation for low and medium-carbon microalloyed steels but studies on this standpoint for high-carbon microalloyed steels are limited. Hence, the findings of the current study could enlighten the design of new high-carbon microalloyed steels with optimized microstructures and mechanical properties, leading to improved performance and sustainable applications.

3.2 Experimental Procedures

In the present study, the chemistry of three forged bars containing high carbon, such as HC0, HC1 and HC9, is mentioned in Table 3.1. At first, the bars were completely homogenized prior to hot rolling (HR) with finish rolling temperature (FRT) of 1000°C and exposed to air cooling (AC). Dilatometry was employed to investigate phase transformations, including austenite decomposition and martensite formation, in steel specimens. Cylindrical samples (71 mm × 6 mm) were subjected to dilatometry tests using a Gleeble 3800 simulator. Temperature control was achieved through a K-type thermocouple fixed to the specimen surface. This setup enabled the precise measurement of critical transformation temperatures, specifically Ac_1, Ac_3, and M_s, providing valuable insights into the material's microstructural evolution.

Small samples (10 mm × 10 mm) were cut from the HR steel, and mirror-polished, scratch-free surfaces were prepared *via* typical metallographic specimen preparation techniques. The mirror-polished surfaces were etched in 2% nital solution to visualize the microstructures in an optical microscope (OM) and scanning electron microscope (SEM). A high-resolution energy-filtered transmission electron microscope (TEM) was also utilized to witness the nanostructures with better resolution. Before TEM investigation, 3 mm diameter disc samples were taken from these thin foils and consequently electropolished using a solution of 90% CH_3COOH and 10% $HClO_4$. X-ray diffraction studies are conducted in a machine incorporated with a Cu tube (Cuk_α) at 40 kV voltage and 40 mA current, and a systematic scanning was carried out with a step of 0.02°/s. The data files were then analyzed in DIFFRAC.TOPAS software and the

Table 3.1 The chemistry of the three experimental steels (wt.%).

Steel	C	Mn	Si	Nb	S	P	Fe
HC0	0.66	0.67	0.20	-	0.006	0.014	Balance
HC1	0.66	0.67	0.20	0.030	0.006	0.014	Balance
HC9	0.68	1.72	2.20	0.036	0.008	0.012	Balance

volume fractions of RA (V_γ) was obtained by using the Rietveld refinement method.

The austenite grain size of all samples subjected to direct water quenching from the austenitizing temperature was estimated from 30 to 40 optical micrographs by using the linear intercept method. Reliable Image J and Axiovision 4.8 Image analysis software were used to compute the average volume fractions of pearlite in HRAC samples for all three investigated steels, and 25 to 30 OM micrographs were chosen for this purpose. The pearlite nodule size (NS), colony size (PS) and interlamellar spacing (λ) were measured by using the linear intercept method from 30 to 40 high-resolution FESEM and TEM micrographs, respectively.

The Rockwell hardness (HRC) was determined using a calibrated hardness tester. Consequently, dog bone-shaped flat tensile specimens having 25 mm gauge length were machined following the ASTM standard [13]. Uniaxial tensile tests were carried out in an Instron 5900R instrument with a precision strain measuring extensometer attached to it, and the cross-head velocity was kept at 0.5 mm/min.

3.3 Results and Discussion

3.3.1 Cleanliness Level

Figure 3.1 shows various types of inclusions in the investigated steels. Figures 3.1(a) and (b) show the OM microstructures of these inclusions, whereas Figures 3.1(c)-(f) reveal the morphologies and corresponding EDS spectra and elemental mapping of these inclusions. Two types of inclusions are visible in Figure 3.1. The morphology and EDS spectrum in Figures 3.1(a) and (d), respectively, confirms it is Type-A MnS inclusion having length of 54 μm whereas, from the morphology, EDS spectrum and elemental mapping of the other one, Figures 3.1(b) and (c), correspondingly, confirms it is a Type-B SiO_2-Al_2O_3-CaO type oxide inclusion having length of 27 μm. It is also evident from Figure 3.1 that the former one is an elongated type, whereas the latter one is present in nearly a globular shape. It is well known that in free-cutting grade steels, FeS or (Fe, Mn)S are unwanted owing to their low melting temperatures (<1000°C). These varieties result in hot-shortness as well as crack formation in the course of HR and therefore, need to be altered to favorable Mn-rich (Mn, Fe)S or MnS having higher melting temperatures. It has been reported earlier that the occurrence of Type-A inclusions may be considered as a product of cooling or solidification, whereas the Type-B inclusions are associated

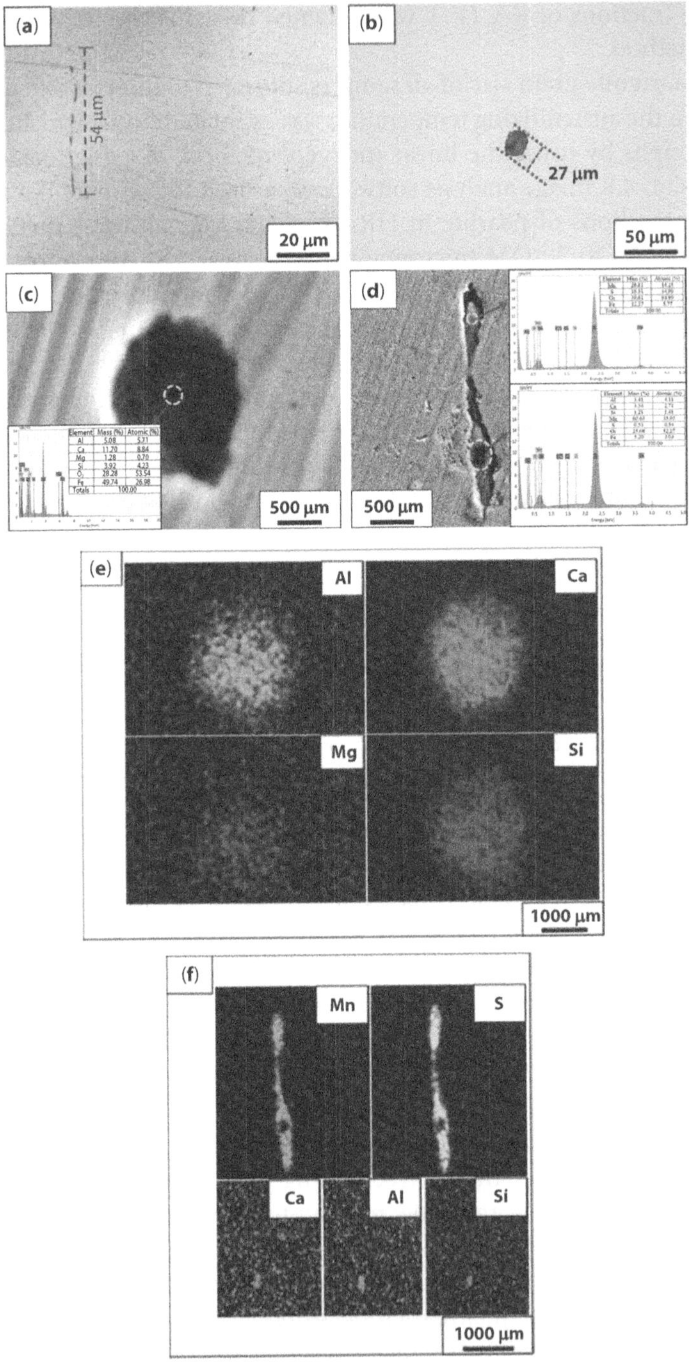

Figure 3.1 Types of inclusions present in the investigated steels.

with steelmaking, and it is more likely to come from slag [14, 15]. It has also been reported that Type-B inclusions usually have a low melting point, and even at 1300°C, most of them remain in liquid form.

The cleanliness of the steels was measured under 100× magnification in a 200 mm^2 area and was categorized according to their shape, size, and distribution in accordance with IS 4163:2004 standard [16]. As mentioned in this standard, inclusions are mainly of five types, i.e., A: deformable sulfide, B: non-deformable aluminate, C: deformable silicate, D: non-deformable globular oxide and DS: single globular type. According to this standard, the inclusions as evident in Figures 3.1(a) and (b) fall under A0.5 type thin series and B0.5 type thick series. However, it cannot be overruled that while investigating about the nature of these inclusions in a more extensive way as shown by the SEM micrographs, EDS spectrum and elemental mapping analysis, it has been observed that significant numbers of oxides and complex oxide inclusions i.e., containing more than one oxide and duplex i.e., oxide and sulfide inclusions were also present in the steels as evident in Figures 3.1(c) to (f). Therefore, the present study on the high-carbon steel witnessed complex inclusions which could not be classified into just one group, as mentioned in the existing standards. Some of the inclusions were composites which consisted of manganese sulfide along with different oxides. It is pertinent that these observations are in good agreement with earlier research carried out by M. Faraji et. al. on the cleanliness level of a high-carbon clean steel [17]. Therefore, from the above discussion, it can be concluded that the cleanliness level of the studied steels is within the acceptable limit according to the standards followed by the steel industries.

3.3.2 Continuous Cooling Transformation (CCT) Diagrams Obtained from JMatPro® Software

Figures 3.2(a) to (c) represent the CCT diagrams estimated from JMatPro® software for HC0, HC1 and HC9 steels. It is notable that apart from the chemical compositions of each studied steel, quenching temperature (200°C) and austenite grain size (AGS) have also been given as input to estimate the CCT diagrams by using JMatPro® software. The prior austenite grain sizes (PAGS) were calculated from the optical microstructures of directly water-quenched samples from the soaking temperature of 1200°C for each investigated steel. The measured values of PAGS were 56±8, 40±5 and 36±10 micrometers (μm) for HC0, HC1 and HC9 steels, respectively. It is imperative that although the cooling rate (CR) of 5°C/s crossed the nose of the ferrite (F), pearlite (P) and bainite start (B_s) curves

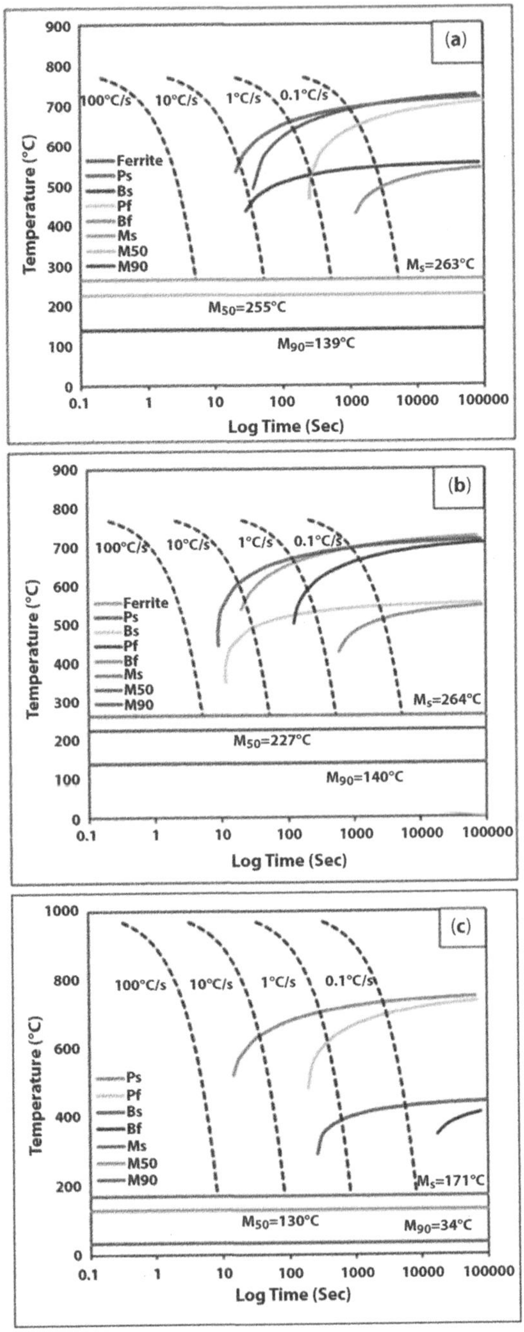

Figure 3.2 CCT diagrams estimated from JMatPro® software for (a) HC0, (b) HC1 and (c) HC9 steels.

in Figures 3.2(a) and (b) but in case of HC9 steel the bainite nose has been shifted and hence 5°C/s CR has not reach the B_s as evident from Figure 3.2(c). Therefore, after HR and at a CR of 5°C/s, the possible microstructure will be predominantly F and P for both HC0 and HC1 steels. Whereas, in the case of HC9 steel, after HR and air cooling at 5°C/s, the resultant microstructure is expected to be a two-phase mixture comprising primarily M with a certain amount of P, as revealed in Figure 3.2(c). Conversely, it is also factual that B will form in HC9 steel at comparatively slow cooling rates of 1°C/s or less, as evident from the CCT curve in Figure 3.2(c). This may be attributed to the fact that the joint addition of Mn and Si in HC9 steel shifts the noses of transformation curves to the right side, as it is well known that Mn and Si act as austenite stabiliser and graphitiser, respectively [18–25]. It is also evident from Figure 3.2(c) that the predicted M_s for HC9 steel is very low as compared to HC0 and HC1 alloys. In this connection, E. Erisir have already reported that in the JMatPro® software, the Kirkaldy and Venugopalan model is generally used for the estimation of M_s temperature, in which the presence of alloying elements like Si is not often considered [25]. However, in the present study, Si content is higher (2.2 wt.%) for HC9 steel. Therefore, it cannot be overruled that the estimated and measured M_s may vary significantly for HC9 steel.

3.3.3 Dilatometry and Phase Transformation Temperatures

Figure 3.3 represents the dilatometry results for all three investigated steels. It is noteworthy that for all the cases, both the heating and cooling rates were maintained at 5°C/s. The critical phase transformation temperatures, Ac_1 and Ac_3, were determined from dilatometric curves, where Ac_1 marks the onset of austenite transformation (deviation from linear expansion) and Ac_3 signifies the completion of austenitisation (return to linear expansion behavior upon reaching a fully austenitic state). It is visible from Figure 3.3(a) and Figure 3.3(b) that addition of Nb results in minor changes in austenite start (Ac_1) and finish (Ac_3) temperatures whereas, it significantly lowers both the pearlite start (P_s) and finish (P_f) temperatures in the order of 33°C and 5°C, respectively. On the other hand, for HC9 steel, where the Mn and Si contents are significantly higher than both HC0 and HC1 steels, in addition to the suppression of pearlite transformation, bainite and martensite transformations are also triggered under the same CR, as evident in Figure 3.3(c). It is noticeable that Figures 3.3(a) and (b) clearly show a distinct hump in the cooling curves between P_s and P_f temperatures, resulting from recalescence-induced temperature rise in both steels. In a preceding research, I. Dey *et al.* [26], have conveyed that the temperature rise due to

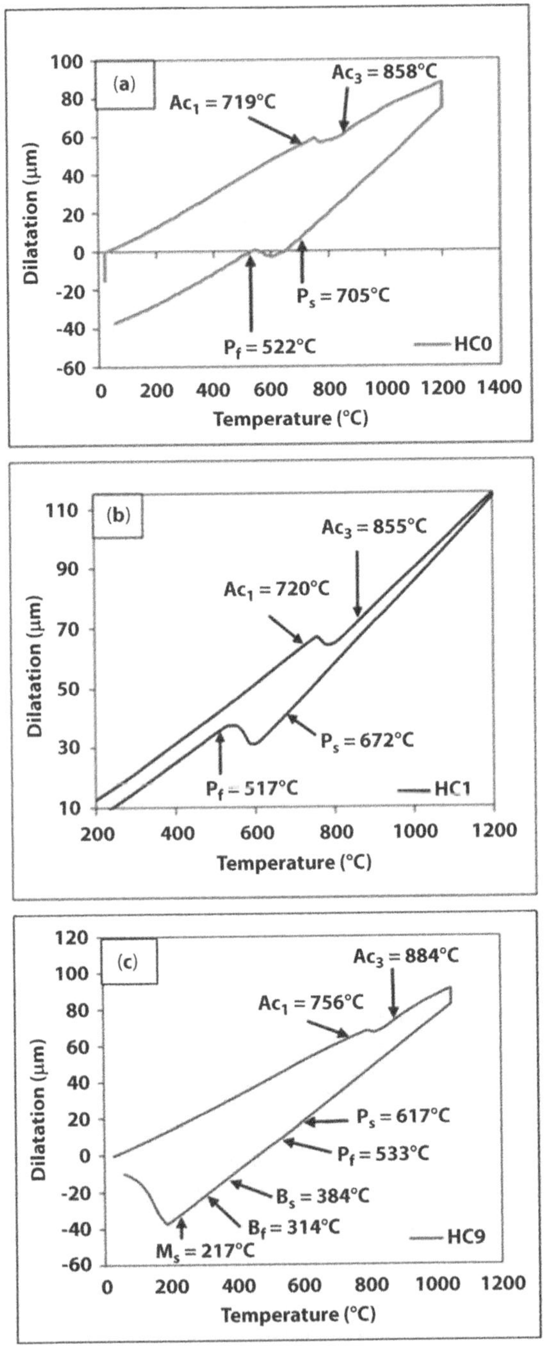

Figure 3.3 Dilatometry results of (a) HC0, (b) HC1, and (c) HC9 steels.

recalescence increases with increasing CR, which is an indication of substantial pearlite formation from the parent austenite. They have also witnessed that Nb, being an F stabilizer, reduces the enthalpy change at each available CR and thereby results in microstructure refinement. However, the negligible temperature increase caused by recalescence in Figure 3.3(c) indicates that the P is significantly less in HC9 steel as compared to HC0 and HC1 steels under the same CR, which may be due to the considerable alloying elements of Mn and Si as discussed in the preceding sections. Therefore, it is obvious that during cooling after the P transformation, the rest untransformed austenite can only transform to phases like B or M, as apparent in Figure 3.3(c). In this context, the bainite start (B_s) and finish (B_f) as well as martensite start (M_s) temperatures in Figure 3.3(c) were determined from the dilatation curves where a faint change in the slope of the curve takes place after the P transformation region.

3.3.4 Optical and SEM Microstructures

Figure 3.4 shows microstructures of HRAC samples of the three investigated steels. Figures 3.4(a) and (b) display the OM and FESEM micrographs of HC0, whereas Figures 3.4(c) and (d) show OM and FESEM micrographs of HC1 steels. Figures 3.4(e) and (f) reveal the OM and SEM micrographs of HC9 steel. The rolling and transverse axes are also shown in the optical micrographs, and the normal axis is kept parallel to the observer plane. The measured HRC values are given in each micrograph. Figures 3.4(a) to (d) reveal that a grain boundary ferrite and pearlite (shown by arrows), which correlates well with the CCT plots in Figures 3.2(a) and (b). It is pertinent that the basic difference between resolved and unresolved pearlite is that, for unresolved pearlite, the pearlite interlamellar spacing (λ) is finer than that of resolved pearlite. Figures 3.4(e) and (f) display that hot rolled HC9 steel leads to a complex microstructure of martensite, pearlite, with marginal retained austenite. Here, the greyish contrast in Figure 3.4(e) reveals martensite. The maximum value of 66 HRC in Figures 3.4(e) and (f) also confirms the presence of a hard phase like martensite/bainite in HC9 steel. The PAGBs are also perceptible in Figure 3.4(e). However, retained austenite in Figure 3.4(f) (denoted by red arrow) may be associated with the fact that Mn, as an austenite stabilizer and Si, as a graphitizer, will stabilize some amount of RA and also restrict the harmful carbide precipitation [18–22].

Figure 3.5 shows a bar plot displaying the quantities of phases in all the current steels. Here, B, RA, PTM, FM, P and F represent bainite, retained austenite, pretempered/preformed martensite, fresh martensite, pearlite

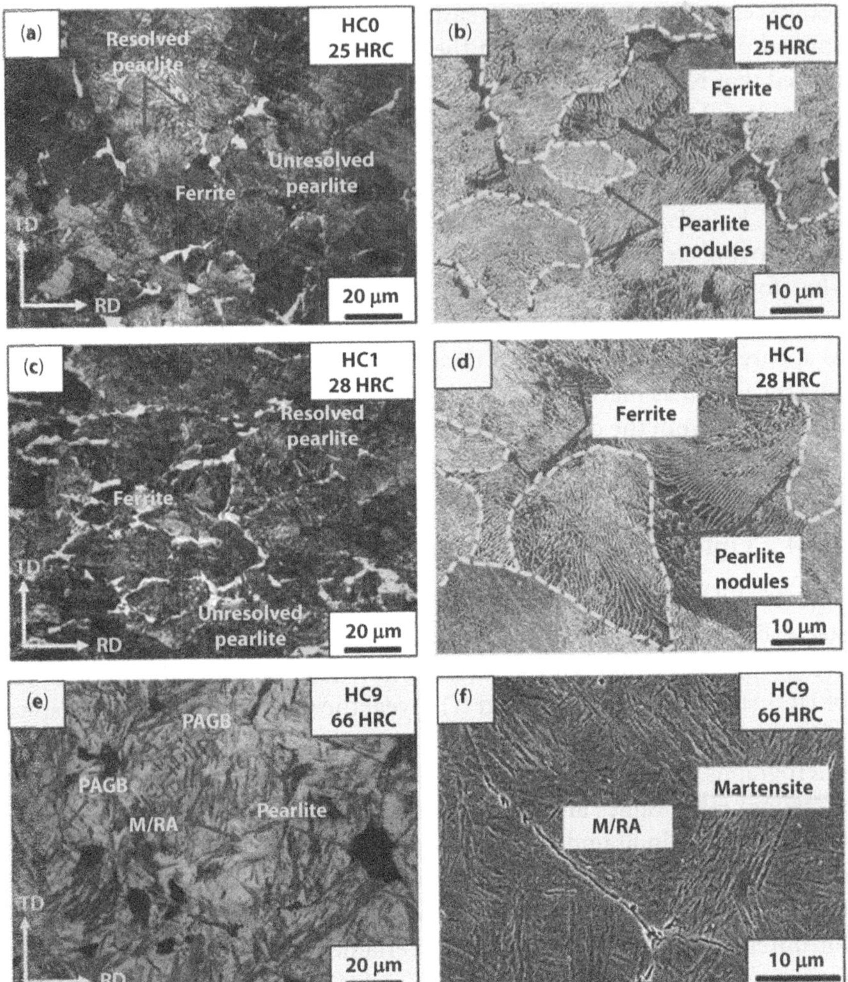

Figure 3.4 OM and SEM micrographs of HRAC samples (a) & (b) HC0, (c) & (d) HC1, and (e) & (f) HC9 steels.

and ferrite, respectively. It discloses the higher amount of F (9 vol.%) for HC1 steel, due to the addition of Nb, a F stabilizer [23, 24]. The quantities of F and P for HC0, HC1 and HC9 steels were calculated by image analysis and Image J software, whereas the other constituents of HC9 steel were estimated from available empirical equations [27].

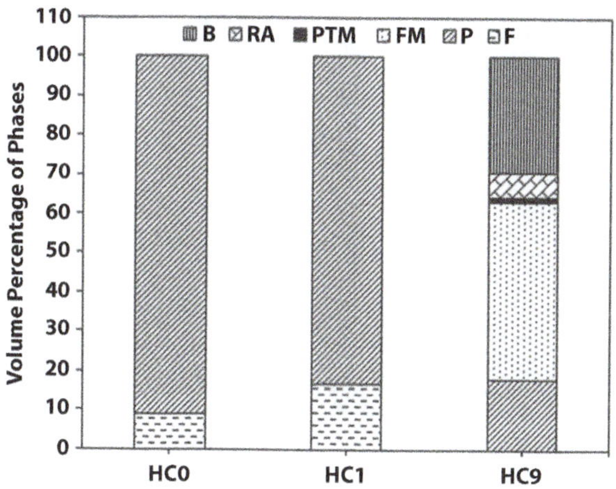

Figure 3.5 Plot showing the quantities of the constituting phases of the investigated steels.

3.3.5 TEM Microstructures

Figure 3.6 shows the TEM micrographs of (a) & (b) HC0-HRAC, (c) & (d) HC1-HRAC and (e) & (f) HC9-HRAC steels. The corresponding EDS spectra signifying the presence of ferrite-cementite phase as well as NbC precipitates for HC0, HC1 and HC9 steels are shown in the insets of Figures 3.6(b), (d) and (f), respectively. Table 3.2 summarizes AGS, NS, PS, λ, CT, width of MA islands and C content in RA (C_γ) of the investigated steels. The ferrite matrix and cementite peaks are visible in Figures 3.6(a) and (c). It is also evident from Figure 3.6(a) vis-à-vis Figure 3.6(c) that Nb addition in HC1 steel leads to a refinement of pearlite interlamellar spacing (λ) in the order of about 80 nm than HC0 steel. A region comprising degenerated/pseudo pearlite is also shown with the yellow dashed circle in Figure 3.6(c), which may result from insufficient C diffusion during cooling [26, 28]. It is known that Nb depresses the phase transition and thereby reduces NS, PS, λ and CT values, as evident in Table 3.2. A very fine NbC precipitate of size ≈70-90 nm is shown in Figure 3.6(d). The corresponding EDS spectrum indicates NbC precipitate formation. It is pertinent that Nb influences the C diffusion and thereby contributes to the thinning of Fe_3C, Figure 3.6(c) and Table 3.2. On the other hand, Figure 3.6(e) reveals the existence of lath martensite (LM) and fine interlath retained austenite (RA) (dark regions) type microstructures with lath widths of 0.2-0.9 μm and 25-90 nm, respectively. Figure 3.6(e) reveals a high dislocation density in C-depleted martensite laths, characterized by dislocation tangles (indicated by red arrows). Secondary twin martensite with high carbon

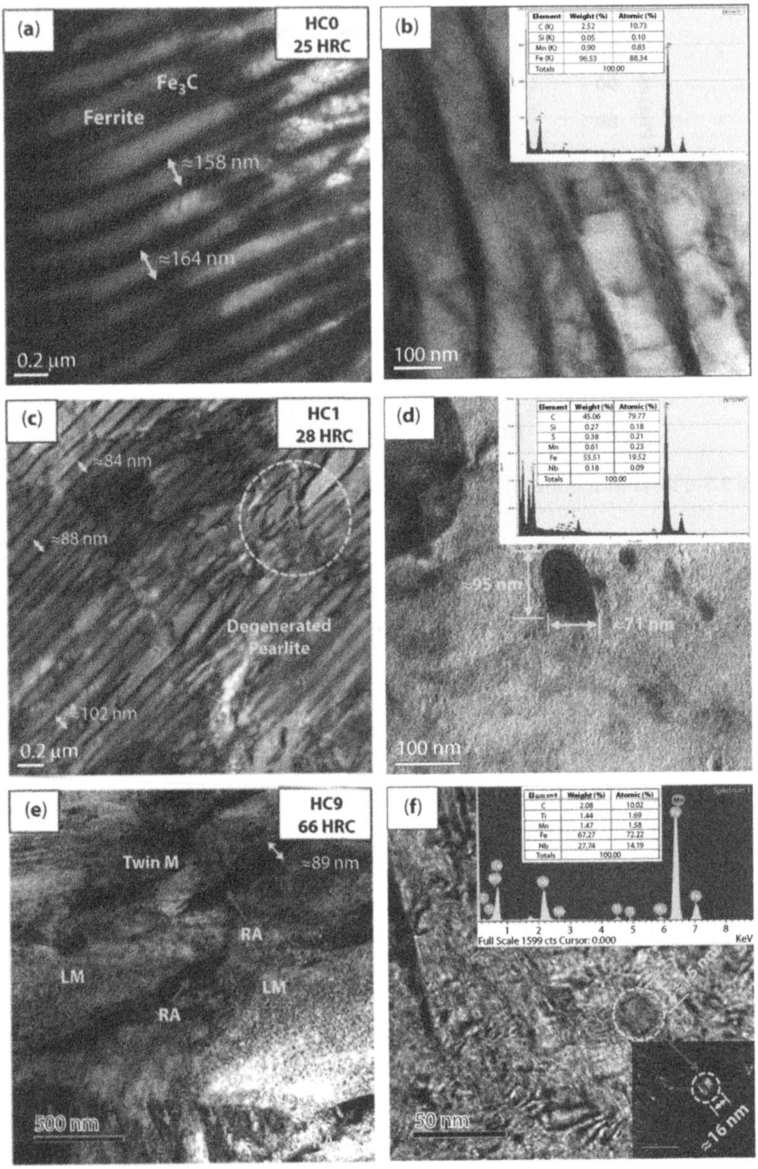

Figure 3.6 TEM micrographs of (a) & (b) HC0-HRAC, (c) & (d) HC1-HRAC and (e) & (f) HC9-HRAC steels.

content is identified (indicated by arrows in Figure 3.6(e)). The twinned M can enhance both hardness and strength [27]. Moreover, NbC precipitate (15-20 nm) is also witnessed in HC9 steel in the HRAC condition, as shown in the inset of the TEM dark field image in Figure 3.6(f). The EDS

Table 3.2 AGS, NS, PS, λ, CT, width of RA laths and C content in RA (C_γ) of the steels.

Sample code	AGS (μm)	NS (μm)	PS (μm)	λ (nm)	CT (nm)	Width of RA laths (nm)	C content in RA (C_γ)
HC0-AC	60±8	48±7	10±6	157±9	54±10	-	-
HC1-AC	47±8	26±7	7±6	92±9	42±10	-	-
HC9-AC	38±8	-	-	-	-	25-90	0.82±0.05

spectrum in Figure 3.6(f) suggests the occurrence of NbC precipitate. In this context, the relationship of the product of solute concentrations ([M].[X]) with temperature (T) from the solubility product equation designates the consistent rise of NbC with the drop of T, which contains the formation of precipitate inside γ or M matrix [27, 29, 30]. Therefore, the occurrence of relatively finer NbC precipitates (15 nm) in HC9 steel than HC1 steel is most preferably associated with its kinetics of formation at lower temperatures on a finer transformation product matrix as a result of effective alloying additions, as obvious in Figure 3.6(d) in relation to Figure 3.6(f).

The lattice parameter of austenite was computed by using Bragg's law (Equation (3.1)) and the planar equation (Equation (3.2)), whereas C content in RA was estimated from Equation (3.3).

$$2d_{hkl}\sin\theta_{hkl} = \lambda \tag{3.1}$$

$$\frac{1}{d_{hkl}^2} = \frac{h^2+k^2+l^2}{a^2} \tag{3.2}$$

$$C_\gamma = \frac{a_\gamma - 3.547}{0.046} \tag{3.3}$$

Where, d_{hkl} is inter-planar distance; θ is Bragg angle; λ is the wavelength of the incident beam; a is the lattice parameter, and hkl are the Miller indices.

3.3.6 Tensile Properties

Figure 3.7 represents the experimental tensile test results in a bar chart. Table 3.3 summarizes the values of yield strength (YS), ultimate tensile strength (UTS), yield ratio (YR), uniform elongation (UEL) and total elongation (TEL) of the investigated steels. It can be observed from Figure 3.7 and Table 3.3 that the YS and UTS values are higher for HC1 steel of around 14 MPa and 59 MPa, respectively, than HC0 steel which can be correlated to Nb microalloying which refines the AGS (13 μm), NS (22 μm), PS (3 μm) and λ (65 nm) by depressing the pearlite transformation and also contributes to the precipitation strengthening through fine NbC precipitates, as evident in Figures 3.3(a), 3.3(b), 3.6(a), 3.6(c) and Table 3.2 and the current findings are consistent with the previous research [31]. Nevertheless, 1710 MPa YS and 1986 MPa UTS, for HC9 steel, as evident in Figure 3.7 and Table 3.3, may be attributed to both C-enriched M as a consequence of direct air cooling to room temperature after hot rolling as well as very fine NbC precipitates (Figure 3.6(f)) which considerably refines the transformation products (like M/B) and contribute to the precipitation strengthening. In this connection, it cannot be overruled that despite such high strengths, the values of UEL and TEL for HC9 steel decrease marginally in the order of 3-4% than both HC0 and HC1 steels which may be associated with the presence of 6% of RA in the final microstructure as evident in Figure 3.6(e). In this connection, it is well documented that with increasing loads, RA yields first during tensile testing and sooner or later,

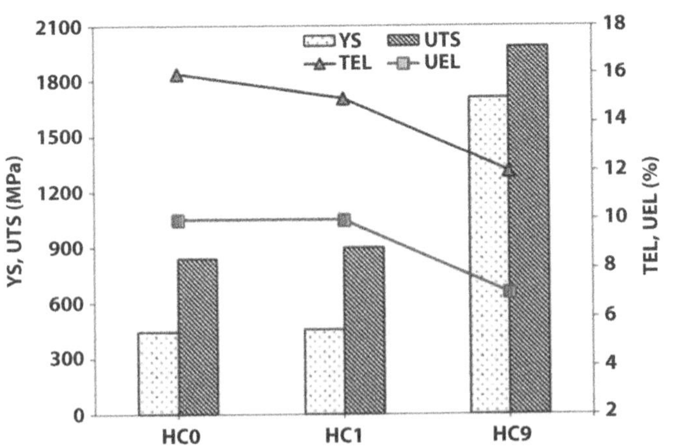

Figure 3.7 Bar chart showing the tensile properties of the investigated steels.

Table 3.3 Summarized tensile test results of the investigated steels.

Sample condition	YS (MPa)	UTS (MPa)	UEL (%)	TEL (%)	UTS×TEL (MPa%)	$YR = \frac{YS}{UTS}$
HC0-HRAC	446	839	10	16	13424	0.53
HC1-HRAC	460	898	10	15	13470	0.51
HC9-HRAC	1710	1986	7	12	23832	0.86

at higher loads, it tends to transfer a portion of the load to the corresponding B or M phases [27, 32].

This study demonstrates that the investigated steels display good balance of tensile properties, as revealed by their microstructure-mechanical property relationships, which is obtained by suitable chemistry design and process parameter selections. Hence, it can be said that these steels fall under the category of third (3rd) generation advanced high strength steels and are suitable for various applications including automotive wires, wires for constructions, railroad industries for both rails stock and wheel materials, tee rails, piano wires, tire cords and tire bids, suspension bridge cables, dies, tools, springs, fasteners etc. The required range of carbon in the chemistry of the wire rods for all these aforementioned applications is 0.60 to 0.85 wt.%, which suits well with the compositions of the three currently investigated high carbon steels (HC0, HC1 and HC9). Furthermore, all these aforesaid applications demand ultimate tensile strength values in the range 920-1290 MPa. Therefore, it can be concluded that the investigated three high-carbon steels are capable of providing such types of high-strength and ductility values and hence are suitable for various applications according to industry norms.

3.4 Conclusions

The significant outcomes of this investigation are summarized below:

1. The prediction made by CCT curves obtained from JMatPro® software and the results obtained from dilatometry are in good agreement in phase transition temperature determination and microstructure predictions for HC0 and HC1 steels. However, the little inconsistency in determining the

martensite start temperature (M_s) and the other phase constituents for HC9 steel may be elucidated by the limitation of the model in JMatPro® software, with certain restrictions in the selection of the alloying element of Si.
2. Hot rolled HC0 and HC1 steels yield microstructure comprising ferrite (F) and pearlite (P), with HC1 steel exhibiting refined microstructural features due to the addition of Nb. On the contrary, for HC9 steel, it leads to the formation of a mixed microstructure comprising martensite, grain boundary pearlite with some amount of retained austenite.
3. HC9 steel exhibits the optimal balance of UTS and TEL at 23,832 MPa%, surpassing HC1 (13,470 MPa%) and HC0 (13,424 MPa%) steels after hot rolling and air cooling, likely due to the higher fraction of martensite (M), retained austenite (RA), pearlite (P), and fine NbC precipitates. However, HC0 steel exhibits the lowest hardness (25 HRC), yield strength (446 MPa), and ultimate tensile strength (839 MPa) due to its microstructure dominated by softer phases, specifically ferrite (F) and pearlite (P). The improvement of hardness (3 HRC), YS (14 MPa) and UTS (59 MPa) values in HC1 steel compared with HC0 steel is linked to the decreased size of the AGS, NS, PS, λ and CT values due to lower phase transformation temperatures for Nb addition and precipitation strengthening by fine NbC precipitates.
4. The currently investigated three high-carbon hot-rolled and air-cooled steels offer a notable balance of tensile properties. The specific properties and characteristics of these steels can be customized for diverse industrial applications.

AI disclosure statement: The author confirms that no artificial intelligence (AI)-generated content or AI-assisted writing tools were used in the preparation of this chapter (including summaries and the abstract). All text, analysis, and conclusions are the result of the author's original work.

References

1. Mohrbacher, H., High-performance steels for sustainable manufacturing of vehicles, in: *Green and sustainable manufacturing of advanced material*, M. Singh, T. Ohji, R. Asthana (Eds.), pp. 135–163, Elsevier Pub., Netherlands, 2016.

2. Hosseini, N., Forouzan, F., Vuorinen, E., *In-situ* microstructural evolution during quenching and partitioning of a high-carbon steel by high-temperature X-ray diffraction. *Mater. Today Commun.*, 31, 103503, 2022.
3. Hossain, R., Pahlevani, F., Sahajwalla, V., Effect of small addition of Cr on stability of retained austenite in high carbon steel. *Mater. Charact.*, 125, 114, 2017.
4. Tomita, Y., Development of mechanical properties of structural high-carbon low-alloy steels through modified heat treatment. *J. Mater. Sci.*, 24, 1357, 1989.
5. Babasafari, Z., Pan, A.V., Pahlevani, F., Kong, C., Sahajwalla, V., du Toit, M., Dippenaar, R., Effect of silicon and partitioning temperature on the microstructure and mechanical properties of high-carbon steel in a quenching and partitioning heat treatment. *J. Mater. Sci.*, 56, 15423, 2021.
6. Rodriguez-Páez, J., A microstructure based constitutive model for pearlite, Doctoral dissertation, pp. 1–233, Universidad de Navarra, Pamplona, Spain, 2023.
7. Fu, L.M., Wang, H.R., Wang, W., Shan, A.D., Austenite grain growth prediction coupling with drag and pinning effects in low carbon Nb microalloyed steels. *Mater. Sci. Technol.*, 27, 6, 996, 2011.
8. Karmakar, A., Kundu, S., Roy, S., Neogy, S., Srivastava, D., Chakrabarti, D., Effect of microalloying elements on austenite grain growth in Nb-Ti and Nb-V steels. *Mater. Sci. Technol.*, 30, 6, 653, 2014.
9. Eskandari, M., Bonab, M.A.M., Szpunar, J.A., Evolution of the microstructure and texture of X70 pipeline steel during cold-rolling and annealing treatments. *Mater. Des.*, 90, 618, 2016.
10. Nishioka, K. and Ichikawa, K., Progress in thermomechanical control of steel plates and their commercialization. *Sci. Technol. Adv. Mater.*, 13, 23001, 2012.
11. Dutta, B. and Sellars, C.M., Effect of composition and process variables on Nb(C, N) precipitation in niobium microalloyed austenite. *Mater. Sci. Technol.*, 3, 197, 1987.
12. Javaheri, V., Khodaie, N., Kaijalainen, A., Porter, D., Effect of niobium and phase transformation temperature on the microstructure and texture of a novel 0.40% C thermomechanically processed steel. *Mater. Charact.*, 142, 295, 2018.
13. ASTM E8/E8M-16: Standard test methods for tension testing of metallic materials, ASTM International, Conshohocken, PA, USA, 2016.
14. Deng, Z., Cheng, L., Chen, L., Zhu, M., Effect of refractory on nonmetallic inclusions in Si-Mn killed steel. *Steel Res. Int.*, 90, 1900268, 2019.
15. Liu, Y., Zhang, L., Duan, H., Zhang, Y., Luo, Y., Conejo, A.N., Extraction, thermodynamic analysis, and precipitation mechanism of MnS-TiN complex inclusions in low-sulfur steels. *Metall. Mater. Trans. A*, 47A, 3015, 2016.

16. IS 4163:2004: Steel - Determination of content of nonmetallic inclusions-micrographic method using standard diagrams (Third Revision), Bureau of Indian Standards, New Delhi, India, 2004.
17. Faraji, M., Wilcox, D.P., Thackray, R., Howe, A.A., Todd, I., Tsakiropoulos, P., Quantitative characterization of inclusions in continuously cast high-carbon steel. *Metall. Mater. Trans. B*, 46, 2490, 2015.
18. Banerjee, K. and Venugopalan, T., Development of hypoeutectoid graphitic steel for wires. *Mater. Sci. Technol.*, 24, 10, 1174, 2008.
19. Speer, J.G. and Bhadeshia, H.K.D.H., Microstructure evolution in irons and steels: a tribute to David V. Edmonds. *Int. Heat Treat. Surf. Eng.*, 4, 2, 62, 2010.
20. Vuorinen, E.P., Ozugurler, A.G., Ion, J.C., Eriksson, K., Somani, M.C., Karjalainen, L.P., Allain, S., Caballero, F.G., Hot forming of ultra-fine-grained multiphase steel products using press hardening combined with quenching and partitioning process. *Metals*, 9, 357, 2019.
21. Klueh, R.L., Maziasz, P.J., Lee, E.H., Manganese as an austenite stabilizer in Fe-Cr-Mn-C steels. *Mater. Sci. Eng., A*, 102, 115, 1988.
22. Moor, E.D., Matlock, D.K., Speer, J.G., Merwin, M.J., Austenite stabilization through manganese enrichment. *Scr. Mater.*, 64, 185, 2011.
23. Wang, H.H., Wang, J., Tong, Z., Hodgson, P.D., Wan, X.L., Wu, K.M., Wei, R., Timokhina, I., Characterization of Nb interface segregation during welding thermal cycle in microalloyed steel by atom probe tomography. *Metall. Mater. Trans. A*, 49A, 6224, 2018.
24. Kuzucu, V., Aksoy, M., Korkut, M.H., The effect of strong carbide-forming elements such as Mo, Ti, V and Nb on the microstructure of ferritic stainless steel. *J. Mater. Process. Technol.*, 82, 165, 1998.
25. Erisir, E., Ayhan, I.I., Guney, C., Alan, E., Durger, N.B., Un, S., Microstructure and phase transformations in high-strength bainitic forging steel. *J. Mater. Eng. Perform.*, 30, 3458, 2021.
26. Dey, I., Ghosh, S.K., Saha, R., Effects of cooling rate and strain rate on phase transformation, microstructure and mechanical behaviour of thermomechanically processed pearlitic steel. *J. Mater. Res. Technol.*, 8, 3, 2685, 2019.
27. Dey, I., Saha, R., Mahato, B., Ghosh, M., Ghosh, S.K., Effect of quenching and partitioning on microstructure and mechanical properties of high-carbon Nb microalloyed steel. *Metall. Mater. Trans. A*, 55, 8, 2736, 2024.
28. Dey, I., Chandra, S., Saha, R., Ghosh, S.K., Effect of Nb micro-alloying on microstructure and properties of thermo-mechanically processed high carbon pearlitic steel. *Mater. Charact.*, 140, 45, 2018.
29. Gladman, T., *The Physical Metallurgy of Microalloyed Steels*, ISBN 0910716812, pp. 81–135, Institute of Materials, London, UK, 1997.
30. Wang, X.D., Xu, W.Z., Guo, Z.H., Wang, L., Rong, Y.H., Carbide characterization in a Nb-microalloyed advanced ultrahigh strength steel after quenching–partitioning–tempering process. *Mater. Sci. Eng., A*, 527, 3373, 2010.

31. Dey, I., Saha, R., Ghosh, S.K., Influence of microalloying and isothermal treatment on microstructure and mechanical properties of high carbon steel. *Met. Mater. Int.*, 28, 1662, 2022.
32. Gao, G., Zhang, H., Tan, Z., Liu, W., Bai, B., A carbide-free bainite/martensite/austenite triplex steel with enhanced mechanical properties treated by a novel quenching–partitioning–tempering process. *Mater. Sci. Eng., A*, 559, 165, 2013.

4

High-Performance Steels for Sustainable Manufacturing of Vehicles

Ankita Bhattacharya[1], Ghanshyam Das[1] and Swarup Kumar Ghosh[2*]

[1]Department of Materials and Metallurgical Engineering, National Institute of Advanced Manufacturing Technology (NIAMT), Ranchi, Jharkhand, India
[2]Department of Metallurgy and Materials Engineering, Indian Institute of Engineering Science and Technology, Shibpur, Howrah, West Bengal, India

Abstract

The growing demand for lightweight, safe, and environmentally sustainable vehicles has accelerated the adoption of high-performance steels in automotive manufacturing. This chapter explores the critical role of advanced high-strength steels, high-strength low-alloy steels, ultra-high-strength steels, and galvanized steels in achieving sustainable mobility. Key properties such as high strength, excellent formability, crashworthiness, and corrosion resistance enable the design of lighter yet safer vehicle structures, directly improving fuel efficiency and reducing CO_2 emissions. Applications of high-performance steels span body-in-white structures, safety components, chassis systems, and electric vehicle battery enclosures. Despite challenges such as hydrogen embrittlement and formability limitations, advances in alloy development, surface engineering, and smart manufacturing are driving future growth. The chapter also highlights the role of recyclable, low-carbon steels and green manufacturing practices in enhancing sustainability. Ultimately, high-performance steels remain indispensable for enabling lighter, safer, and more eco-friendly vehicles, underpinning the automotive industry's path toward carbon neutrality and sustainable innovation.

Keywords: High-performance steel, manufacturing, sustainability, vehicle, lightweight, safety

**Corresponding author*: skghosh@metal.iiests.ac.in

Gurudas Mandal, Swarup Kumar Ghosh, Snehanshu Pal and Sandip Kunar (eds.) Sustainable Manufacturing of Advanced Materials, (65–92) © 2026 Scrivener Publishing LLC

4.1 Introduction

4.1.1 Importance of Sustainability in Vehicle Manufacturing

Sustainability is now a core focus in vehicle manufacturing, driven by environmental, regulatory, and economic factors [1–3]. High-performance steels are vital, enabling lower emissions, regulatory compliance, and cost-efficiency. Manufacturers reduce carbon footprints by using lightweight materials and energy-efficient processes, and support a circular economy through recycling. These efforts also cut vehicle weight, improve fuel efficiency, and lower costs, enhancing both competitiveness and brand value while future-proofing against evolving market demands.

4.1.2 Requirements in Car Body Design

Designing a car's Body-in-White (BIW) requires balancing strength, safety, durability, efficiency, and aesthetics. Structurally, the BIW must offer high static and dynamic stiffness for stable handling and comfort, achieved mainly through optimizng geometry rather than material changes [4]. For crash performance, critical areas need steels with high yield and tensile strength, as well as excellent work-hardening ability to absorb impact and protect passengers. Fatigue resistance is also important since components like the chassis must withstand repeated stresses, making high-strength steels ideal. Additionally, materials must be easy to form, weld, and assemble while delivering a high-quality finish and supporting the vehicle's design appeal [4, 5].

4.1.3 Role of High-Performance Steel in Sustainable Vehicle Manufacturing

Traditionally, low-carbon steels were favored in vehicle manufacturing for their low cost and easy formability. However, growing demands for crash safety, fuel efficiency, and lower emissions have driven the shift toward high-performance steels. The Ultralight Steel Auto Body (ULSAB) project demonstrated that lighter, stronger, and cost-effective structures were achievable, paving the way for the widespread use of Advanced and Ultra High-Strength Steels (AHSS/UHSS) [4, 6]. The reasons for the widespread adoption of high-performance steels are given below:

(i) Safety: Better energy absorption and deformation resistance protect passengers in crashes [7].

(ii) Weight reduction: Higher strength enables thinner, lighter parts, improving fuel economy [7].
(iii) Formability: Modern AHSS offers strength with flexibility for complex designs [8].
(iv) Crash performance: Superior behavior under high strain rates enhances impact resistance [9].

4.2 Properties and Characteristics of High-Performance Steels

High-performance steels used in automotive applications broadly include:

- Advanced High Strength Steels (AHSS)
- High-Strength Low Alloy (HSLA) Steels
- Ultra High-Strength Steels (UHSS)
- Galvanized High Strength Steels

Each category addresses specific performance, cost, and manufacturing needs, and they collectively support the industry's goals of improving safety, reducing emissions, and enhancing sustainability.

4.2.1 Advanced High-Strength Steels

In today's automotive industry, sustainability, safety, and cost-efficiency drive material choices [9, 10]. While lightweight materials like Al, Mg, and composites offer high strength-to-weight ratios, their high cost, manufacturing complexity, and recycling challenges hinder large-scale use. Conventional steels are affordable and easy to process, but lack the strength-ductility balance needed for modern vehicles [8]. To bridge this gap, the steel industry developed Advanced High-Strength Steels (AHSS), which provide high strength, crash resistance, and weight savings, all while remaining compatible with existing manufacturing methods [11]. AHSS also offers excellent recyclability and lower lifecycle emissions, making it a key material in the pursuit of sustainable, high-performance vehicle design.

The evolution of Advanced High-Strength Steels (AHSS) spans three generations, each addressing the limitations of the last [11]. First-generation AHSS, such as Dual-Phase (DP), Complex Phase (CP), and TRIP steel, offer a major leap in strength over conventional steels, using martensitic or bainitic phases. However, their limited ductility kept the tensile strength-elongation (TS × El) product below 25,000 MPa%, restricting formability [4, 11–13]. Second-generation AHSS aimed to overcome this by using fully austenitic structures, such as TWIP steels, with high manganese content (15–30%). These achieved exceptional strength and formability (TS × El > 50,000 MPa%), but they were costly, hard to weld, and prone to delayed cracking, limiting their widespread use [11, 14]. Third-generation AHSS strikes a balance, combining good strength and ductility with lower alloying costs. Innovations like Quenching and Partitioning (Q&P) and Medium Manganese steels target TS × El values above 25,000 MPa%, while remaining compatible with mass production [11, 14, 15].

4.2.2 Characteristics of Some AHSS Steels

4.2.2.1 Dual Phase Steel

Dual-phase (DP) steels are widely used in the automotive industry due to their superior strength–formability balance, ease of production, and weldability [11, 16]. They offer high specific strength, excellent ductility, continuous yielding, and better work hardening than conventional HSLA steels [17]. DP steels are key materials among first-generation AHSS, designed with a soft ferritic matrix and dispersed hard martensitic islands [12, 18]. Their typical composition (in wt.%) includes 0.05–0.20% C, 1.2–1.6% Mn, and 0.3–0.6% silicon, with microalloying elements like Ti, V, and Nb for grain refinement and strength enhancement [19, 20].

Figure 4.1 represents the schematic diagram showing the commonly used heat treatment schedule of DP steel. Produced *via* full (Route A) or intercritical annealing (Route B) followed by rapid quenching, DP steels develop 10–40% martensite within a ferritic matrix [11]. However, Route B shown in Figure 4.1 results in lower tensile strength and higher ductility of the products as compared to Route A [11]. A typical microstructure of DP steel is shown in Figure 4.2. Figure 4.2 (a) shows the optical micrograph of DP steel comprising bright ferrite grain and dark martensite island, and high-resolution SEM micrograph (Figure 4.2(b)) highlights martensite's characteristic island structure. This structure yields high tensile strength (400–1000 MPa) and good ductility (10–30%) [21, 22]. The soft-phase ferrite allows easy initial deformation, while hard-phase martensite provides resistance, enabling high work-hardening rates and smooth yielding without a defined yield point. The various strength–ductility combinations in DP microstructures can be tailored by adjusting phase fractions, phase distribution, carbon content in the hard phase, and grain size [23, 24].

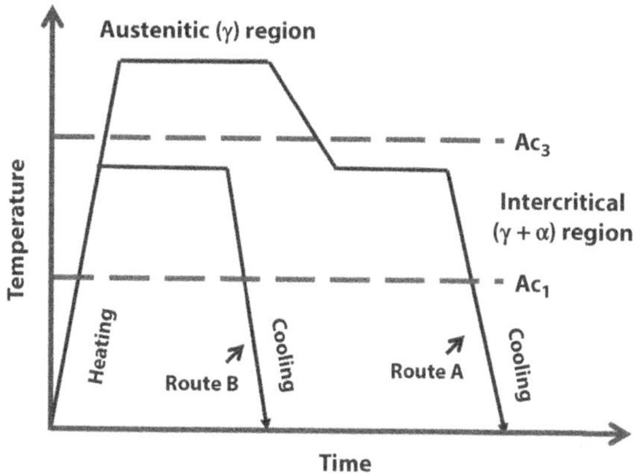

Figure 4.1 Schematic diagram showing the heat treatment schedule of DP steel.

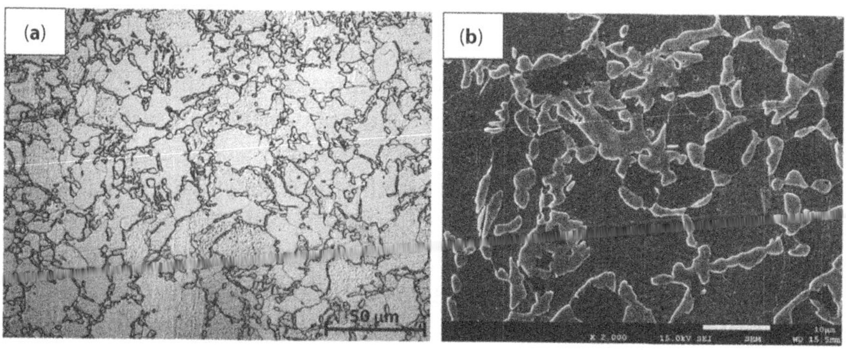

Figure 4.2 (a) Optical micrograph of dual-phase steel showing bright ferrite grain and dark martensite island, and (b) SEM micrograph highlighting martensite's characteristic island structure.

DP steels offer excellent weldability and strength-to-weight ratios, though higher martensite levels can reduce ductility and hole expansion [11, 12]. They are commonly used in automotive structural parts, closures, and crash management systems [11, 25].

4.2.2.2 Transformation-Induced Plasticity Steel

Transformation-induced plasticity (TRIP) steels are ideal for lightweight car body construction, offering excellent strength, ductility, and formability that enhance both performance and safety [8]. TRIP steels enhance strength and ductility through the transformation of retained austenite into martensite during deformation, a mechanism known as the TRIP effect [12, 26]. They typically contain 0.20–0.25% C, ~1.5% Mn, Si or Al to suppress carbide formation, and sometimes P to aid hardenability and carbon partitioning [15, 27].

Their microstructure consists of a ferritic matrix with bainite, martensite, and 7–15% metastable retained austenite (RA) [12, 28]. This retained austenite transforms under strain, enhancing work-hardening and delaying necking for improved uniform elongation. The variation in retained austenite content in TRIP steel takes place with enhancement in tensile strain. The TRIP effect is influenced by the volume and stability of RA, its C content, grain size, and alloying elements [11, 29]. The higher C increases RA stability, while smaller grains reduce it due to early phase transformation. The larger grains favor twinning, delaying transformation and enhancing the TRIP effect.

Manufacturing involves full austenitization, inter-critical annealing, deformation, and controlled cooling to stabilize austenite, making processing complex [11, 30]. The conventional processing route of TRIP steels is shown in Figure 4.3. Deformation in the inter-critical temperature range accelerates the austenite (γ) to ferrite (α) transformation, enriching the remaining γ phase with C and stabilising it. This also increases bainite nucleation while slowing its growth, producing smaller bainite plates that further enrich the γ phase. TRIP steels achieve tensile strengths over 1200 MPa with ~14% uniform elongation. Due to such high uniform elongation and work hardening, TRIP steel retains significant deformation capacity after forming [10]. This stored mechanical energy enhances its ability to absorb impact, making it especially effective in crash situations compared

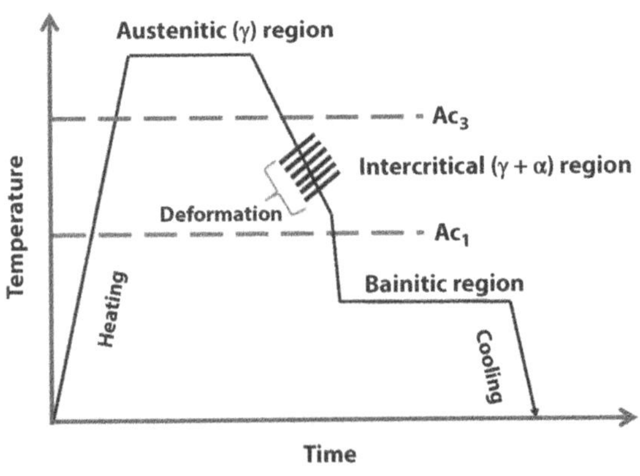

Figure 4.3 Schematic illustration of a thermomechanical processing route of TRIP steel.

to other AHSS grades [9]. TRIP steels are used in vehicles for their excellent strength and ductility, making them ideal for lightweight, safe car body components like B-pillars, sills, and crash boxes. However, high C content reduces weldability and increases processing time, limiting widespread use [11].

4.2.2.3 *Twinning-Induced Plasticity Steel*

To enable weight reduction without sacrificing safety and allow room-temperature stamping, steelmakers in the 1990s developed fully austenitic twinning-induced plasticity (TWIP) steels [7]. TWIP steels are high-Mn (15–25%) austenitic steels designed to remain fully austenitic at room temperature [11]. This structure promotes mechanical twinning during deformation, which greatly enhances strain hardening by impeding dislocation movement [10, 31]. Alloying elements typically lower the stacking fault energy (SFE), promoting twinning and improving ductility. Small additions of Si and Al optimize SFE, ensuring effective twinning without phase transformation [32]. Figure 4.4(a) shows a mixed ferrite and austenite microstructure in dual-phase TWIP steel. The tint-etched micrograph in Figure 4.4(b) reveals the phase distribution, i.e., light pink is austenite and the yellow is ferrite [33]. The annealing twins are visible in both micrographs. With tensile strengths over 1000 MPa and elongations up to 50%, TWIP steels allow the cold forming of complex shapes [34]. However, their high alloy content leads to elevated costs, welding and processing difficulties, delayed fracture risks, and recycling challenges [11, 14, 35].

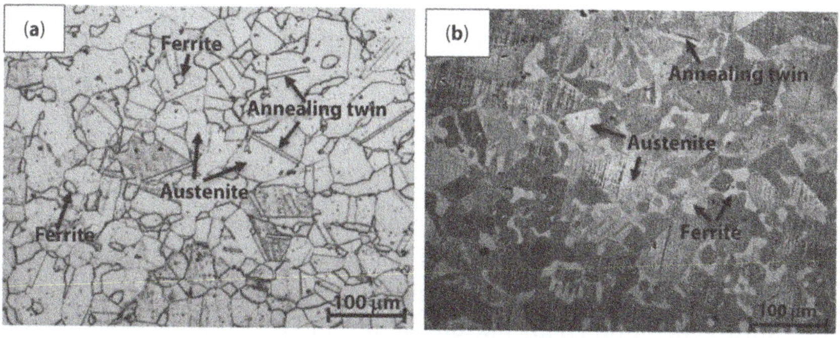

Figure 4.4 (a) Optical micrograph of dual-phase TWIP steel showing the ferrite-austenite structure and (b) tint-etched optical micrograph highlighting ferrite along the grain boundary of the austenite matrix.

Despite this, their superior crash energy absorption makes them ideal for safety-critical components like door beams and side impact bars.

4.2.2.4 *Complex Phase Steel*

Recent advances in phase transformation kinetics have driven the development of multiphase steels, where the combination of phases like ferrite, martensite, bainite, and retained austenite provides application-specific properties [11, 36]. Such steels are known as complex phase (CP) steels within the AHSS family. CP steels achieve high strength by combining hard phases like bainite and martensite with ferrite while limiting retained austenite. Microalloying with niobium, titanium, and vanadium enables precipitation strengthening and grain refinement [11, 36]. Processed through austenitization and controlled cooling or isothermal holding, CP steels form a fine-grained, multi-phase structure. Bainite is the dominant phase and plays a key role in determining its properties. Factors like austenitizing temperature, cooling rate, isothermal holding, and microalloying elements strongly affect its formation and the resulting steel microstructure [11, 37]. CP steels have emerged as a key solution for achieving high strength levels above 800 MPa, offering a unique combination of mechanical properties [11, 38]. Compared to DP and conventional high-strength steels, CP steels provide higher yield strength along with improved bendability and hole expansion capacity at similar strength levels. Although they don't match the formability of TRIP steels, CP steels offer a better overall balance of strength, formability, weldability, and bake hardenability, often at a lower alloying cost. Their excellent energy absorption makes them valuable in crash-critical applications [39]. As a result, CP steels are increasingly used in vital automotive components such as B-pillars, bumper systems, anti-intrusion beams, side members, and suspension parts, helping drive the development of safer, lighter vehicles [11].

4.2.2.5 *Quenching and Partitioning Steel*

TRIP steels are limited by their low RA content due to restricted carbon levels, resulting in only moderate strength and limited strain hardening. While alloying elements like carbon, manganese, and nickel can improve austenite stability, they raise costs and harm weldability. To overcome this, third-generation AHSS like quenching and partitioning (Q&P) steels use tailored thermal treatments to stabilize RA by altering its composition and lowering its martensite start (M_s) temperature, rather than relying on chemical composition alone [10, 40]. Q&P steels are advanced

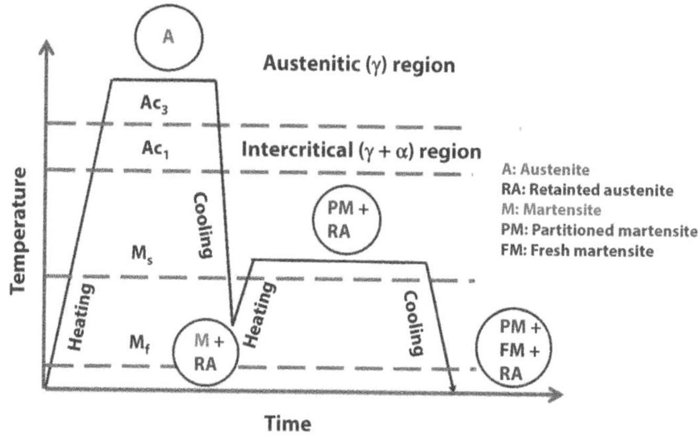

Figure 4.5 Schematic illustration of the Q&P schedule and evolution of the multiphase microstructure.

third-generation AHSS developed to stabilize RA efficiently without relying on high alloy content. Their typical composition includes 0.15–0.4% C, 1.5–2.5% Mn, and additions of Si or Al to prevent carbide formation during processing [10, 11]. The Q&P process involves a controlled partial quench to a temperature between the martensite start (M_s) and finish (M_f) points, followed by a partitioning step at that temperature or slightly higher [10, 41]. During this step, C diffuses from the freshly formed martensite into the remaining austenite, stabilizing it at room temperature. A simplified schematic diagram showing the Q&P heat treatment schedule is represented in Figure 4.5.

The final microstructure consists of tempered martensite, retained austenite, and occasionally bainite, delivering a strong balance of high tensile strength (1000–1500 MPa) and moderate ductility (~20% elongation) [11]. This combination provides excellent formability and superior crash energy absorption, making Q&P steels well-suited for structural reinforcements in automotive applications. However, the process demands precise thermal control, and weldability remains a challenge, requiring further development for widespread industrial use [7].

4.2.2.6 Medium Mn Steel

Medium Mn steels are an emerging class of third-generation AHSS, developed to balance high performance with cost efficiency [42]. Medium Mn steels are favored for their excellent formability, lower cost, and process-friendly composition. They avoid casting issues seen in high-Al steels as well as the complex processing demands of Q&P steels, making them more efficient for industrial use [43]. They contain intermediate Mn levels (3–12%) and low C content (0.05–0.3%), along with minor additions of Si or Al to control phase transformation and prevent carbide formation [44, 45]. These steels leverage both the TRIP and TWIP effects by stabilizing a significant amount of ultrafine-grained retained austenite within a refined ferrite or martensite matrix.

Their processing typically involves intercritical annealing (IA) within a temperature range where both ferrite and austenite coexist. Medium Mn steels feature a distinct multi-phase microstructure of ultra-fine ferrite and/or martensite along with a significant amount of (up to ~ 50%) retained austenite, formed through the reverse transformation of martensite [42, 46]. RA fraction and total elongation are found to first increase, then decrease with an increase in IA temperature and time. The high IA leads to the formation of unstable coarse austenite grains, while the low IA forms stable fine austenite grains [47]. Both extremes reduce toughness; therefore, the annealing temperature and time need to be optimized to achieve the desired volume fraction and stability of RA, which directly influences mechanical behavior.

Medium Mn steels demonstrate exceptional mechanical properties, combining tensile strengths up to 1400 MPa with elongations as high as 50% [10], owing to the presence of abundant RA in the microstructure. Strain hardening and UTS increase with higher annealing temperatures, while uniform elongation peaks at intermediate temperatures. Stable austenite slows transformation, while unstable austenite leads to a ferrite–austenite–martensite microstructure, resembling DP steel. Compared to TWIP steels, they offer similar strength-ductility performance at a lower cost due to reduced Mn content. They also show better compatibility with hot forming, allowing for precise dimensional control and reduced springback. Despite these advantages, medium Mn steels require strict control over processing conditions, particularly during annealing, and may present weldability concerns at higher alloy levels. Nevertheless, their superior energy absorption, lightweight potential, and cost-effectiveness make them well-suited for automotive structural components demanding both strength and formability.

4.2.3 Benefits and Limitations of Advanced High-Strength Steels

AHSS has significantly impacted the automotive industry, offering clear benefits across all vehicle types [11, 48]. Small cars gain better crash protection, while larger vehicles become more fuel-efficient. A key advantage of AHSS is that it can be processed using existing methods for mild or conventional high-strength steels, but in thinner gauges, reducing weight without sacrificing strength. Techniques like roll forming allow for the efficient production of complex shapes with minimal tool wear. Compared to Al, AHSS delivers the same strength at a lower cost and thickness, and it excels in forming intricate profiles. In structural applications, Al must be up to three times thicker than AHSS to achieve equivalent strength.

Still, AHSS presents some forming challenges [49]. While not introducing new issues, it intensifies existing ones, such as higher press forces, increased tool wear, and greater springback. Its complex microstructure complicates material testing and consistency, often requiring new methods and closer quality control. The forming operations can lead to early fractures, not always predicted by conventional metrics. Additionally, increased friction and heat during forming raise tool maintenance costs, and tool coatings, lubricants, and process parameters must be carefully selected. Despite these difficulties, AHSS remains a preferred choice due to its performance, cost-efficiency, and manufacturability.

4.3 High Strength Low Alloy Steel

High Strength Low Alloy (HSLA) steels gained prominence in the automotive industry during the oil crisis, when reducing vehicle weight became crucial for improving fuel efficiency [50]. Unlike traditional carbon steels, which added bulk to meet strength needs, HSLA steels offered higher strength at lower thicknesses, enabling lighter, more fuel-efficient vehicles [51]. Today, their relevance extends beyond weight reduction. HSLA steels support advanced safety systems, offer excellent formability and weldability, enhance crash performance, and provide superior corrosion resistance, making them vital to modern, sustainable vehicle manufacturing.

4.3.1 Characteristics of HSLA Steel

HSLA steels are specifically designed by carefully controlling their chemical composition, typically with C content ranging from 0.05% to 0.25% [52]. This low C level ensures the steels maintain good ductility and weldability, essential characteristics for automotive manufacturing [53]. Their strength (YS up to 1000 MPa) is not derived from C alone but from the addition of microalloying elements like Nb, V, and Ti. These elements promote the formation of fine carbide and nitride precipitates, which strengthen the steel by obstructing dislocation movement within the microstructure.

The microstructure of HSLA steels is mainly ferritic, with a fine-grained structure that contributes significantly to their mechanical properties. Hot-rolled HSLA microalloyed steels are produced using controlled thermo-mechanical rolling and, sometimes, cold rolling with heat treatment [54]. Microalloying elements, due to their varying solubilities in austenite, play key roles during thermomechanical processing (TMP). TMP represents the three stages of the controlled rolling process in the recrystallization region, non-recrystallization region and two-phase ($\gamma+\alpha$) region, dealing with the corresponding microstructural changes due to deformation [55]. TiN controls grain coarsening, while NbC and Nb(C, N) control the non-recrystallization (T_{NR}) and transformation temperatures. VC and VN enhance precipitation strengthening during the austenite to ferrite transformation. T_{NR} is crucial for controlling microstructure by ensuring uniformity in the transfer bar and enabling significant deformation to create nucleation sites for ferrite through intercrystalline defects. This process refines the austenite grain structure, transforming it into fine-grained ferrite with possible bainite formation depending on cooling rates.

Cold-rolled HSLA steels are derived from hot-rolled products and further processed at room temperature, often followed by annealing. This results in improved surface finish, dimensional accuracy, and a more uniform microstructure [56]. These steels combine high yield strengths (500–600 MPa), tensile strengths up to 730 MPa, and elongations over 20%, with a typical hardness around 200 HV. They also offer excellent weldability and resistance to brittle fracture, making them ideal for structural automotive components such as chassis parts, side rails, suspension systems, wheels, and brake discs. Their strength-to-weight advantage supports lighter, safer, and more fuel-efficient vehicle designs. While they typically exhibit slightly lower strength than hot-rolled steels, cold-rolled HSLA steels offer superior formability, essential for intricate stamping operations in body-in-white applications [57]. Their smoother surfaces make them suitable for aesthetic panels and interior components. Applications include structural reinforcements like center pillars and inner body locks, where strength, ductility, and appearance must be balanced. Their enhanced formability and impact resistance make them particularly valuable in safety-critical zones.

The exceptional mechanical properties of HSLA steels arise from a combination of strengthening mechanisms working synergistically [57, 58]. Through extensive grain refinement, grain boundary strengthening is the most significant, improving yield strength while preserving ductility.

Precipitation strengthening, enabled by fine carbides and nitrides of niobium, vanadium, and titanium, further restricts dislocation motion. Solid solution strengthening adds moderate resistance by distorting the ferritic lattice with minor alloying elements, and dislocation strengthening increases during thermomechanical processing as deformation raises dislocation density. These mechanisms collectively deliver the strength, ductility, and toughness essential for vehicle structures facing impact and dynamic loads.

HSLA steels offer key benefits that support sustainable automotive design [57]. Their high strength-to-weight ratio allows for lighter vehicles without sacrificing structural integrity, leading to better fuel economy and lower emissions. They also exhibit excellent weldability and formability, making them suitable for complex, lightweight components in modern vehicle architectures. Some grades provide enhanced corrosion resistance, reducing the need for coatings and extending vehicle life. However, these advantages come with trade-offs. Higher strength can make machining more challenging and increase tool wear, raising production costs. Certain compositions may also require advanced welding methods to avoid defects like hydrogen-induced cracking. Additionally, reduced ductility in some grades may limit formability, requiring careful material selection during design.

4.4 Ultra-High Strength Steel

The automotive industry must balance improved crash safety with better fuel efficiency. Traditional safety measures increased vehicle weight, reducing fuel economy. Ultra-high strength steel (UHSS) solves this by enabling lighter, thinner components without sacrificing strength, enhancing both safety and efficiency. As global crash and emissions standards tighten, UHSS is increasingly used in key areas like pillars, door beams, and bumpers, making it essential for building safer, more sustainable vehicles [59]. UHSS derive their exceptional properties from carefully engineered alloy compositions and microstructures [60, 61]. Based on C–Si–Mn systems, their design balances strength, ductility, and manufacturability. C strengthens the steel but must be optimized to preserve ductility and weldability. Mn enhances hardenability and lowers transformation temperatures, while Si promotes RA by suppressing carbide formation. Microalloying with elements like Nb, Ti, and V refines the grain structure and enhances strength through precipitation hardening. Figure 4.6(a) of the SEM micrograph reveals a minor amount of lower bainite and a higher amount of lath

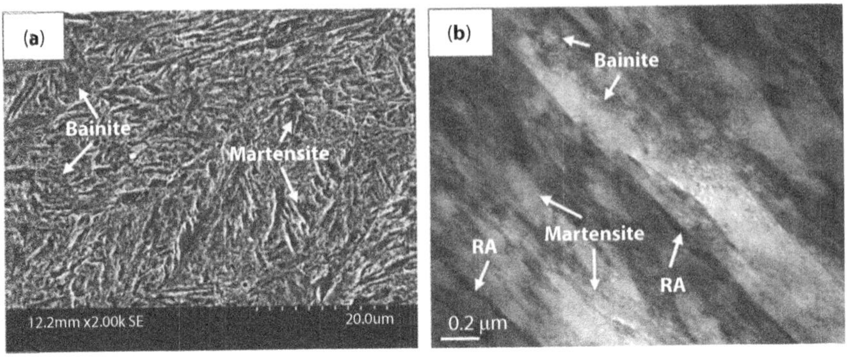

Figure 4.6 (a) SEM micrograph and (b) TEM micrograph of a thermomechanical processed water-quenched microalloyed UHSS steel.

martensite obtained after water quenching. Figure 4.6(b) shows the bright field TEM image of the dislocated lath martensite with an average lath of 400 nm and the dark (arrowed) interlath retained austenite (RA) films. The resulting microstructure, consisting mainly of fine lath martensite, with bainite or retained austenite, is further optimized through processes like Quenching and Partitioning (Q&P), which improve ductility *via* the TRIP effect. This design ensures not only high strength but also superior formability and crash performance. Recent advances in ultrahigh-strength steels (UHSS) include cold-formable sheets with over 980 MPa strength and hot-stamped parts reaching 1,470 MPa, used in vehicle structures [62]. While combining Q&P and hot stamping can boost ductility, the process complexity and thermal control challenges limit industrial use. Hydroforming and roll forming are emerging alternatives, with roll forming favored for its ability to shape high-strength steels without excessive plastic deformation, requiring materials with high yield-to-tensile strength ratios and excellent bendability [63]. The synergy between alloy design and microstructural control gives UHSS its outstanding mechanical performance, with tensile strengths of 980–1700 MPa and elongations up to 16–17% in cold-rolled martensitic grades [64]. While higher strength can reduce formability and weldability, especially in high-C steels, local ductility depends heavily on microstructural uniformity. Optimizing phase distribution is key to balancing strength, global formability, and local deformation. As a result, UHSS enables the design of lighter, stronger, and safer vehicle structures. The properties of UHSS in automotive components depend heavily on the chosen forming method, which must balance performance, complexity, and cost. Cold forming is economical and ideal for high-volume production of simple parts, using low-alloy martensitic steels. However, limited formability restricts its use to less complex shapes. Hot stamping improves formability by shaping UHSS at high temperatures in its austenitic state, followed by in-die quenching to form martensite. It enables complex, high-strength parts with precise geometry but requires tight thermal control, increasing cost and cycle time. Roll forming excels in producing long, straight components like beams, using a series of rollers for gradual shaping. It offers high efficiency and material use but is best suited for simple, prismatic geometries. Cold-formed martensitic steel shows strong potential in components like battery shells and frames, while hot-stamped parts are widely applied in reinforced areas, including A/B-pillars, door rings, sills, bumpers, roof rails, and underbody structures [60, 65].

UHSS offers major benefits in vehicle design, combining a high strength-to-weight ratio for weight reduction and fuel efficiency with excellent crash energy absorption for enhanced safety [60, 66]. Its weldability also supports integration into existing manufacturing systems. However, limited formability in high-strength grades restricts use in complex shapes, often requiring hot stamping. Additionally, hydrogen embrittlement, especially in galvannealed components, poses a risk of delayed brittle failure [60]. Hydrogen embrittlement occurs when atomic hydrogen accumulates at microstructural defects in steel, causing premature failure. In UHSS, this is intensified by internal stresses in the martensitic structure and fine grains. Hydrogen sources include environmental exposure, corrosion, and manufacturing processes like pickling and galvanizing. To prevent embrittlement, strategies include minimizing hydrogen uptake during processing, applying post-forming bake-out treatments, optimizing alloy compositions, and controlling galvanizing conditions. Protective coatings also help reduce susceptibility, ensuring UHSS components maintain durability in automotive applications.

4.5 Galvanised Steel

Galvanization is essential for automotive components due to the harsh conditions on vehicles face, e.g., exposure to moisture, road salts, dirt, and temperature fluctuations, all of which accelerate corrosion [67]. This not only affects aesthetics but also compromises structural integrity, safety, and resale value. With design features like collision sensors limiting traditional protective methods, galvanized steel offers an ideal solution [68]. Its sacrificial zinc coating corrodes first, protecting the underlying steel and enhancing component longevity. Additionally, its durability and recyclability align with the growing demand for sustainability in automotive design. Two main galvanization processes are used in automotive applications [69].

Hot-dip galvanizing involves immersing steel in molten zinc at around 455°C, creating a durable, corrosion-resistant coating through zinc-iron alloy layers. It provides excellent protection, especially in harsh environments, and is cost-effective for high coating weights. However, the process can reduce the strength of ultra-high-strength steels and lead to surface imperfections, such as spangle formation, affecting aesthetics. It also complicates spot welding due to the thicker zinc layer. Electro-galvanizing, on the other hand, applies a thin, uniform zinc coating *via* electroplating at lower temperatures. It preserves the steel's strength, is ideal for ultra-high-strength materials, and offers a smooth surface perfect for painted automotive panels. While electro-galvanizing excels in appearance and formability, it offers lower corrosion resistance compared to hot-dip galvanizing and is less economical for parts requiring heavy-duty protection due to its higher cost and thinner coating. The choice between hot-dip and electro-galvanized steel is influenced by several critical factors, as summarized in Table 4.1.

Thus, for applications requiring high corrosion resistance and where minor strength loss is acceptable, hot-dip galvanizing is preferred. For outer body panels demanding a pristine appearance and retained strength, electro-galvanized steel is the material of choice. In vehicle manufacturing, different types of galvanized steel are chosen based on specific needs [67]:

Table 4.1 Comparison of characteristics between hot-dip and electro-galvanised steels.

Requirement	Hot-dip galvanized	Electro-galvanized
Mechanical strength	Slightly reduced	Fully retained
Coating weight	High and economical	Thin and costly at high weight
Spot welding	Challenging (high current)	Easier but still requires care
Formability	Good with surface treatment	Excellent for deep drawing
Surface quality	Moderate (needs additional smoothing)	High-quality (ideal for painting)

(i) **One-sided galvanized steel (GI-1S):** Coated with zinc on one side, this is used for parts where only the interior is exposed to corrosion, like hoods and doors. The advantages are cost-effective and sufficient protection for internal surfaces. However, it requires additional protection for external surfaces.

(ii) **One-and-a-half side galvanized steel (GI-1½S):** The exterior side is coated with a zinc-iron alloy (galvanneal), and the interior has a zinc coating. Used in parts like doors and fenders, where appearance and corrosion resistance are important. The advantages are excellent paintability and corrosion protection on visible surfaces. However, it is more complex and costly to manufacture.

(iii) **Two-sided galvanized steel (GI-2S):** Zinc-coated on both sides for comprehensive protection, typically used in exposed structural components like floor pans and truck beds.

These steels are full of corrosion-resistant for critical areas. However, the demerits are added weight, which may slightly reduce fuel efficiency.

(iv) **Galvannealed steel (GA):** Produced through hot-dip galvanizing followed by annealing, this steel features a zinc-iron alloy coating on both sides. It's ideal for parts requiring good paint adhesion and moderate formability. The advantages are excellent paint adhesion and improved weldability. The disadvantages are slightly lower ductility and formability compared to GIs.

Galvanized steels offer a balanced mix of mechanical properties ideal for vehicle manufacturing [69]:

- Strength: High tensile strength ensures safety and structural integrity.
- Ductility: Suitable for forming complex shapes, especially in electro-galvanized and galvannealed grades.
- Formability: Enhanced by treatments and lubricants; hot-dip steels form well, while electro-galvanized excels in deep drawing.
- Corrosion Resistance: Excellent, particularly in hot-dip steels with heavier coatings.

- Surface Finish: Critical for painted panels; electro-galvanized and galvannealed steels provide superior results.

Galvanized steels offer several key advantages for vehicle manufacturing [70]. Their exceptional corrosion resistance greatly extends the lifespan of components and vehicles, while their durability ensures minimal maintenance over time. Highly recyclable, they support sustainable production practices. Though initial costs may be higher, their long-term performance makes them cost-effective. Additionally, they adapt well to forming, welding, and painting, enhancing manufacturing flexibility. Despite their many benefits, galvanized steels come with a few limitations [71]. The added weight from zinc coatings can slightly reduce vehicle efficiency, while surface imperfections like spangles or wrinkles may impact visual quality if not properly managed. Spot welding also demands precise process control to ensure strong joints. Additionally, galvanized steels are generally more expensive upfront than uncoated alternatives, and the galvanizing process itself can pose environmental concerns if emissions aren't effectively controlled.

4.6 Applications of High-Performance Steels

The integration of high-performance steels has revolutionized the way vehicles are built, influencing both their structural framework and safety performance. The various parts of a car body are made of high-performance steels [72].

4.6.1 Body and Chassis Structures

At the heart of any vehicle lies its body and chassis, where strength, stiffness, and durability are crucial. High-performance steels have become indispensable in these areas. In the Body-in-White (BIW) phase, AHSS and UHSS ensure the necessary rigidity and crash protection while enabling weight reduction [59, 60]. Components such as roof rails, side sills, B-pillars, and cross-members make extensive use of these steels. Material selection is based on performance needs: DP steels provide a balance of strength and ductility; TRIP and CP steels are used where high energy absorption during crashes is required; and hot-stamped martensitic steels are applied in the most critical safety zones, like the B-pillars [11]. Similarly, the chassis relies on high-performance steels for long-term durability and crash resistance. Hot-rolled HSLA steels are typically chosen for frame rails, suspension arms, and engine cradles due to their strength and cost-effectiveness [57]. Where higher crash resistance and lower weight are required, cold-formed martensitic UHSS steels are utilized, ensuring the vehicle remains both robust and lightweight [59, 60].

4.6.2 Safety-Related Applications

Beyond structural performance, high-performance steels are vital in enhancing occupant protection during accidents. Door beams, usually made from DP steels or martensitic UHSS, play a key role in preventing intrusion during side impacts [59, 60]. In seat structures, martensitic and high-strength DP steels ensure that seats maintain their position during collisions, aiding effective airbag deployment and minimizing occupant movement. Roof bows and pillars, often formed from hot-stamped boron steels, provide critical rollover protection by preserving the occupant's survival space when the vehicle is subjected to severe crushing forces.

4.6.3 Other Applications

The move toward electric vehicles and new safety requirements has expanded the applications of high-performance steels beyond traditional structures. Battery enclosures in electric vehicles use cold formed UHSS and hot-stamped martensitic steels to protect against collisions and underfloor hazards while keeping weight low to maintain battery range [59, 60]. Bumpers and crash boxes, essential for absorbing impact energy in collisions, are made from cold-rolled martensitic UHSS and roll-formed DP/CP steels [21, 25]. Similarly, battery mounting frames and cross members employ Q&P and medium-Mn steels, combining high strength with the flexibility needed to handle dynamic forces and thermal changes.

4.7 Benefits of High-Performance Steels

The widespread adoption of high-performance steels offers several critical advantages that align with evolving automotive priorities [7, 8].

4.7.1 Weight Reduction and Fuel Efficiency

Lightweighting remains a key objective in vehicle design, directly impacting fuel economy, emissions, and driving dynamics [7]. High-performance steels allow thinner, lighter components without sacrificing strength.

Research indicates that reducing a vehicle's weight by 30% can lead to an approximate 30% increase in fuel efficiency [73]. In electric vehicles, lighter structures also contribute to extended battery range, enhancing overall vehicle performance.

4.7.2 Improved Safety and Crashworthiness

Safety remains a fundamental requirement, and high-performance steels greatly enhance crash protection [7]. Components made from UHSS maintain cabin integrity under impact, while steels like DP and TRIP provide controlled deformation to absorb energy progressively. Advanced steels with the TWIP effect further improve crash energy management, making modern vehicles safer than ever before.

4.7.3 Sustainability

The environmental benefits of high-performance steels are equally significant [3]. Steel is one of the most recycled materials globally, with recycling rates exceeding 90% for automotive steel. Recycling not only conserves resources but also uses about 25% of the energy needed for primary steel production, reducing overall CO_2 emissions. Newer steelmaking methods, such as Electric Arc Furnace (EAF) production and hydrogen-based processes, further enhance the sustainability profile of high-performance steels [1, 3].

4.8 Challenges and Opportunities

While the advantages of high-performance steels are numerous, the path to widespread adoption comes with important challenges and emerging opportunities. The use of high-performance steels can involve higher material costs due to the addition of expensive alloying elements [11]. Specialized processing techniques, such as precise thermal treatments and advanced coating processes, add further complexity [10, 11]. In manufacturing, challenges like reduced formability, springback, tool wear, and difficulties in welding UHSS require advanced solutions [59]. Hydrogen embrittlement also poses a risk, particularly in steels exceeding 1200 MPa strength, necessitating careful management during production and service [60]. In addition, material variability and slower updates in industry standards can complicate quality control and part certification, making consistent production more challenging.

Despite these challenges, the potential for innovation remains strong. The development of third-generation AHSS and nano-structured steels promises better performance at lower costs [11, 74]. Forming innovations, such as indirect hot stamping and hybrid forming techniques, are improving manufacturability [59, 75]. At the same time, smart manufacturing technologies, including simulations and real-time quality monitoring, are making it easier to predict and control complex production processes. The rise of sustainable steelmaking, particularly through hydrogen-based reduction and closed-loop recycling, will further support the environmental goals of the automotive industry [1, 3].

4.9 Future Trends and Developments

The future of high-performance steels lies in continuous material innovation and sustainable manufacturing practices. Third-generation AHSS aims to bridge the gap between high strength and high ductility, using advanced treatments like Q&P and medium-Mn steel technologies [10, 11, 44, 45]. These steels are set to become essential for lightweight, crash-resistant vehicle structures. Hot-formable and ultra-thin steel parts are becoming more common, driven by new forming technologies that allow for stronger yet lighter components [59]. Improvements in surface coatings, such as Zn-Al-Mg layers, are enhancing corrosion resistance while minimizing the risk of hydrogen embrittlement [69]. On the manufacturing side, the push toward green steel production through hydrogen-based methods and expanded use of Electric Arc Furnaces is reducing the carbon footprint of steelmaking. Vehicle designs are also evolving to prioritize recyclability, supporting a closed-loop materials economy [3]. Finally, as strength levels continue to rise, addressing challenges like weldability, hydrogen embrittlement, and cost management will remain key [11, 60]. The use of simulations will be crucial in optimizing material selection, forming, and joining processes, ensuring that high-performance steels continue to meet the demands of the next generation of vehicles [75].

4.10 Conclusions

High-performance steels are indispensable in modern vehicle manufacturing, offering an unmatched combination of lightweight capabilities, superior crashworthiness, manufacturing flexibility, and environmental sustainability. Despite challenges like cost and formability complexities, the continuous evolution of AHSS, HSLA, and UHSS technologies holds vast potential for future automotive design. AHSS offers exceptional strength, formability, and crashworthiness, making it ideal for critical vehicle components. HSLA provides a balance of strength, weldability, and formability, making it suitable for a wide range of automotive applications. UHSS exhibits extremely high strength-to-weight ratios, enabling significant weight reduction while maintaining structural integrity. These steels will enable the development of new vehicle designs and structures, driving innovation in the automotive industry. Their role in achieving safe, efficient, and sustainable transportation systems cannot be overstated. As material science and manufacturing techniques advance, high-performance steels will remain at the heart of greener, safer, and more innovative mobility solutions.

AI disclosure statement: The author confirms that no artificial intelligence (AI)-generated content or AI-assisted writing tools were used in the preparation of this chapter (including summaries and the abstract). All text, analysis, and conclusions are the result of the author's original work.

References

1. Broadbent, C., Steel's recyclability: demonstrating the benefits of recycling steel to achieve a circular economy. *Int. J. Life Cycle Assess.*, 21, 1658, 2016.
2. Conejo, A.N., Birat, J.P., Dutta, A., A review of the current environmental challenges of the steel industry and its value chain. *J. Environ. Manage.*, 259, 109782, 2020.
3. Suer, J., Traverso, M., Jäger, N., Review of life cycle assessments for steel and environmental analysis of future steel production scenarios. *Sustainability*, 14, 14131, 2022.
4. Hilditch, T.B., de Souza, T., Hodgson, P.D., 2-Properties and automotive applications of advanced high-strength steels (AHSS), in: *Welding and joining of advanced high strength steels*, pp. 9–28, Woodhead Publishing, Cambridge, UK, 2015.
5. Malen, D.E., *Fundamentals of automobile body structure design*, 2nd edition, SAE International, Warrendale, PA, USA, 2011.
6. van Schaik, M. and Martin, D., ULSAB advanced vehicle concepts-the latest steel demonstration for automotive, SAE Technical Paper, 2000.
7. Bouaziz, O., Zurob, H., Huang, M., Driving force and logic of development of advanced high strength steels for automotive applications. *Steel Res. Int.*, 84, 937, 2013.

8. Kuziak, R., Kawalla, R., Waengler, S., Advanced high strength steels for automotive industry. *Arch. Civ. Mech. Eng.*, 8, 103, 2008.
9. Galán, J., Samek, L., Verleysen, P., Verbeken, K., Houbaert, Y., Advanced high strength steels for automotive industry. *Rev. Metal.*, 48, 2, 118, 2012.
10. Schmitt, J.H. and Iung, T., New developments of advanced high-strength steels for automotive applications. *C.R. Phys.*, 19, 8, 641, 2018.
11. Nanda, T., Singh, V., Singh, V., Chakraborty, A., Sharma, S., Third generation of advanced high-strength steels: Processing routes and properties. *Proc. Inst. Mech. Eng., Part L: J. Mater.: Des. Appl.*, 233, 2, 209–238, 2019.
12. Fonstein, N., *Advanced High Strength Sheet Steels: Physical Metallurgy, Design, Processing, and Properties*, Springer, Cham, 2015.
13. Matlock, D.K., Speer, J.G., De Moor, E., Gibbs, P.J., Recent developments in advanced high strength sheet steels for automotive applications: an overview. *International Iron & Steel Symposium*, pp. 1–12, Karabük, Türkiye, 02-04 April, 2012.
14. Kwon, O., Lee, K.Y., Kim, G.S., Chin, K.G., New trends in advanced high strength steel developments for automotive application. *Mater. Sci. Forum*, 638–642, 136, 2010.
15. Matlock, D.K. and Speer, J.G., Processing opportunities for new advanced high-strength sheet steels. *Mater. Manuf. Process.*, 25, 7, 2010.
16. Li, C., Li, Z., Cen, Y., Ma, B., Huo, G., Microstructure and mechanical properties of dual phase strip steel in the overaging process of continuous annealing. *Mater. Sci. Eng., A*, 627, 281, 2015.
17. Bhattacharya, D., Developments in advanced high strength steels, in: *Proc. Jt. Int. Conf. HSLA Steels*, 2005.
18. Ghosh, S.K. and Yadav, M., Effects of intercritical annealing on microstructure and mechanical properties of low C-Mn steels. *Metallogr. Microstruct. Anal.*, 13, 817, 2024.
19. Adamczyk, J. and Grajcar, A., Effect of heat treatment conditions on the structure and mechanical properties of DP-type steel. *J. Achiev. Mater. Manuf. Eng.*, 17, 305, 2006.
20. Manoj, M.K., Pancholi, V., Nath, S.K., Mechanical properties and fracture behavior of medium carbon dual phase steels. *Int. J. Res. Advent Technol.*, 2, 4, 2443, 2014.
21. Mohrbacher, H., Modern steel grades and advanced steel semi-products for automotive body applications. *Metal*, 24, 1, 2005.
22. Rocha, R.O., Melo, T.M.F., Pereloma, E.V., Santos, D.B., Microstructural evolution at the initial stages of continuous annealing of cold rolled dual-phase steel. *Mater. Sci. Eng., A*, 391, 296, 2005.
23. Hüper, T., Endo, S., Ishikawa, N., Osawa, K., Effect of volume fraction of constituent phases on the stress-strain relationship of dual phase steels. *ISIJ Int.*, 39, 288, 1999.

24. Park, K.-T., Han, S.Y., Ahn, B.D., Shin, D.H., Lee, Y.K., Um, K.K., Ultrafine grained dual phase steel fabricated by equal channel angular pressing and subsequent intercritical annealing. *Scr. Mater.*, 51, 9, 909, 2004.
25. Wang, W. and Wei, X., The effect of martensite volume and distribution on shear fracture propagation of 600-1000 MPa dual phase sheet steels in the process of deep drawing. *Int. J. Mech. Sci.*, 67, 100, 2013.
26. Burgold, A., Kuna, M., Prüger, S., Material forces in consideration of phase transformation in TRIP-steel. *Procedia Mater. Sci.*, 3, 461, 2014.
27. Grajcar, A., Kuziak, R., Zalecki, W., Third generation of AHSS with increased fraction of retained austenite for the automotive industry. *Arch. Civ. Mech. Eng.*, 12, 334, 2012.
28. Ghosh, S.K. and Chattopadhyay, P.P., Phase evolution and mechanical behavior of 0.36 wt% C high strength TRIP-assisted steel. *Steel Res. Int.*, 83, 12, 1163, 2012.
29. Wang, M.-M., Tasan, C.C., Ponge, D., Kostka, A., Raabe, D., Smaller is less stable: Size effects on twinning vs. transformation of reverted austenite in TRIP-maraging steels. *Acta Mater.*, 79, 268, 2014.
30. Luo, H., Dong, H., Huang, M., Effect of intercritical annealing on the Lüders strains of medium Mn transformation-induced plasticity steels. *Mater. Des.*, 83, 42, 2015.
31. Bouaziz, O. and Guelton, N., Modelling of TWIP effect on work-hardening. *Mater. Sci. Eng., A*, 319–321, 246, 2001.
32. Duan, X., Wang, D., Wang, K., Han, F., Twinning behaviour of TWIP steel studied by Taylor factor analysis. *Philos. Mag. Lett.*, 93, 316, 2013.
33. Tewary, N.K., Ghosh, S.K., Chatterjee, S., Ghosh, A., Deformation and annealing behaviour of dual phase TWIP steel from the perspective of residual stress, faults, microstructures and mechanical properties. *Mater. Sci. Eng., A*, 733, 43, 2018.
34. Imandoust, A., Zarei-Hanzaki, A., Heshmati-Manesh, S., Moemeni, S., Changizian, P., Effects of ferrite volume fraction on the tensile deformation characteristics of dual phase twinning induced plasticity steel. *Mater. Des.*, 53, 99, 2014.
35. Kim, S.K., Kim, G., Chin, K.G., Development of high manganese TWIP steel with 980 MPa tensile strength, in: *Proc. Int. Conf. New Dev. Adv. High-Strength Sheet Steels*, p. 249, AIST, Orlando, FL, 2008.
36. Mallick, P.K., *Materials, design and manufacturing for lightweight vehicles*, 2nd edition, Woodhead Publishing, Cambridge, UK, 2021.
37. Karelova, A., Hairer, F., Krempaszky, C., Werner, E., Hebesberger, T., Pichler, A., Influence of the overaging temperature on microstructure and mechanical properties of complex-phase bainitic steel. *J. Mater. Sci. Technol.*, 2, 129, 2009.
38. Bhattacharya, D., Microalloyed steels for the automotive industry. *Tecnol. Metal. Mater. Min.*, 11, 4, 371, 2014.

39. Singh, N.K., Cadoni, E., Singha, M.K., Gupta, N.K., Quasi-static and dynamic tensile behavior of CP800 steel. *Mech. Adv. Mater. Struct.*, 21, 7, 531, 2014.
40. Speer, J.G., Edmonds, D.V., Rizzo, F.C., Matlock, D.K., Partitioning of carbon from supersaturated plates of ferrite, with application to steel processing and fundamentals of the bainite transformation. *Curr. Opin. Solid State Mater. Sci.*, 8, 219, 2004.
41. Dey, I., Saha, R., Mahato, B., Ghosh, M., Ghosh, S.K., Effect of quenching and partitioning on microstructure and mechanical properties of high-carbon Nb microalloyed steel. *Metall. Mater. Trans. A*, 55A, 2736, 2024.
42. Suh, D.-W. and Kim, S.-J., Medium Mn transformation-induced plasticity steels: Recent progress and challenges. *Scr. Mater.*, 126, 63, 2017.
43. Speer, J., Matlock, D.K., De Cooman, B.C., Schroth, J.G., Carbon partitioning into austenite after martensite transformation. *Acta Mater.*, 51, 9, 2611, 2003.
44. Bhattacharya, A., Biswal, S., Barik, R.K., Mahato, B., Ghosh, M., Mitra, R., Chakrabarti, D., Comparative interplay of C and Mn on austenite stabilization and low temperature impact toughness of low C medium Mn steels. *Mater. Charact.*, 208, 113658, 2024.
45. Bhattacharya, A., Bokinala, P.K., Mitra, R., Chakrabarti, D., Relative effect of C and Mn on strength-toughness of medium Mn steels. *Mater. Sci. Technol.*, 35, 55, 2019.
46. Benzing, J.T., Da Silva, A.K., Morsdorf, L., Bentley, J., Ponge, D., Dutta, A., Han, J., McBride, J.R., Van Leer, B., Gault, B., Raabe, D., Wittig, J.E., Multi-scale characterization of austenite reversion and martensite recovery in a cold-rolled medium-Mn steel. *Acta Mater.*, 166, 512, 2019.
47. Cai, Z.H., Ding, H., Misra, R.D.K., Ying, Z.Y., Austenite stability and deformation behavior in a cold-rolled transformation-induced plasticity steel with medium manganese content. *Acta Mater.*, 84, 229, 2015.
48. Baluch, N., Udin, Z.M., Abdullah, C.S., Advanced High Strength Steel in Auto Industry: an Overview. *Eng. Technol. Appl. Sci. Res.*, 4, 4, 686, 2014.
49. Dykeman, J., Hoydick, D., Link, T., Mitsuji, H., Material property and formability characterization of various types of high strength dual phase steel, SAE Technical Paper 2009-01-0974, 2009.
50. Dinda, S., JA, D., AS, K., Using microalloyed steels to reduce weight of automotive parts, 1975.
51. Ferry, B.N. and Paliwoda, E.J., Formability and high strength low alloy steels. *Can. Metall. Q.*, 13, 2, 415, 1974.
52. Totten, G.E., *Steel heat treatment: Metallurgy and technologies*, CRC Press, Taylor & Francis Group, Boca Raton, FL, USA, 2007.
53. Avner, S.H., *Introduction to Physical Metallurgy*, 2nd edition, McGraw-Hill Book Company, Singapore, 1974.
54. Davis, J.R., *Alloying: understanding the basics*, ASM International, Materials Park, OH, USA, 2001.
55. Tanaka, T., Controlled rolling of steel plate and strip. *Int. Met. Rev.*, 26, 1, 185, 1981.

56. Liu, Q., Zhou, Q., Venezuela, J., Zhang, M., Wang, J., Atrens, A., A review of the influence of hydrogen on the mechanical properties of DP, TRIP, and TWIP advanced high-strength steels for auto construction. *Corros. Rev.*, 34, 3, 127, 2016.
57. Garcia, C.I., 6-High strength low alloyed (HSLA) steels, in: *Automotive Steels: Design, Metallurgy, Processing and Applications*, pp. 145–167, Woodhead Publishing, Cambridge, UK, 2017.
58. Petch, N.J., The cleavage strength of polycrystals. *J. Iron. Steel Inst.*, 174, 25, 1953.
59. Chen, H., Zhao, L., Lu, S., Lin, Z., Wen, T., Chen, Z., Progress and Perspective of Ultra-High-Strength Martensitic Steels for Automobile. *Metals*, 12, 2184, 2022.
60. Bandyopadhyay, P.S., Kundu, S., Ghosh, S.K., Chatterjee, S., Structure and properties of a low-carbon, microalloyed, ultra-high-strength steel. *Metall. Mater. Trans. A*, 42A, 1051, 2011.
61. Kobayashi, J., Nakajima, Y., Sugimoto, K.I., Effects of cooling rate on impact toughness of an ultrahigh-strength TRIP-aided martensitic steel. *Adv. Mater. Res.*, 922, 366, 2014.
62. Ronin, F., Variant-Flexible Steel Forming in Automotive Production. *2nd European Practice Conf.*, Bad Nauheim, Frankfurt, 2002.
63. Lengauer, G., Lightweight components through custom rollforming in the automotive industry. *International Conference SCT 2005*, Wiesbaden, Germany, 5-10 June, 2005.
64. Fujita, N., Matsumura, K.I., Tomokiyo, T., Kusumi, K., Nonaka, T., Present situation and future trend of ultra-high strength steel sheets for auto-body, Nippon Steel Technical Report, pp. 99–103, 2013.
65. Lu, H., Zhao, Y., Feng, Y., Ma, M., Side, A., Liu, Y., Guo, A., Progress and prospect for development and application of microalloying press-hardening steel. *Mater. Mech. Eng.*, 44, 1, 2020.
66. Senuma, T., Hot stamping steel, in: *Encyclopedia of Materials: Metals and Alloys*, pp. 26–36, Elsevier, Oxford, UK, 2022.
67. Bologna, D.J., Corrosion resistant materials and body paint systems for automotive applications, SAE Technical Paper 862015, 1986.
68. Porter, F.C. and Walden, G.K., Paper 13: The Use of Zinc Coated Steels to Prevent Body Corrosion. *Proc. Inst. Mech. Eng. Conf. Proc.*, 182, 10, 51–60, 1967.
69. Dallin, G.W. and Director, G.C., Galvanizing-2012, Contin. Hot-Dip Galvaniz. Prod. Galvinfo Center-A Progr. Int. Zinc Assoc. URL (available http//www.Galvinfo.Com/GalvInfo_2012_Course_Docs/Galvanizing%20 20120831.Pdf), 2012.
70. Yoichi, T. and Kazuhiro, A., Hot-dip galvanized steel sheet with excellent surface quality for automotive outer panels. *JFE Tech. Rep.*, 4, 55–60, 2004.

71. Koumura, H., Kato, C., Mochizuki, K., Morito, N., Thermodynamic and morphological investigations of dross in molten Zn bath. *Tetsu-to-Hagane*, 81, 8, 815, 1995.
72. Karbasian, H. and Tekkaya, A.E., A review on hot stamping. *J. Mater. Process. Technol.*, 210, 15, 2103, 2010.
73. Borcherts, R.H., Stadler, H.L., Brehob, W.M., Auiler, J.E., Improvements in automotive fuel economy. *Energy*, 3, 4, 439, 1978.
74. Branagan, D.J., Parsons, C.S., Machrowicz, T.V., Frerichs, A.E., Meacham, B.E., Cheng, S., Sergueeva, A.V., 3rd Generation AHSS; Mechanistic responses enabling cold deformation during stamping. *Mater. Sci. Forum*, 941, 542, 2018.
75. Raabe, D., Sun, B., Kwiatkowski Da Silva, A., Gault, B., Yen, H.W., Sedighiani, K., Thoudden Sukumar, P., Souza Filho, I.R., Katnagallu, S., Jägle, E., Kürnsteiner, P., Kusampudi, N., Stephenson, L., Herbig, M., Liebscher, C.H., Springer, H., Zaefferer, S., Shah, V., Wong, S.L., Baron, C., Diehl, M., Roters, F., Ponge, D., Current challenges and opportunities in microstructure-related properties of advanced high-strength steels. *Metall. Mater. Trans. A*, 51, 5517, 2020.

5

Sustainable Manufacturing of Ferritic Stainless Steel: Experimental and Review

Swarnendu Das[1], Nishant Kumar[2], Piyush Kumar[3], R. Rejeesh[1], A. Ghosh[3] and S. Patra[1]*

[1]*Metallurgical Engg., Indian Institute of Technology (BHU), Varanasi, Uttar Pradesh, India*
[2]*R&D Center, Jindal Stainless Limited, Jajpur, Odisha, India*
[3]*Metallurgical Engg. & Materials Science, Indian Institute of Technology, Simrol, Indore, Madhya Pradesh, India*

Abstract

The current study investigates a sustainable way of manufacturing the ferritic stainless steel. The ferritic stainless steels mainly suffer from the ridging and drawability during manufacturing. To obtain the high deep drawability and ridging resistance, the ferritic stainless steel undergoes a number of processes including close control of alloying, continuous casting, hot rolling, annealing, cold rolling and final annealing process. The process control to get better ridging resistance and drawability is very critical in the casting, hot rolling, cold rolling and annealing part. In this study, four different process routes were explored to identify the best route suited for ferritic stainless steel of 409L variety. The ultimate aim was to identify if any process can be removed so that energy can be saved and sustainability can be achieved. The study suggests that the ferritic stainless steel can be processed without heavy cold rolling in Z mill after hot rolling. The EMS can eliminate the columnar grain and can eliminate the need of offline heavy cold rolling.

***Keywords*:** Ferritic stainless steel, cold rolling, annealing, EMS

**Corresponding author*: sudipta.met@iitbhu.ac.in

Gurudas Mandal, Swarup Kumar Ghosh, Snehanshu Pal and Sandip Kunar (eds.) Sustainable Manufacturing of Advanced Materials, (93–110) © 2026 Scrivener Publishing LLC

5.1 Introduction

Ferritic stainless steel (FSS) is widely utilized in numerous applications, such as cooking utensils, kitchenware, elevators, automobiles, and white goods. This results from its exceptional corrosion resistance, elevated heat conductivity, and improved surface polish. FSS serves as a cost-efficient substitute for the more expensive, nickel-containing austenitic stainless steels [1–4]. However, a major drawback is the emergence of ridging after cold forming processes, which significantly affects its surface appearance and demands intensive polishing.

Ridging refers to the unwanted formation of surface undulations that arise when the material undergoes stretching or deep drawing. These undulations create alternating peaks and valleys on opposite surfaces without altering the sheet's overall thickness, typically with ridge depths ranging between 20 and 50μm [1, 2, 5–17]. While earlier studies linked this to the segregation of alloying elements such as carbon, chromium, and molybdenum, it is now widely accepted that ridging stems from anisotropic plastic deformation, which results from alternating bands of varied crystallographic textures [1, 2, 5–7, 13–17].

Numerous studies have highlighted the impact of crystallographic texture on ridging behavior. It is understood that ridging occurs due to non-uniform shear strain distribution (specifically the γ23 component) brought about by the banded arrangement of texture components in conjunction with rolling direction (RD) [1, 2, 13, 15–17]. These texture bands display varying resistance to thinning (i.e., thickness strain ε_{ZZ}) when subjected to uniaxial tensile testing along the RD [1, 13–16].

Takechi and colleagues earlier described ridging on the basis of the formation of ferrite grain bands oriented in the ⟨011⟩ direction parallel to the RD, which causes fluctuations in the γ-23 shear strain during tensile deformation along the RD. Electron backscatter diffraction (EBSD) analysis on the rolling plane further revealed that stronger intensity of similar orientations results in more pronounced ridging [18–20].

To fully understand the ridging phenomenon in FSS, one must also consider the influence of ferrite grain size alongside crystallographic texture [5, 21, 22]. Studies have shown that cold-rolled (CR) and annealed (A) sheets of low-carbon ferritic and ferritic stainless steels can exhibit a range of grain sizes, each associated with distinct crystallographic textures [9, 14, 21–28]. When these grains of different sizes cluster or form bands, the material can undergo uneven yielding during tensile stress, consistent with the Hall–Petch relationship. Additionally, variations in grain size can lead to localized differences in the Taylor factor, contributing further to non-uniform plastic deformation and promoting ridging.

Therefore, to effectively prevent ridging during the industrial processing of FSS, it is essential to gain a comprehensive understanding of its origins. This requires a combined evaluation of both crystallographic texture and variations in grain size [29–31].

5.2 Remedial for Ridging

Employing a low finish rolling temperature (FRT) in HR facilitates the development of shear bands inside coarse grains, which leads to the formation of the recrystallization of fine ferrite grains exhibiting a favorable ⟨111⟩|| ND orientation during hot band annealing. Intermediate annealing in the process of cold rolling [1, 2, 10, 32, 33]. Cross-rolling at a 45° angle relative to the normal direction (ND) enhances the randomization of texture components [21, 34]. Two stage cold-rolling with intermediate annealing processes to eliminate cube fiber solidification texture and attain a homogeneous grain structure with advantageous γ-fibre texture [8, 19, 35–40].

Preventing the development of a columnar structure of grains during solidification [5–10, 13–16], by utilizing substantial roughing deformation with extended inter-pass durations attain full recrystallization of grains [1].

This study examined how differential speed rolling affects both the ridging resistance and the evolution of crystallographic texture in ultra-purified 17% Cr ferritic stainless steel, with the goal of identifying the most effective method for enhancing surface quality. Three cold rolling techniques were employed: standard (conventional) cold rolling and two asymmetric rolling processes with varying speed ratios.

In the conventional cold rolling process, noticeable ridging was observed, primarily due to the formation of grain clusters with similar crystallographic orientations arranged in band-like structures. In contrast, applying differential speed rolling led to a reduction in ridging depth, with the effect becoming more significant at higher speed ratios. The findings indicate that severe ridging is closely associated with the presence of banded clusters of γ-fiber grains, specifically those oriented with ⟨111⟩ parallel to the normal direction (ND). Differential speed rolling introduces shear bands that facilitate recrystallization within both grain interiors and at grain boundaries. These shear bands disrupt the continuity of similarly oriented grain clusters, promoting a more randomized texture and thereby enhancing the material's surface quality.

5.3 Experimental Details

The present investigation used industrial grade 409L ferritic stainless steel. All the as-cast slabs were produced from the identical temperatures, utilizing the same melt composition and casting conditions. Casting route involved the use of electromagnetic stirrer (EMS). The chemical composition of the investigated steel is given in Table 5.1.

Schematic diagram in Figure 5.1 shows the process route applied in all three studied samples. The samples were taken in different stages and name of the samples are written under bracket in bold font.

HAPL-Z-CAPL: First one is the cast material is hot rolled to 5.3mm thickness (sample name **HR-5.3mm**) with the finish rolling temperature of 850°C, then 14% online cold rolled and annealed at 990°C (**4.5mm HR-annealed**), and after that cold rolled 75% in an offline Z mill and finally annealed at 925°C (**HAPL-Z-CAPL**). Total cold reduction is 77% considering intermediate online intermediate cold rolling and final offline cold rolling.
Bell-Z-CAPL: The second one is that the cast material was hot rolled at 5.3mm thickness (**HR-5.3mm**) with the finish rolling temperature of 850°C, and then bell annealed at 800 °C for 3hours (**5.3mm bell annealed**), cold rolled 75% in an offline Z mill and finally annealed at 890°C in a continuous annealing linc (**Bell-Z-CAPL**). Total cold reduction is 77% in a single step.
HAPL-CAPL: The third one is hot rolled to 3.3mm thickness with the finish rolling temperature of 800°C (**HR-3.3mm**), and then cold rolled 21% in online annealing line and with annealing at 990°C (**2.6mm HR+annealed**), and after that it was cold rolled again 54% and finally annealed at 925°C (**HAPL+CAPL**). Total cold reduction here is 60% considering intermediate cold rolling and final cold rolling.

The samples were taken in each step of processing to understand comprehensively the microstructure, texture at different stages. The samples were cut in the cross section of the sample and metallographically polished according to the standard methods. Then optical microstructures were taken using Leica microscope. The electropolishing of the samples

Table 5.1 Chemical composition of the investigated steel.

%C	%Mn	%S	%P	%Si	%Cr	%Ti	N(ppm)
0.0075	0.42	0.001	0.035	0.44	11.28	0.23	80

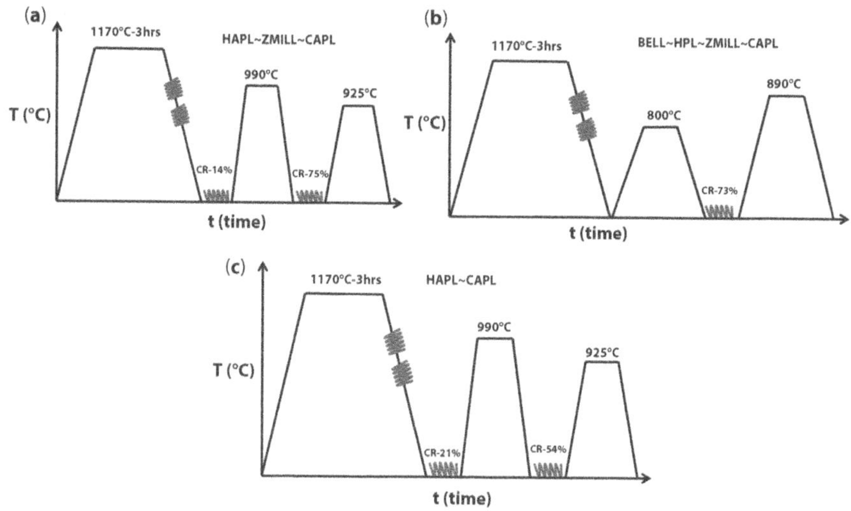

Figure 5.1 The process route applied to the investigated three samples.

were carried out using 9:1 ratio of methanol and perchloric acid. EBSD studies were carried out in the samples of each processing stage such as hot rolled, annealed and cold rolled annaealed sample in a Zeiss scanning electron microscope with EBSD attachment. TSL OIM software was used for analyzing the data obtained from EBSD. A minimum misorientation of 2° was established for the EBSD (Electron Backscatter Diffraction) analysis. Boundaries exhibiting misorientations between 2° and 15° were classified as low-angle grain boundaries (LAGB), also referred to as 'sub-boundaries', while those exceeding 15° were designated as high-angle grain boundaries (HAGB) or 'grain boundaries. The spatial distribution of crystallographic orientations throughout the scanned region was displayed using orientation distribution functions (ODFs) in the $\Phi 2 = 45°$ section of Euler space. The $\Phi 2 = 45°$ section prominently reveals the key texture components and fibers commonly formed in BCC (body-centered cubic) materials, such as ferritic steel [27].

Flat tensile specimens were fabricated in accordance with ASTM E8 guidelines and tested at room temperature (25°C) using a 10-ton Instron® 8862 servo-electric testing machine. The strain rate applied was 6.6×10^{-4}s, and a 50 mm gauge length extensometer was employed to obtain precise strain measurements.

To evaluate normal anisotropy ($\bar{R}$-value), samples were prepared at 0°, 45°, and 90° angles relative to the rolling direction (RD) of the sheet, following ASTM E517. Two specimens were tested according to orientation to ensure measurement repeatability. Strips measuring 35 mm in width and 200 mm in length were cut with their length aligned along the RD of cold-rolled and annealed sheets, following acid pickling to remove surface scale. These strips underwent 20% tensile deformation along RD, and the ridging severity was assessed using a Mitutoyo® Surftest SJ-210 portable surface roughness tester. Measurements were conducted with a stylus of 2 μm tip radius, 0.75 mN depth force, and a 5 mm travel length—five measurements were taken per sample for a total of 25 mm.

Standard tensile samples were also extracted from the transverse direction of the cold-rolled and annealed sheets. These were prepared in accordance with ASTM E8 standards and tested at ambient temperature (25°C) using the Instron® 8862 machine, with a crosshead speed of 0.5 mm/min.

5.4 Results

The continuous casting with electromagnetic stirrer can lead to formation of more than 50% equiaxed grain structure as observed in the optical microstructure where the chilling zone appeared in the surface of the slabs with very fine grains and then columnar grains started to form up to 15% of the slab surface and sometime continued little bit more towards the center. The equiaxed zone formed about 12-14% of the length towards the center regions and then a small strip of 10mm thickness of columnar zone appeared, but then, finally the central equiaxed zone with about 45-50% appeared.

5.4.1 Microstructure and Texture of Hot-Rolled and Annealed Structures

The as-cast samples (Figure 5.2) were soaked at 1170°C and finished about 850°C for 5.3mm samples and 800°C for 3.3mm samples. Both the microstructures have been given in Figure 5.3. The optical microstructure shows completely through thickness grain structure in the hot rolled 3.3mm samples (Figures 5.3a, c) and 5.3mm sample (Figures 5.3b, d). The surface and subsurface region shows recrystallized grains with more refinement in the grain size, but the center region shows elongated grain structure with <100> cube texture (Figures 5.3e, 5.3f). The 5.3mm hot rolled sample shows more band width of the uncrystallized columnar grain in the center region of the sample. The surface and sub surface recrystallization were also comparatively more in 3.3mm sample. The presence of more high

Figure 5.2 As cast microstructure of the studied heat after continuous casting.

angle boundaries suggest that the microstructure is more annealed in the 3.3mm sample (Figures 5.3e, 5.3f). The Φ2=45°ODF of both the sample suggests that the 5.3mm sample is having more stronger (18.3 times random) rotated cube fiber texture along with alpha and week gamma fiber. In 3.3mm sample comparatively weaker rotated cube and alpha texture was observed (with 8.5 times of random) and gamma fiber was not observed (Figures 5.3g, 5.3h). Ideal orientations and fibres shown in Figure 5.4.

5.4.2 Hot Rolled and Annealed Microstructure

Figures 5.5a & 5.5d represents the hot rolled and annealed sample shows IPF map and optical microstructure of 4.5mm sample. The microstructure is well recrystallized with coarser grain size about 60-70μm in the sample. Figures 5.5b & 5.5e shows the IPF map and optical microstructure of 2.6mm sample which is much finer as it was rolled at a lower finishing temperature and more reduction at low temperature helps in the grain refinement. The intermediate 21% reduction also helps in the grain refinement to about ~50μm. The bell annealing at 800C again lead to grain coarsening and the addition cold rolling of hot band annealing was missing in the structure. The grain size obtained was ~100 μm. The ODF Φ2 = 45° map (Figure 5.6) shows that the bell annealed sample is much stronger texture 11.98 times

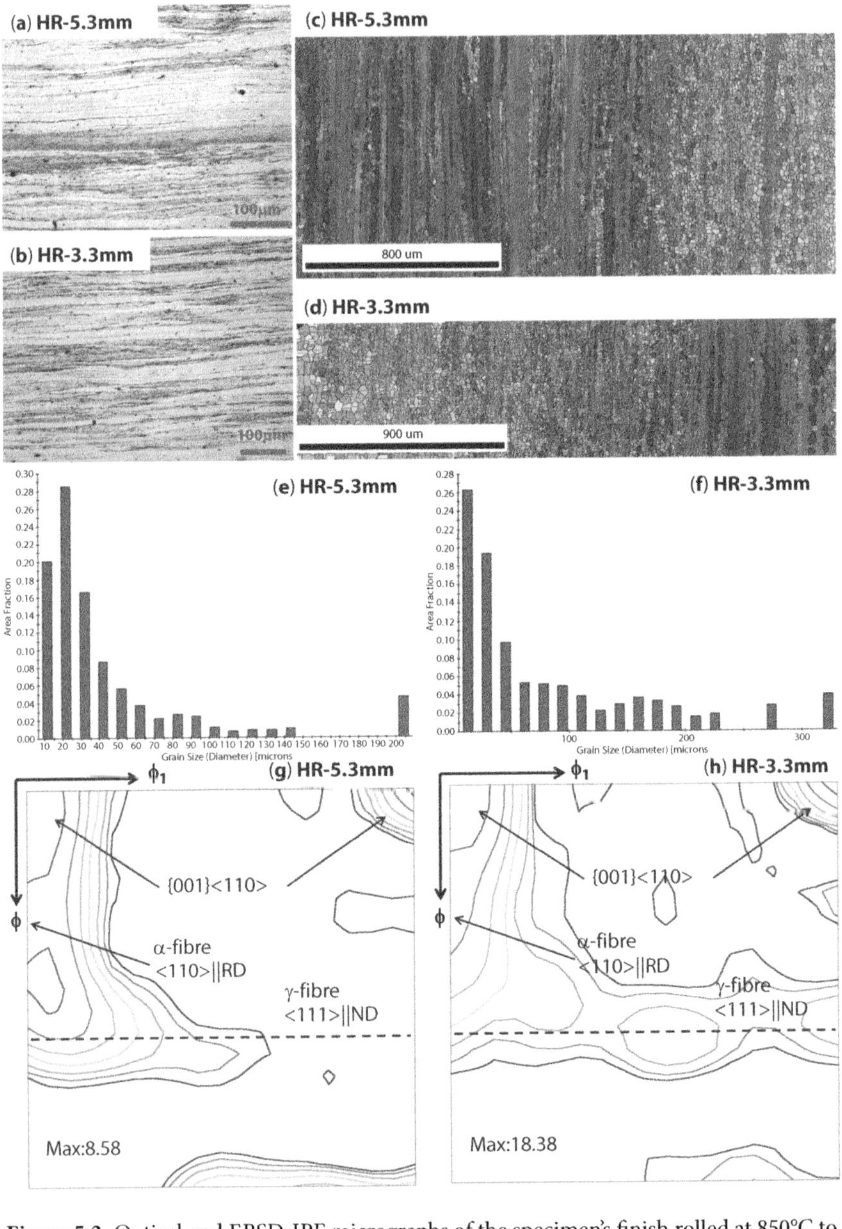

Figure 5.3 Optical and EBSD-IPF micrographs of the specimen's finish rolled at 850°C to (a, c) 5.3mm and (b, d) 800°C for 3.3mm respectively. (e, f) Respective relative grain size distribution and (g, h) ODF Maps.

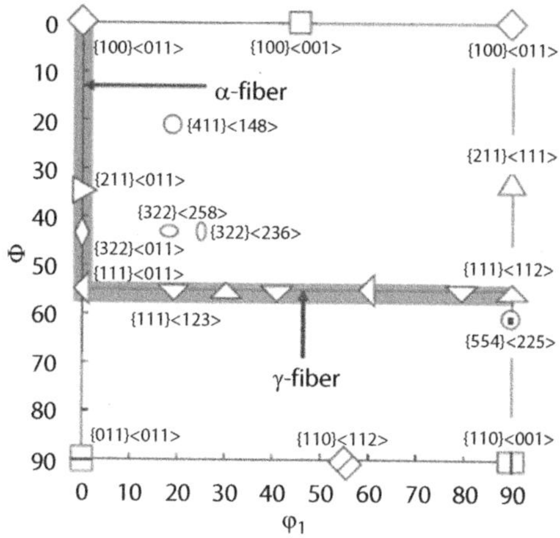

Figure 5.4 Location of ideal orientations and fibers.

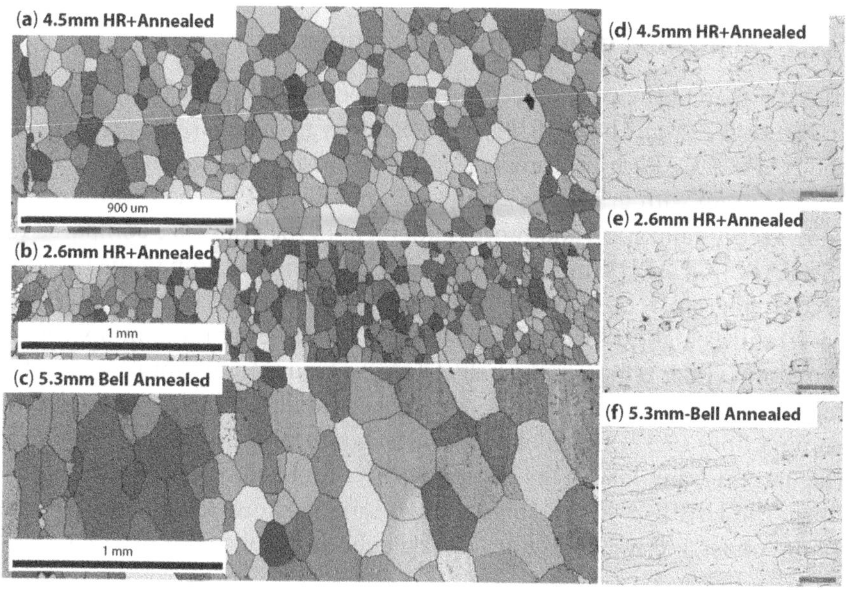

Figure 5.5 Hot rolled and annealed sample of (a, d) 4.5mm, (b, e) 2.6mm and (c, f) 5.3mm bell annealed sample along with optical microstructure.

of random than the other two samples. The bell annealed sample (Figure 5.6b) shows that both the strong alpha fiber and gamma fiber orientation. Whereas 4.5mm hot band annealed sample (Figure 5.6a) shows rotated cube, alpha and [554]<225> orientation. The 2.6mm hot band annealed

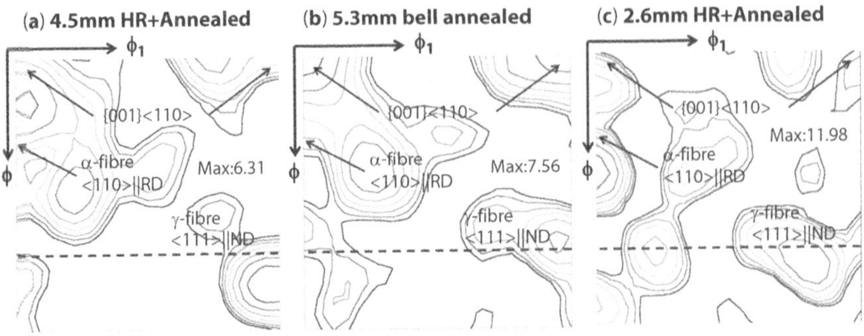

Figure 5.6 ODF maps of (a) 4.5mm HR + annealed, (b) 5.3mm bell annealed, (c) 2.6 mm HR + annealed.

sample (Figure 5.6c) shows that rotated cube, [334]<113> texture along with [554]<225> texture component.

5.4.3 Cold Rolled and Annealed Samples

The IPF map of the cold rolled and annealed samples have been provided in Figures 5.7 (a-c). The ODF $\Phi2 = 45°$ section is given below. The grain size of the HAPL-Z-CAPL route sample (Figures 5.7 (d-f)) shows complete recrystallized grain with grain size about ~45μm. The HAPL-CAPL sample also shows recrystallization and more refinement in the grain size compared to the HAPL-Z-CAPL sample. The Bell-Z-CAPL shows comparatively more recovered grain as observed in Figure 5.7b. The $\Phi2 = 45°$ ODF (Figures 5.7 (d-f)) shows highest textural intensity of 11.39 times of random texture in the gamma fiber region in the bell-Z-CAPL (Figure 5.7e) sample with little bit weaker component in the rotated cube and alpha fiber. The moderate gamma fiber (9.37) times of random) was observed in the HAPL-CAPL (Figure 5.7f) sample with the lower intensity of rotated cube, cube and alpha fiber. The HAPL-Z-CAPL (Figure 5.7e) sample shows lowest intensity of gamma fiber and weakest cube, rotated cube and alpha fiber orientation.

5.4.4 Mechanical and Ridging Behavior of the Samples

The tensile results of the samples have been given in Table 5.2. The ridging and cupping behavior of the sample has been reported in Table 5.2. The Bell-HAPL_Z-CAPL sample shows comparatively higher ridging value (Rz) and the HAPL-Z-CAPL sample shows lowest ridging value and HAPL-CAPL sample was comparatively moderate ridging was observed. The r-bar values are the highest in case of bell-HAPL-Z-CAPL sample

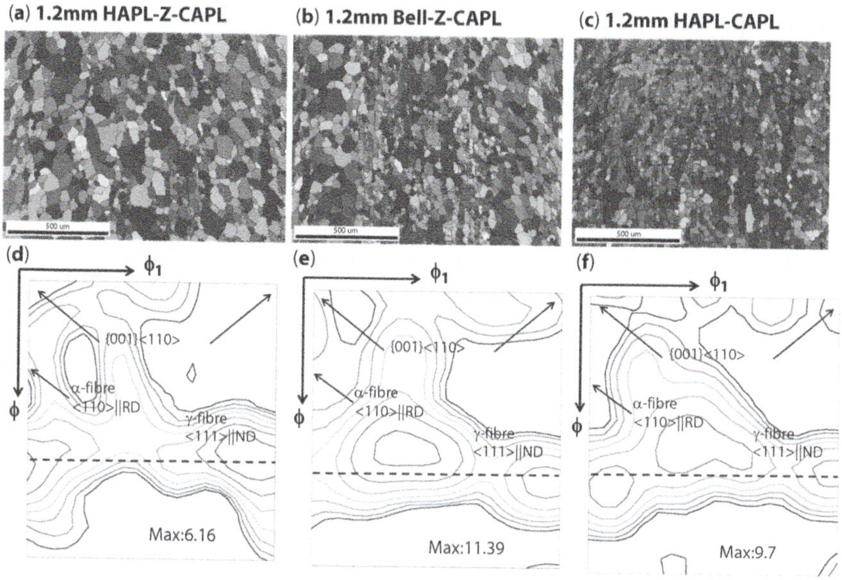

Figure 5.7 (a-c) EBSD-IPF map of the specimens investigated and corresponding (d-f) ODF Φ2 = 45° section.

Table 5.2 Tensile results of the investigated steels.

Process route	Ridging (Rz)	Cupping (MM)	YS	UTS	%El	R-bar
HAPL-ZMILL-CAPL	9	12.75	220	404	34	1.6
BELL-ZMILL-CAPL	13	12.75	227	414	37	1.85
HAPLCAPL	11	12.75	230	410	35	1.65

and comparatively lesser value of r-bar was observed in the HAPL-CAPL and HAPL-Z-CAPL sample. The cupping value was similar for all three samples.

5.5 Discussion

5.5.1 Effect of Processing on Microstructure and Texture Evolution

Three different routes were investigated to produce the 409L ferritic stainless steel with the same starting microstructure and the same cold rolled annealed finishing thickness. The hot rolling reduction and finishing temperature was varied in two samples and one was hot rolled to 5.3mm with finishing temperature of 5.3mm, and another was hot rolled to 3.2mm with finishing temperature of 3.2mm. Both coils were annealed and cold rolled to 1.2mm thick sheet with intermediate annealing and final rolling and annealing. The purpose was to understand the effect of low cold rolling reduction on the drawing and ridging behavior. It was clearly observed that both the sample is almost having equivalent ridging resistance and drawability. The microstructural development suggests that the 3.2 mm hot rolling can lead to finer grain size with more recrystallization and more shear band with lesser amount of harmful cube texture which can modify the cube texture and can increase the gamma fiber orientation have an impact on ridging and drawability. On the other hand, the cold rolled annealed microstructure was also having finer structure in the HAPL-Z-CAPL case and the texture intensity of the sample was also reduced. The HAPL-Z-CAPL sample underwent high finish rolling temperature with coarser hot rolled annealed microstructure which decreases the intensity of γ-fiber and other orientation as well. The bell annealed sample, on the other hand, having higher harmful texture in the hot band annealed microstructure. The reason behind the harmful cube texture is the low temperature annealing. The low temperature annealing for longer time could not convert and rotate the harmful texture in the bell annealed sample. But the cold rolled and annealed sample of Bell-Z-CAPL provides higher intensity gamma fiber along with higher cube fiber orientation also. The higher cold reduction compared to the sample in the final stage can promote the higher intensity of gamma fiber orientation in the bell-Z-CAPL sample.

5.5.2 Effect of Processing and Microstructure, Texture on Ridging Resistance, Mechanical Property and Drawability

The three investigated samples suggested that all the process routes can provide good combination of mechanical, drawability and ridging resistance. Comparatively, best ridging resistance was observed in the HAPL-Z-CAPL sample where the intermediate annealing after cold rolling and large cold reduction decreased the intensity of the gamma fiber texture. The higher finish rolling temperature and low reduction in the lower temperature in hot rolling can provide a low intensity of both favorable gamma and harmful cube texture. So, even though the cold reduction was lesser, the sample provides moderately favorable gamma fiber and lesser harmful cube fiber which actually promoted the similar ridging resistance and drawability in the low cold rolled sample.

Secondly, though there was more harmful cube fiber in the bell annealed sample, but higher cold reduction and the annealing provided a breakdown of the cube fiber and higher reduction promoted more intensity of the gamma fiber which resulted higher drawability in the Bell-Z-CAPL sample. The ridging resistance, however, is little bit more than the other two samples; it may be due to having more intensity of both favorable gamma and harmful cube fiber which promotes anisotropic plastic flow in the sample and ridging value that increases a bit.

5.6 Conclusion and Summary

In general, ferritic stainless steel requires lot of process control to achieve the highest quality in terms of ridging resistance and drawability. Starting from controlling the chemistry especially, low carbon and nitrogen, addition of stabilizer Ti and then casting parameters with EMS to control the columnar grain; the columnar grain is the source of the harmful texture, which is hard to break in the later stage of processing. The control of hot rolling, reduction in low finishing temperature, to impart the shear band alters the harmful cube texture. Then the annealing is used to promote recrystallization and modification of texture, controlling the grain size by breaking the coarse-grain band. Then finally, high cold rolling reduction which promotes favorable texture and also modifies the harmful texture to some extent and finally annealing to get proper recrystallized grains.

The study suggested that lower cold reduction (60%) can also produce good ridging resistance and drawability, which are against the popular belief in the literature of ferritic stainless steel. By reducing the cold reduction, the energy can be saved, and productivity can be improved in a steel mill operation. Thus, sustainability goal can be met following the above benefit.

Secondly, it also proves that the electromagnetic stirrer can provide a high intensity stirring during continuous casting which helps to breakdown the columnar grain which is the origin of harmful cube fiber orientation. This EMS can actually be very helpful for the sustainable operation. Because large size columnar grain is very hard to breakdown, and if present in high quantity it can cause coarse grain band and promotes ridging and also harm the drawability.

The bell annealing provides the opportunity for long term annealing in a ferritic stainless steel at comparatively lower temperature. It was observed that the bell annealing with higher cold reduction is helpful to get the high intensity gamma fiber texture which promotes highest drawability in the samples but with little bit decreasing the ridging resistance.

The mechanical property of the all the samples was, in accordance with the standard requirement of 409L grade and the ductility of high and cupping value is also very good which suggest the produced coils with higher % equiaxed grains can promote formability and good mechanical properties.

AI disclosure statement: The author acknowledges the use of AI-based tools for minor language editing or grammatical correction. However, all intellectual contributions, analyses, and conclusions presented in this chapter are entirely the author's own.

References

1. Patra, S. and Singhal, L.K., Influence of Hot Band Annealing and Cold Rolling on Texture and Ridging of 430 Stainless Steel Containing Aluminum. *Mater. Sci. Appl.*, 04, 01, 70–76, 2013.
2. Zhang, C., Liu, Z., Wang, G., Effects of hot rolled shear bands on formability and surface ridging of an ultra purified 21%Cr ferritic stainless steel. *J. Mater. Process. Technol.*, 211, 6, 1051–1059, 2011.
3. Mallaiah, G., Kumar, A., Ravinder Reddy, P., Madhusudhan Reddy, G., Influence of grain refining elements on mechanical properties of AISI 430 ferritic stainless steel weldments - Taguchi approach. *Mater. Des.*, 36, 443–450, 2012.
4. Ashby M.F. and Jones, D.R.H., Engineering Materials 1-An introduction to properties and applications, 1996.

5. Park, S.H., Kim, K.Y., Lee, Y.D., Park, C.G., Evolution of microstructure and texture associated with ridging in ferritic stainless steels. *Mater. Sci. Forum*, 408–412, II, 1335–1340, 2002.
6. Modak, P., Patra, S., Mitra, R., Chakrabarti, D., Effect of Starting As-cast Structure on the Microstructure–Texture Evolution During Subsequent Processing and Finally Ridging Behavior of Ferritic Stainless Steel. *Metall. Mater. Trans. A Phys. Metall. Mater. Sci.*, 49, 6, 2219–2234, 2018.
7. Siqueira, R.P., Sandim, H.R.Z., Oliveira, T.R., Raabe, D., Composition and orientation effects on the final recrystallization texture of coarse-grained Nb-containing AISI 430 ferritic stainless steels. *Mater. Sci. Eng., A*, 528, 9, 3513–3519, 2011.
8. Patra, S., Ghosh, A., Sood, J., Singhal, L.K., Podder, A.S., Chakrabarti, D., Effect of coarse grain band on the ridging severity of 409L ferritic stainless steel. *Mater. Des.*, 106, 336–348, 2016.
9. Tsuchiyama, T., Hirota, R., Fukunaga, K., Takaki, S., Ridging-free ferritic stainless steel produced through recrystallization of lath martensite. *ISIJ Int.*, 45, 6, 923–929, 2005.
10. Engler, O., Huh, M.Y., Tomé, C.N., Crystal-plasticity analysis of ridging in ferritic stainless steel sheets. *Metall. Mater. Trans. A Phys. Metall. Mater. Sci.*, 36, 11, 3127–3139, 2005.
11. Raabe, D. and Lüucke, K., Textures of ferritic stainless steels. *Mater. Sci. Technol.*, 9, 4, 302–312, 1993.
12. Gao, F., Yu, F., Misra, R.D.K., Zhang, X., Zhang, S., Liu, Z., Microstructure, Texture, and Deep Drawability Under Two Different Cold-Rolling Processes in Ferritic Stainless Steel. *J. Mater. Eng. Perform.*, 24, 10, 3862–3880, 2015.
13. Shin, H.J., An, J.K., Park, S.H., Lee, D.N., The effect of texture on ridging of ferritic stainless steel. *Acta Mater.*, 51, 16, 4693–4706, 2003.
14. Knutsen, R.D. and Wittridge, N.J., Modelling surface ridging in ferritic stainless steel. *Mater. Sci. Technol.*, 18, 11, 1279–1285, 2002.
15. Wu, P.D., Lloyd, D.J., Huang, Y., Correlation of ridging and texture in ferritic stainless steel sheet. *Mater. Sci. Eng., A*, 427, 1–2, 241–245, 2006.
16. Wu, P.D., Jin, H., Shi, Y., Lloyd, D.J., Analysis of ridging in ferritic stainless steel sheet. *Mater. Sci. Eng., A*, 423, 1–2, 300–305, 2006.
17. Fang, X., Wang, W., Guo, H., Qin, C., Zhou, B., Evolutions of texture and grain boundary plane distributions in a ferritic stainless steel. *J. Cent. South Univ.*, 19, 12, 3363–3368, 2012.
18. Brochu, M., Yokota, T., Satoh, S., Analysis of Grain Colonies in Type 430 Ferritic Stainless Steels by Electron Back Scattering Diffraction (EBSD). *Miscellaneous*, 37, 9, 872, 1997.
19. Hayakawa, Y., Szpunar, J.A., Palumbo, G., Lin, P., The role of grain boundary character distribution in Goss texture development in electrical steels. *J. Magn. Magn. Mater.*, 160, 143–144, 1996.
20. Sellars, C.M. and Whiteman, J.A., Recrystallization and grain growth in hot rolling. *Met. Sci.*, 13, 3–4, 187–194, 1979.

21. Samajdar, I., Verlinden, B., Van Houtte, P., Vanderschueren, D., γ-Fibre recrystallization texture in IF-steel: An investigation on the recrystallization mechanisms. *Mater. Sci. Eng., A*, 238, 2, 343–350, 1997.
22. Samajdar, I., Verlinden, B., Kestens, L., Van Houtte, P., Physical parameters related to the developments of recrystallization textures in an ultra low carbon steel. *Acta Mater.*, 47, 1, 55–65, 1998.
23. Martínez-de-Guerenu, A., Arizti, F., Díaz-Fuentes, M., Gutiérrez, I., Recovery during annealing in a cold rolled low carbon steel. Part I: Kinetics and microstructural characterization. *Acta Mater.*, 52, 12, 3657–3664, 2004.
24. Khatirkar, R., Mani Krishna, K.V., Kestens, L.A.I., Petrov, R., Pant, P., Samajdar, I., Strain localizations in ultra low carbon steel: Exploring the role of dislocations. *ISIJ Int.*, 51, 5, 849–856, 2011.
25. Jonas, J.J. and Kestens, L.A.I., Modelling the Effects of Nucleation and Growth on Texture Formation in Commercial Steels. *Grain Growth Polycryst. Mater. II*, 204, 155–168, 1996.
26. Petrov, R.H., Kestens, L.A.I., Verbeken, K., Houbaert, Y., Grain Growth after Intercritical Rolling. *Recryst. Grain Growth*, 467, 305–310, 2004.
27. Verbeken, K. and Kestens, L.A.I., Nucleation of Secondary Recrystallization in Ultra Low Carbon Steel. *Textures Mater. - ICOTOM 14*, 495, 1189–1194, 2005.
28. Yazawa, Y., Ozaki, Y., Kato, Y., Furukimi, O., Development of ferritic stainless steel sheets with excellent deep drawability by {1 1 1} recrystallization texture control. *JSAE Rev.*, 24, 4, 483–488, 2003.
29. Amuda, M.O.H. and Mridha, S., Grain refinement and hardness distribution in cryogenically cooled ferritic stainless steel welds. *Mater. Des.*, 47, 365–371, 2013.
30. Zhang, C., Zhang, L.W., Liu, Z.Y., Improvement of surface ridging resistance of an ultra-purified ferritic stainless steel by optimizing hot rolling condition. *Acta Metall. Sin. (English Lett.*, 29, 6, 561–567, 2016.
31. Dieter, G.E., Mechanical metallurgy, 2011.
32. Tsuji, N., Tsuzak, K., Mak, T., Effect of Initial Orientation on the Recrystallization Behavior of Solidified columnar Crystals in a 19% Cr Ferritic Stainless Steel. *ISIJ Int.*, 33, 783–792, 1993.
33. Hayakawa, Y., Muraki, M., Szpunar, J.A., The changes of grain boundary character distribution during the secondary recrystallization of electrical steel. *Acta Mater.*, 46, 3, 1063–1073, 1998.
34. Lee, J.H., Park, S.H., Huh, M.Y., Modification of the Recrystallization Texture by Means of Cross Rolling in Ferritic Stainless Steel Sheets. *Des. Process. Prop. Adv. Eng. Mater.*, 449, 113–116, 2004.
35. Huh, M.Y. and Engler, O., Effect of intermediate annealing on texture, formability and ridging of 17%Cr ferritic stainless steel sheet. *Mater. Sci. Eng., A*, 308, 1–2, 74–87, 2001.
36. Davies, E.R., Introduction to texture analysis, 2008.

37. Ray, R.K. and Jonas, J.J., Transformation textures in steels. *Int. Mater. Rev.*, 35, 1, 1–36, 1990.
38. Hamada, J., Matsumoto, Y., Fudanoki, F., Maeda, S., Effect of Initial Solidified Structure on Ridging Phenomenon and Texture in Type 430 Ferritic Stainless Steel Sheets. *ISIJ Int.*, 43, 12, 1989–1998, 2003.
39. Li, R., Fu, B., Wang, Y., Li, J., Dong, T., Li, G., Zhang, G., Liu, J., Effect of Cold-Rolling Reduction on Recrystallization Microstructure, Texture and Corrosion Properties of the X2CrNi12 Ferritic Stainless Steel. *Materials (Basel)*, 15, 19, 2022.
40. Zhang, X., Fan, L., Xu, Y., Li, J., Xiao, X., Jiang, L., Texture, microstructure and mechanical properties of aluminum modified ultra-pure 429 ferritic stainless steels. *Mater. Des.*, 89, 626–635, 2016.

6

Sustainable Steel Processing Techniques: Minimising Waste and Energy Consumption

Muralidhar Yadav[1] **and Swarup Kumar Ghosh**[2]*

[1]*School of Materials Science and Engineering, Nanyang Technological University, Jurong West, Singapore*
[2]*Department of Metallurgy and Materials Engineering, Indian Institute of Engineering Science and Technology, Shibpur, Howrah, West Bengal, India*

Abstract

This chapter highlights the pivotal role of sustainable steel processing in mitigating the environmental impact of conventional steel production. It provides an in-depth analysis of cutting-edge technologies, including electric arc furnace steelmaking, waste heat recovery-integrated basic oxygen furnace systems, and direct reduced iron processes, designed to minimize environmental impact. Real-world examples of sustainable steel production are presented through case studies of industrial stalwarts, e.g., ArcelorMittal, SSAB, Nucor, and Tata Steel, highlighting accomplishments of sustainable practices. Emerging technologies and future directions such as hydrogen-based steelmaking, carbon capture, utilization and storage, and AI-driven digitalization are discussed as key transformative tools towards a sustainable low-carbon steel industry. Considering innovative technologies, policy support, and sustainable business practices, this chapter offers a comprehensive framework for driving the steel industry's transition to a more environmentally friendly and energy-efficient future. This chapter illustrates how the steel industry can adopt advanced technologies, regulatory alignment, and circular economy practices to reduce its environmental footprint, energy consumption, and waste, while fostering a more sustainable, low-carbon and resource-responsible future. This roadmap outlines a strategic approach for the steel industry to reduce its environmental footprint and contribute to a low-carbon future.

**Corresponding author*: skghosh@metal.iiests.ac.in

Gurudas Mandal, Swarup Kumar Ghosh, Snehanshu Pal and Sandip Kunar (eds.) Sustainable Manufacturing of Advanced Materials, (111–136) © 2026 Scrivener Publishing LLC

Keywords: Green steel manufacturing, hydrogen-based direct reduction, electric melting furnace, carbon management technologies, clean steel innovations

6.1 Introduction

The steel industry is considered the backbone of modern society as it provides indispensable materials for construction, transportation, energy, and other industries [1]. The steel industry, which is a foundational pillar of modern infrastructure, with an annual production of over 1.8 billion metric tons of steel, plays a vital role in economic growth, but is also a significant source of global greenhouse gas (GHG) emissions, accountable for around 8% of worldwide emissions. Conventional steel production processes, like the Blast Furnace-Basic Oxygen Furnace (BF-BOF) method, rely heavily on carbon-intensive processes that consume immense energy and generate a significant amount of waste, including slag, dust, and CO_2 emissions. The growing demand for steel, especially in developing economies, augments the ecological challenges associated with its production. As governments implement stringent environmental regulations and pursue climate goals, the demand for sustainable steelmaking practices has become more critical than ever. The development of sustainable steel production methods is vital for reducing environmental damage, conserving resources, and safeguarding the industry's future prosperity.

The environmental consequences of traditional steel production are diverse and widespread. The use of fossil fuels, especially coal, in blast furnaces and basic oxygen furnaces (BOFs) leads to substantial carbon dioxide (CO_2) emissions, a foremost contributor to global warming [2]. The sourcing and treatment of essential materials like iron ore and coal contribute to habitat destruction, deforestation, and depletion of resources. Additionally, significant volumes of by-products, including slag, dust, and sludge, pose notable environmental and logistical challenges for the steel industry [3].

In the face of aggravating environmental challenges, the steel industry is recognizing the need for a more sustainable approach, prompting a major transformation [4], which involves a fundamental reconsideration of steel production, incorporating cleaner technologies, resource efficiency, and circular economy principles. The objective of sustainable steel processing is to break the link between steel production and environmental degradation, enabling the industry to meet global demand while reducing its ecological footprint.

This chapter provides a detailed look at sustainable steel processing by highlighting a broad spectrum of techniques, ranging from traditional to state-of-the-art approaches. These techniques offer numerous pathways to conserve energy, minimize waste generation and carbon footprint. Furthermore, the chapter will discuss approaches to minimizing waste, promoting energy efficiency, and employing pioneering methods like additive manufacturing to foster sustainability in the steel industry. Hands-on examples and best practices will be showcased to demonstrate the successful application and impact of sustainable steel processing techniques.

6.2 Importance of Sustainable Steel Processing

Sustainability principles are widely applied across various disciplines, including engineering, manufacturing, and design. It involves finding a balance between economic, social, and environmental needs to ensure a healthy and thriving planet for future generations. Figure 6.1 illustrates the key sustainability components of steel manufacturing processes [5]. Sustainable steel processing refers to implementing innovative solutions and approaches aimed at minimizing waste, optimizing energy efficiency, and reducing carbon emissions. Several key factors behind the growing emphasis on sustainability are listed below [6, 7]:

- **Environmental regulations**: Global agreements like the Paris Agreement and systems like the European Union

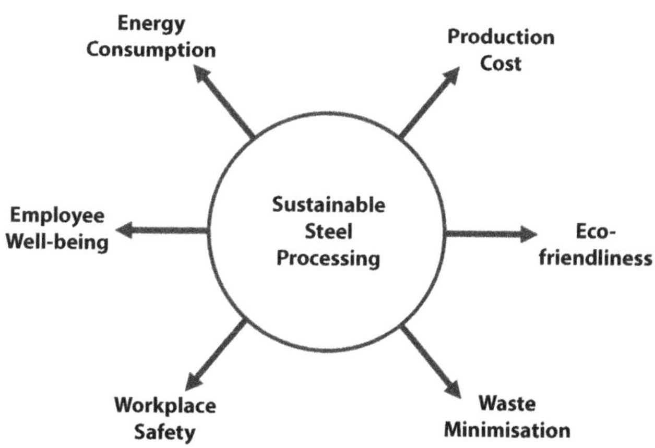

Figure 6.1 Major sustainability elements of steel manufacturing processes.

Emissions Trading System (EU ETS) set emission reduction mandates.

- **Cost savings**: Sustainable practices can yield significant economic benefits, including reduced energy costs, improved resource efficiency, and lower reliance on costly fossil fuels.
- **New technologies**: New steelmaking technologies, including hydrogen-based reduction and electric arc furnace (EAF) steelmaking, provide practical alternatives to traditional methods.
- **Social responsibility initiatives**: As public and investor awareness of sustainability grows, steel producers are under increasing pressure to adopt greener practices.

The chapter investigates sustainable steel processing, waste management, and energy conservation strategies that can help the steel industry transition towards a more environmentally friendly model.

6.3 Sustainable Steel Processing Techniques

Steel production primarily involves the blast furnace (BF) and basic oxygen furnace (BOF) route for iron ore-based production, and the electric arc furnace (EAF) route for scrap-based production. In the BF-BOF process, iron ore is reduced to molten iron using coke and then refined into molten steel through oxygen blowing. The molten steel is then cast into slabs or billets and rolled into finished products [8]. The material flow in steel production is depicted in Figure 6.2. Although the BOF route has been the traditional choice, the EAF route and other emerging technologies provide more environmentally friendly options [9].

6.3.1 Electric Arc Furnace (EAF) Steelmaking

As a sustainable option, the EAF process has gained traction as an alternative to BOF, primarily because of its capacity to use large amounts of scrap steel [9]. Contrary to BOF, which relies on energy-intensive iron ore reduction, EAF uses electric arcs to melt scrap steel, resulting in a considerable reduction in carbon emissions and energy consumption.

- **Process Description**

The EAF process involves loading the furnace with scrap steel and potentially other materials like direct reduced iron (DRI) or pig iron to attain the

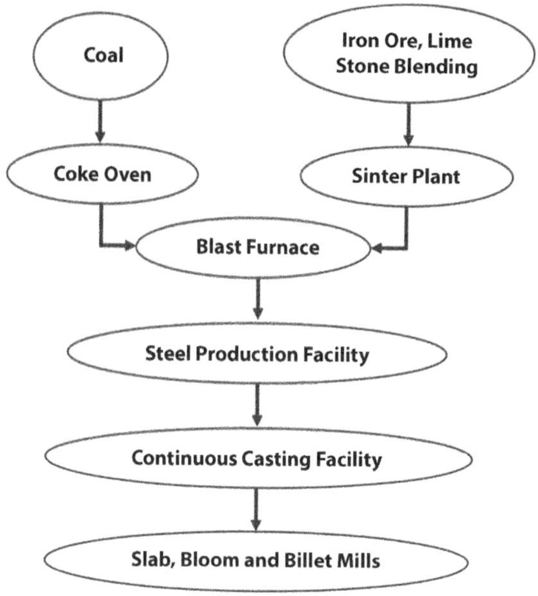

Figure 6.2 Process flow for steel production.

desired steel composition. Next, high-power electrodes are inserted into the furnace, generating an electric arc between the electrodes and the scrap metal. This arc generates intense heat, melting the scrap and any other metallic materials. Lime and dolomite are introduced into the furnace as fluxes to eliminate impurities, resulting in slag that floats on the molten steel. The molten steel is poured out of the furnace and undergoes further refining and processing.

- **Sustainability Advantages**
    - **High Scrap Utilisation:** EAFs are capable of operating with a very high percentage (~ 100%) of scrap steel. This approach lessens the need for iron ore mining and processing, which in turn reduces environmental degradation, including habitat disruption, greenhouse gas emissions, and high energy consumption [10].
    - **Energy Savings:** Compared to BOF, EAF steel production using scrap requires substantially less energy, i.e., around one-third, leading to notable decreases in CO_2 emissions per

ton of steel produced [11]. This energy efficiency is achieved by skipping the energy-intensive step of reducing iron ore.

- **Flexibility in Operations and Capacity:** EAF technology allows for more adaptable production capacity compared to BOFs. By locating mini-mills near scrap sources and markets, transportation-related emissions are minimized, and the ability to respond quickly to local demand is enhanced.

- **Environmental Considerations**

  - **Power Generation Source:** The carbon footprint of EAF steelmaking is closely linked to the generating source of the electricity it uses. Using renewable sources such as solar cells, windmills, or hydroelectricity offers maximum sustainability benefits, which can't be achieved if EAF steelmaking consumes power produced by fossil fuels, potentially making its impact comparable to BOF in some cases.
  - **EAF Dust:** EAF operations give off dust that may contain heavy elements like lead and cadmium, along with several other contaminants. Appropriate dust control and treatment measures using advanced filtration systems to capture the dust particles, followed by safe disposal or recycling of valuable metals like zinc, are essential to redress the environmental impact.
  - **Sound Pollution:** A possible source of noise pollution, EAF operations necessitate the implementation of noise reduction procedures to safeguard workers and the surrounding community. To moderate the elevated noise level, soundproofing measures and stringent operational guidelines can be implemented around the furnace area.

6.3.2 Basic Oxygen Furnace (BOF) with Waste Heat Utilization

The BOF is a leading steel production method, recognized for efficient production of high-quality steel in large quantities [12]. However, it is also an energy-intensive process, relying on the oxidation of impurities in molten iron with oxygen, which releases substantial amounts of heat. Integrating waste heat recovery systems into BOF operations can significantly enhance their energy efficiency and sustainability.

- **Process Description**

The BOF process requires introducing molten iron from a blast furnace and scrap steel into a refractory-lined vessel. Pure oxygen is injected through a water-cooled lance, and sparking chemical reactions oxidize and eliminate impurities such as carbon, silicon, manganese and phosphorus. These reactions generate a large amount of heat, raising the temperature of the mixture to around 1600-1700°C. Traditionally, much of this heat is released into the atmosphere as hot gases.

- **Waste Heat Utilization Systems**

Waste heat utilization systems make use of the thermal energy contained in the off gases from the BOF process and utilize it for various purposes:

  - **Steam Generation:** The captured heat can be converted into steam in a waste heat boiler system. This steam can then be used for electricity production *via* steam turbines or for heating purposes within the steel plant, reducing the need for additional fuel combustion and associated emissions [13].
  - **Preheating:** Waste heat can be used to preheat incoming air for combustion or raw materials, such as scrap steel, increasing the overall energy efficiency of the BOF process and reducing fuel consumption.
  - **Additional Uses:** Recovered heat can be utilized in a range of industrial processes, either within the steel plant or in other industries, depending on the energy demand and system design. Recovered heat could be applied to district heating or driving absorption chillers for air conditioning.

- **Sustainability Benefits**

Several benefits that utilization of waste heat recovery in BOF steelmaking provides:

  - **Lower energy usage:** A steel plant's overall energy usage can be reduced by using waste heat, resulting in lesser fuel consumption and greenhouse gas emissions [12]. The reduction in fuel consumption results in significant cost savings for the steel manufacturer.
  - **Enhanced Energy performance:** The application of waste heat recovery in the BOF process boosts thermal efficiency, ensuring more effective energy and less waste.

- **Cost Savings:** Waste heat recovery and reuse can harvest significant economic benefits for steel producers by reducing energy expenses and creating opportunities for energy sales.

- **Challenges and Considerations**

  - **System Design and Integration:** The installation of waste heat recovery systems in existing BOF facilities can be a complex and expensive process. For effective recovery of waste heat, a detailed assessment of off-gas characteristics, heat exchange selection, and plant output is essential.
  - **Waste Heat Characteristics:** The variability in waste heat characteristics throughout the BOF process requires advanced systems capable of dynamic response to variation in temperature and flow rate, ensuring the operation remains consistent and efficient.
  - **Fouling and Corrosion:** Containing dust and corrosive substances, BOF off-gases can seriously affect heat exchangers, which include reduced performance and increased requirements for maintenance due to fouling and corrosion.

6.3.3 Direct Reduced Iron (DRI) Process

The DRI process offers another option for the traditional blast furnace process for steelmaking raw materials [10]. In contrast to blast furnaces that rely on coal, the DRI process uses natural gas or hydrogen as reducing agents to transform iron ore into metallic iron in a solid state [13].

- **Process Description**

In the DRI process, lumpy iron ore or pellets are fed into a reactor where it is exposed to a reducing gas in a hot environment (800-1050°C). The reducing gas, which can be syngas produced from natural gas (a blend of carbon monoxide and hydrogen) or pure hydrogen, removes oxygen from the iron ore, producing metallic iron, designated as DRI or sponge iron. There are several variations of the DRI process, including shaft furnaces (e.g., Midrex and HYL processes) and fluidized bed reactors.

- **Sustainability Advantages**

  - **Reduced Greenhouse Gas Emissions:** In natural gas-based reduction, the DRI method can produce less greenhouse gas emissions compared to blast furnaces, primarily due to the lower carbon content of natural gas [14]. However, the extent of emissions reduction depends on the efficiency of the process and any additional measures taken to mitigate methane emissions, which can occur during natural gas production and transportation.

- **Potential for Hydrogen Integration:** The DRI process is particularly well-suited for integration with hydrogen as a reducing agent. Using sustainable hydrogen, generated from renewable sources including solar, wind, and hydropower, can drastically reduce or even eliminate CO_2 emissions from ironmaking, offering a pathway to near-zero emissions in steel production.
- **High-Quality Iron:** The DRI process produces high-purity iron with a low residual element content, which is beneficial for EAF steelmaking, as it can improve steel quality, enhance EAF productivity, and reduce the need for extensive refining.

- **Environmental Considerations**

  - **Natural Gas Dependence:** The sustainability of current DRI processes largely depends on the availability and environmental impact of natural gas. While natural gas is cleaner than coal, it is still a fossil fuel, and its extraction and combustion contribute to greenhouse gas emissions. Transitioning to alternative reducing agents, especially hydrogen, is critical for long-term sustainability.
  - **Hydrogen Production:** The environmental benefits of hydrogen-based DRI are contingent on the production method of hydrogen. Green hydrogen, produced through electrolysis using renewable electricity, is essential for achieving near-zero emissions. Using traditional hydrogen production, such as grey hydrogen from natural gas and brown hydrogen from coal, would negate the environmental benefits and perpetuate reliance on fossil fuels.
  - **Carbon Capture and Storage (CCS):** Even with natural gas, integrating CCS technologies with the DRI process can further lower carbon emissions, particularly in the interim period before widespread green hydrogen adoption. CCS involves capturing and storing process-related CO_2 emissions underground to reduce atmospheric emissions.

6.3.4 Hydrogen Steelmaking

Hydrogen steelmaking represents a potentially revolutionary approach to decarbonizing the steel industry [15]. By using hydrogen instead of fossil fuels like coal and natural gas as the reducing agent, this technology offers the prospect of near-zero CO_2 emissions in primary steel production, fundamentally transforming the environmental footprint of steelmaking.

- **Process Description**

Hydrogen steelmaking involves using hydrogen gas (H_2) to reduce iron ore (iron oxide, Fe_2O_3) to metallic iron (Fe). The core chemical reaction is:

$$Fe_2O_3 \text{ (iron ore)} + 3H_2 \text{ (hydrogen)} \rightarrow 2Fe \text{ (iron)} + 3H_2O \text{ (water)}$$

In this process, the by-product is water vapor (H_2O) instead of CO_2, which is the primary emission from traditional blast furnaces. This fundamental shift in the chemical reaction eliminates the major source of greenhouse gas emissions in conventional steelmaking.

- **Sustainability Benefits**

    - **Drastic Reduction in CO_2 Emissions:** The most significant advantage of hydrogen steelmaking is the potential to eliminate almost all direct CO_2 emissions from the ironmaking stage, a major contributor to the steel industry's carbon footprint [16]. The ability of hydrogen steelmaking to significantly reduce emissions makes it a key enabler of the steel industry's journey towards carbon neutrality.
    - **Suitability for DRI:** DRI plants can be readily adapted to use hydrogen, making the DRI process a prime candidate for hydrogen-based steel production. The alignment of DRI technology with hydrogen steelmaking facilitates easier adoption and integration into current operations.
    - **Closed-Loop Potential:** The water vapor generated as a by-product can be captured and recycled back into the hydrogen production process, thereby supporting a circular economy model and boosting sustainability.

- **Challenges and Considerations**

    - **Green Hydrogen Production:** The sustainability of hydrogen steelmaking is entirely linked to the origin of hydrogen.

Clean hydrogen, produced by electrolysis with renewable electricity from solar, wind, and hydropower, is key to significantly reducing emissions. Relying on fossil fuel-based hydrogen production, such as grey hydrogen from natural gas and brown hydrogen from coal, would negate the environmental benefits and perpetuate reliance on fossil fuels [17].

- **Infrastructure Upgrades:** The adoption of hydrogen steelmaking depends on transforming existing infrastructure and securing a stable hydrogen supply chain. This includes:
  - Major facility redesign and operational adjustments are required to support hydrogen steelmaking, regardless of whether new plants are built or existing ones are modified.
  - Establishing a reliable system for delivering hydrogen to steelmaking facilities, including refuelling stations and handling equipment. Hydrogen is a low-density gas, which poses challenges for efficient storage and transportation.
  - The establishment of safety protocols and standards for handling hydrogen, which is a flammable gas, requires careful management to prevent accidents.

- **Economic Viability:** The cost of green hydrogen production exceeds that of fossil fuel-based hydrogen production at present. Lowering green hydrogen costs is key to scaling up hydrogen steelmaking. This requires:
  - Continued advancements in renewable energy technologies to reduce electricity expenses, enhancing the cost competitiveness of green hydrogen production.
  - Improvements in the efficiency and scalability of electrolysis technologies to increase hydrogen production rates and reduce capital costs.
  - Government initiatives and funding to accelerate the growth of green hydrogen and hydrogen steelmaking.

- **Technological Maturity:** While the concept of hydrogen steelmaking is promising, the technology is still under development and requires further research, pilot projects, and scale-up efforts to demonstrate its technical and economic feasibility at an industrial scale.

6.4 Emerging Steelmaking Technologies

Beyond the established and developing technologies mentioned above, research and development are ongoing to explore even more innovative and potentially transformative steelmaking processes that could further enhance sustainability.

- **Molten Oxide Electrolysis:** This process uses electrolysis to directly reduce iron oxide in a molten oxide bath, potentially simplifying the steelmaking process and reducing energy consumption [18]. This technology aims to eliminate the need for reducing gases altogether, potentially leading to a more energy-efficient and emissions-free route to iron production.
- **Biomass-Based Steelmaking:** Utilizing biomass, such as wood, agricultural residues, or other organic matter, as a reducing agent or energy source offers the potential to decrease dependence on fossil fuels [19]. However, the sustainability of biomass-based steelmaking depends on responsible biomass sourcing and management practices to avoid deforestation, land conversion and conflicts with food production.
- **Carbon Capture and Utilization (CCU):** While not strictly a steelmaking process itself, CCU technologies can play a role in mitigating CO_2 emissions from traditional steelmaking. CCU entails capturing and utilizing CO_2 emissions from steel production to create valuable commodities like fuels, chemicals, or construction materials. However, the overall impact of CCU on emissions reduction is complex and depends on the specific application of the captured CO_2 and the energy sources used in the conversion process.

6.5 Waste Minimization and Recycling

As populations grow and development expands, waste generation increases. Effective waste management and prevention are crucial to mitigate environmental and health impacts. This involves planning and implementing comprehensive programs for collection, transportation, disposal, and recycling/reuse [20]. A sustainable waste management process, as shown in Figure 6.3, involves recovering valuable resources from waste. Products are either reused or recycled into new items or converted into green energy to meet human needs. The effective sustainable waste management requires consideration of social, economic, and regulatory factors. In addition to adopting cleaner production technologies, effective waste minimization and recycling strategies are essential for steel production and promoting a closed-loop system in steel production [21]. The steel industry generates various waste streams, but many of these can be valuable resources if properly managed and processed.

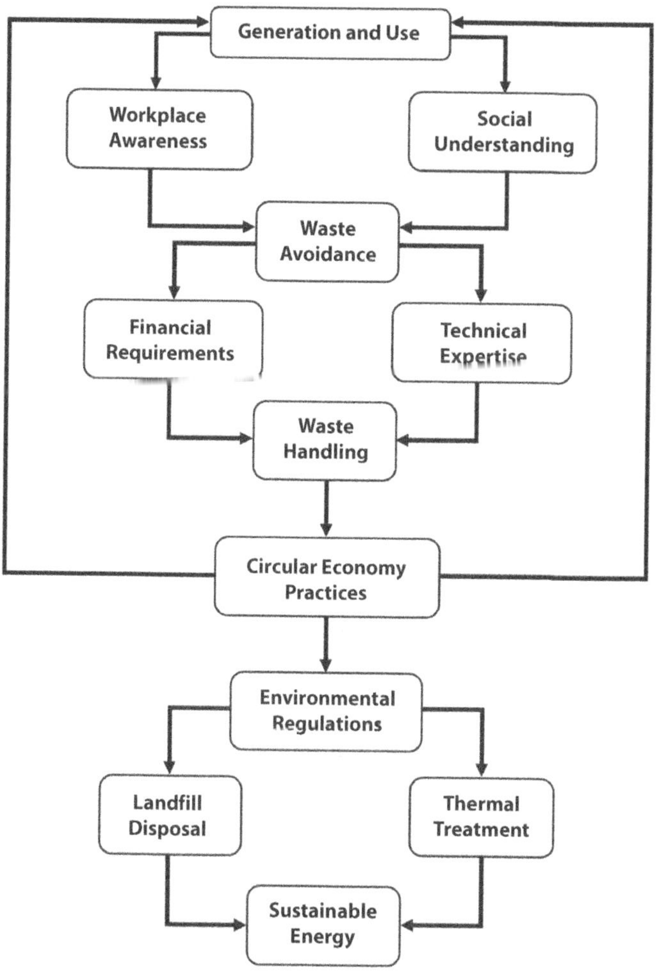

Figure 6.3 Process diagram of waste management.

6.5.1 Recycling of Steel Slag

Slag, a by-product of steel production in BOF and EAF processes, is composed of various oxides, silicates, and non-metallic materials [22]. Slag can be transformed from a waste product to a valuable resource and thereby offering numerous benefits for the environment and the economy.

- **Types of Slag**
  - **BOF Slag:** BOF slag, a byproduct of the basic oxygen furnace process, typically contains high levels of calcium oxide (CaO) and iron oxide (FeO), as well as smaller quantities of silicon dioxide (SiO_2), magnesium oxide (MgO), and aluminum oxide (Al_2O_3).
  - **EAF Slag:** EAF slag composition changes based on scrap material and steel grade. It can exhibit higher metal concentrations, notably iron, manganese, and chromium, in contrast to BOF slag.

- **Recycling Applications**
  - **Cement Manufacturing:** Slag serves as a valuable resource in cement production. Slag replaces a portion of the clinker, which is considered the most energy-intensive component of cement. As clinker manufacturing is a major emission source, using supplementary materials in place of clinker decreases the energy footprint and CO_2 output of cement production [23].
  - **Construction Aggregates:** After processing, slag can be utilized in construction, including roads, buildings, and other infrastructure as a sustainable aggregate material. This improves the concrete's properties, leading to enhanced strength, durability, and protection against chemical degradation. In addition, slag can be used in asphalt mixtures for pavements and highways.
  - **Soil Management:** Slag, utilized as a soil additive to enhance soil health, can mitigate acidity and deliver beneficial nutrients such as calcium, magnesium, and silicon. Given the possibility of heavy metal contamination, the composition of slags must be carefully examined to ensure it meets safety standards and does not pose a risk.

  - **Additional Uses:** Slags have widespread applications beyond construction, ranging from landfill capping, mine reclamation, glass and ceramic production, to metal recovery.

- **Environmental and Economic Benefits**

  - **Waste Minimisation:** Recycling of slag leads to conserving landfill space and reducing pollution, such as leachate contamination of groundwater, by substantially reducing the waste sent to landfills.
  - **Natural Resource Preservation:** Reliance on virgin materials like limestone, clay, and aggregates can decrease if slag is utilized, preserving these limited resources and minimizing the environmental footprint of extraction and processing.
  - **Reduced Energy Consumption:** Compared to conventional clinker production from limestone, incorporating slag into cement production lowers power consumption.
  - **Financial Benefit:** Transformation of slag from a waste material into a valuable product can provide economic benefits for steel producers and reduce expenses for waste disposal.

- **Challenges and Considerations**

  - **Slag Processing:** Slag undergoes various techniques like crushing, grinding, and separation before its utilization in various applications. The specific application of slag dictates the necessary processing steps and level of treatment.
  - **Slag Composition:** The composition of slag differs based on steel type, raw materials, and production methods. This variation determines its suitability for different uses, and blending or other treatments might be necessary for specific requirements.
  - **Ecological Issues:** Inappropriate slag disposal and handling can cause environmental problems, such as air and water pollution from dust and heavy metal contamination. Therefore, safe storage, handling, and processing procedures are crucial to avert or minimize these risks.

6.5.2 Scrap Recycling

Scrap steel, a highly valuable resource, can be reused and recycled multiple times with minimal degradation [23]. Recycling of scrap steel offers considerable ecological and economic gains and thereby making it a foundation of sustainable steel production and a critical component of the circular economy in the steel sector.

- **Types of Scrap Steel**
    - **Process Scrap:** Originated from the steel production process, including materials like trimmings, cut-offs, and products that do not meet quality standards. Often high quality, this type of scrap can be quickly recycled back into steel production and thereby minimizing waste.
    - **Industrial Scrap:** This is created during the production of steel-based goods and components, such as in the automotive, construction, and machinery industries. In general, this scrap is of decent quality and can be efficiently collected and processed for recycling.
    - **Used Material Scrap:** This is recovered from used products (e.g., automobiles, appliances, buildings). Recycling this scrap can be more complicated due to the potential for contamination and the need for specialized sorting and processing procedures.

- **Recycling Process**

Steel scrap is aggregated from multiple sources, sorted, and purified to eliminate impurities and contaminants such as coatings and non-ferrous metals. Thereafter, the cleaned scrap is fed into EAFs for melting or added to BOFs as part of the steelmaking process, creating new steel products.

- **Sustainability Benefits**
    - **Lower Energy Use:** Using scrap steel instead of virgin iron ore can slash energy usage by up to 70%, making recycling a highly energy-efficient option [3, 9]. Consequently, this leads to considerable energy savings and a decrease in greenhouse gas emissions, as the steelmaking process burns less fuel.

- **Natural Resource Savings:** Through scrap recycling, the demand for raw materials like iron ore and coal is reduced, which in turn conserves these resources and minimizes the environmental harm caused by their extraction and processing.
- **Minimised Waste Generation:** Through scrap recycling, materials are kept out of landfills, thereby reducing waste management issues and environmental concerns like leachate production and methane emissions from waste decomposition.
- **Lower Emissions and Effluent:** Steel made from scrap generates significantly less pollution, both in air and water, compared to traditional steel production methods. Thus, it decreases pollution from particulate matter, sulfur dioxide, and nitrogen oxides, along with dropping water consumption and wastewater discharge.

- **Economic Importance**

Steel scrap is a valuable material that can be a cost-effective substitute for virgin raw materials in steel production. The scrap steel industry supports a substantial number of jobs and boosts economic activity across collection, processing, and trading segments.

- **Challenges and Considerations**

  - **Scrap Material Quality:** The quality of scrap material can be inconsistent, necessitating contaminant removal for high-quality steel production. Innovative sorting and processing techniques are essential for handling progressively complex scrap streams.
  - **Scrap Supply:** The amount of scrap availability can shift due to economic conditions, recycling rates, and the age of steel products. To optimize the scrap in steel manufacturing, a consistent and dependable scrap supply is essential.
  - **Scrap Processing:** A well-developed and cost-effective scrap processing infrastructure, which includes capabilities for shredding, sorting, cleaning, and enhancing scrap quality to align with steel production requirements, is crucial for optimizing recycling rates.

6.5.3 Dust and Sludge Recycling

Steelmaking operations release waste materials like dust and sludge containing valuable metals and other resources [10]. By reusing and recycling the aforesaid by-products, the steel industry can further reduce its environmental footprint and support a circular economy.

- **Sources of Dust and Sludge**
  - **EAF Dust:** EAF dust can contain valuable metals, such as iron, zinc, lead, and cadmium, and other oxides and compounds. EAF dust, which is categorized as a hazardous waste in many countries because of its heavy metal content, requires cautious management and treatment.
  - **BOF Sludge:** During BOF operations, sludge is produced that is rich in iron oxides, calcium oxide, and other fine particles. Although BOF sludge is not usually considered hazardous, it still necessitates careful handling and appropriate disposal or recycling methods.
  - **Rolling Mill Scale:** Rolling mill scale, generated during the hot rolling process, is composed of iron oxides that break away from the steel surface. Reusing this scale, steel producers can recover valuable iron units and reduce the need for virgin raw materials.
- **Recycling along with Recovery Options**
  - **Metal Recovery:** Dust and sludge can be processed for the extraction and recovery of valuable metals, reducing waste and helping resource efficiency. For example, pyrometallurgical processes can extract zinc from EAF dust, while iron can be recovered from the processing of BOF sludge and rolling mill scale.
  - **Agglomeration:** The materials can be shaped into pellets and briquettes and reused in steel production, limiting waste and offering a valuable source of metals. Recycling those leads to resource conservation, allaying the ecological consequences of steelmaking.
  - **Alternative uses:** The potential for the reuse of dust and sludge in different industries, such as cement manufacturing or the production of building materials, is determined by the

composition. Specific kinds of sludge can be used as valuable raw materials in these applications.

- **Environmental and Economic Benefits**
    - **Waste Reduction:** Recycling dust and sludge reduces the amount of landfill waste, preserves landfill space and checks pollution.
    - **Resource Recovery:** Reducing the need for extraction and processing of raw materials, this approach conserves natural resources and abates environmental degradation.
    - **Economic Value:** A new stream of revenue can be generated for steel producers by recovering metals from dust and sludge, while reducing waste management expenses.
- **Challenges and Considerations**
    - **Hazardous Waste:** EAF dust, considered a hazardous waste, requires specialized procedures and precautions for handling, transportation, and treatment to minimize environmental and medical risks. Regulatory bodies have regularly mandated the use of specialized facilities and processing techniques for the safe treatment and disposal of hazardous waste.
    - **Complex Composition:** The steelmaking process and raw materials used influence the composition of dust and sludge, which can vary widely. The intricate behavior of these materials poses challenges in developing effective and economically sustainable recycling processes.
    - **Processing Costs:** Massive investment in technology and infrastructure is a typical necessity for the processing of dust and sludge, translating to substantial capital and operational expenses.

6.5.4 Energy Efficiency and Conservation

Energy efficiency and conservation are vital for promoting sustainability in the steel sector by reducing energy consumption and greenhouse gas emissions [11]. Steelmaking, which is a highly energy-intensive process, can bring about substantial energy savings and reduced environmental footprint through minor gains in efficiency.

6.5.4.1 *Energy-Efficient Motors and Drives*

Considered as indispensable components in steel plants, electric motors and drives power a variety of machinery such as pumps, fans, compressors, and rolling mills. Installing energy-efficient motors and drives can result in considerable reductions in electricity consumption and related emissions.

- **High-Efficiency Motors:** Replacing standard motors with specially engineered high-efficiency motors can significantly reduce energy consumption.

6.6 Case Studies and Best Practices

Amid the growing demand and mounting pressure to meet rigorous environmental standards, sustainable steel production has become crucial. This section showcases cutting-edge best practices and innovative case studies from industry leaders in sustainable steel production. These examples propose concrete solutions for minimizing waste production and energy consumption and maximizing efficiency and sustainability.

6.6.1 Steel Industry Best Practices

ArcelorMittal's XCarb™ Initiative

A global industry leader, ArcelorMittal, has launched an XCarb™ initiative to decarbonize steelmaking and encourage circular economy practices [24, 25]. The aims of the initiative are to leverage renewable electricity in electric arc furnaces (EAFs) and to advance carbon capture and storage (CCS) technologies. In addition, ArcelorMittal is exploring the use of cleaner alternatives like green hydrogen and biomass for reducing iron ore in its production processes. The XCarb™ innovation fund serves as a pivotal catalyst in promoting low-carbon innovation in the steel industry, encouraging start-ups and technologies to achieve systematic change.

SSAB's HYBRIT Technology

SSAB, in collaboration with LKAB and Vattenfall, is spearheading innovation in the steel sector with a focus on hydrogen-based iron production through an initiative called the Hydrogen Breakthrough Ironmaking Technology (HYBRIT) [26]. Powered by renewable hydrogen, this novel DRI process is reducing the steel industry's dependence on fossil fuels. SSAB revolutionized the concept of decarbonization of the steel industry with the successful production of fossil-free steel in 2021 using HYBRIT, which significantly reduces carbon emissions, setting a benchmark for the integration of hydrogen technology into large-scale metallurgical operations.

6.6.2 Case Studies of Sustainable Steel Processing

Nucor's EAF Expansion Strategy

Nucor Corporation has long campaigned for utilizing electric arc furnaces that recycle scrap steel and consume significantly less energy compared to traditional blast furnaces [27]. With recent investments aimed at expanding EAF capacity, Nucor has demonstrated that EAF-based production is not only environmentally sustainable but also economically competitive. The company utilizes renewable energy sources where possible, and integrates advanced sensor and control systems to optimize energy efficiency. Their initiatives reflect the scalability of EAF technology and its alignment with low-carbon steel production goals.

Tata Steel's Circular Economy Model

Tata Steel has integrated a circular economy approach into its operations by focusing on recycling, resource efficiency, and waste minimization [28]. The company leads the industry in steel recycling rates and employs closed-loop water systems to minimize consumption. Tata Steel's Zero Waste to Landfill policy ensures that by-products such as slag, dust, and sludge are either reused internally or processed for use in other industries, such as cement and construction. Furthermore, Tata Steel's efforts to incorporate life-cycle thinking into product design and development underline the role of sustainable innovation in achieving long-term environmental performance.

6.7 Recommended Future Directions for Sustainable Steelmaking

6.7.1 Hydrogen-Based Steelmaking: Moving towards Fossil-Free Production

Hydrogen-based steelmaking represents one of the most promising pathways for achieving near-zero carbon emissions in the steel industry [29]. Traditional steel production, particularly *via* blast furnaces, is highly dependent on coal for fuel and reduction, resulting in substantial CO_2

emissions. In contrast, hydrogen can serve as a clean alternative reductant in the direct reduction of iron ore. When produced using renewable energy (green hydrogen), this method emits only water vapor instead of carbon dioxide, effectively decoupling steelmaking from fossil fuel dependency.

One of the most notable advancements in this area is the HYBRIT project developed by SSAB, LKAB, and Vattenfall in Sweden, which successfully demonstrated sustainable steel production using hydrogen reduction technology. This technology not only reduces the carbon footprint of steel but also has the potential to create a new standard for clean metallurgy globally. Despite progress, hurdles persist, such as the high cost and scarcity of green hydrogen, the need for infrastructure development, and scaling the technology for widespread commercial use. As these challenges are tackled with policy measures, innovation, and funding, green steel production using hydrogen could become a cornerstone of sustainable industrial development.

6.7.2 Carbon Capture, Utilization, and Storage (CCUS): Capturing and Repurposing CO_2 Emissions

CCUS is a key approach for reducing the steel industry's greenhouse gas emissions, particularly for integrated steel plants that continue to use carbon-intensive methods like blast furnace operations [30]. CCUS involves capturing carbon dioxide emissions generated by industrial operations, moving the captured CO_2, and either loading it into underground subsurface rock formations (carbon storage) or repurposing it into useful products such as chemicals, fuels, or construction materials (carbon utilization).

For steelmakers, CCUS can be retrofitted to existing facilities, offering a practical interim solution while transitioning to low-carbon production methods. ArcelorMittal and other major steel producers are actively exploring CCUS pilots, integrating carbon capture units at steel plants and developing partnerships to create viable carbon markets. Innovations in technology are enhancing the efficiency and affordability of CO_2 capture, while policy instruments such as carbon pricing, tax incentives, and government funding are helping de-risk investment in this field. As CCUS technologies mature, they will play a pivotal role in bridging the gap between current high-emission practices and the net-zero future.

6.7.3 Artificial Intelligence (AI) and Digitalization: Enhancing Efficiency through Predictive Analytics and Automation

AI and digitalization are transforming the steel manufacturing industry by unlocking new levels of efficiency, precision, and sustainability [31–33]. AI-driven manufacturing systems facilitate real-time tracking and predictive insights throughout production, from raw materials to finished products.

These innovations highlight the aspects of energy improvement, predict equipment failure, optimize furnace operations, and deliver benefits including reduced waste, energy savings and improved product consistency. By creating digital twins of physical assets, steelmakers can simulate and optimize processes, pinpoint potential issues, and implement improvements without affecting actual production. The advancement of automation and robotics is making steel production safer and consistent, particularly in high-risk processes like smelting and casting.

Moreover, data-driven insights, cloud computing, and Internet of Things (IoT) are transforming steel production, delivering enhanced efficiency, productivity, and responsiveness. IoT monitoring and real-time data analytics are enabling steel plants to optimize production processes, envisage and prevent downtime, and respond swiftly to ever-changing market and regulatory demands. Along with the steel industry's digital transformation, AI will be instrumental in driving innovation, optimizing resource use, and accelerating the shift towards a more sustainable and low-carbon future. AI's strength in predictive modeling, process optimization, and anomaly detection will aid in uncovering new avenues for boosting efficiency, cutting waste, and reducing emissions, thereby driving the steel industry towards its sustainability targets.

6.8 Conclusions

The key findings from the study are summarized below:

1. Novel sustainable steel processing techniques are transforming the global steel industry, reducing the environmental footprint and enhancing productivity and efficiency. By adopting these advanced strategies, the steel industry can significantly reduce energy consumption, decrease its carbon footprint, and thereby advance sustainability goals.

2. Sustainable practices enable steel manufacturers to diminish their environmental footprint, enhance product quality, lower costs, and stay ahead of the competition in a market that values sustainability.
3. Electric Arc Furnace steelmaking, Basic Oxygen Furnace processes with integrated waste heat recovery, and Direct Reduced Iron production are among the effective methods for reducing emissions and promoting sustainable steel production. These advancements demonstrate a rising trend towards cleaner and more efficient technologies across the industry.
4. The continued evolution of sustainable steelmaking will rely on three critical areas: the adoption of hydrogen-based steel production to eliminate fossil fuel dependence; the widespread implementation of carbon abatement, utilization, and storage systems to manage and utilize captured emissions; and the integration of artificial intelligence and digital tools to optimize processes, predict equipment failures and reduce energy consumption. Together, these directions hold the key to building a resilient, low-carbon future for the steel industry.
5. ArcelorMittal's XCarb™ initiative, SSAB's HYBRIT technology, Nucor's EAF expansion strategy and Tata Steel's circular economy model are considered the best steel processing initiatives for minimizing waste and energy consumption.

AI disclosure statement: The author confirms that no artificial intelligence (AI)-generated content or AI-assisted writing tools were used in the preparation of this chapter (including summaries and the abstract). All text, analysis, and conclusions are the result of the author's original work.

References

1. Na, H., Yuan, Y., Sun, J., Zhang, L., Du, T., Integrative optimization for energy efficiency, CO_2 reduction, and economic gains in the iron and steel industry: A holistic approach. *Resour. Conserv. Recycl.*, 212, 107992, 2025.
2. Chen, Y., Fang, Y., Feng, W., Zhang, Y., Zhao, G.X., How to minimise the carbon emission of steel building products from a cradle-to-site perspective: A systematic review of recent global research. *J. Clean. Prod.*, 368, 133156, 2022.

3. Larsson, M., Wang, C., Dahl., J., Development of a method for analysing energy, environmental and economic efficiency for an integrated steel plant. *Appl. Therm. Eng.*, 26, 1353, 2006.
4. Mallett, A. and Pal, P., Green transformation in the iron and steel industry in India: Rethinking patterns of innovation. *Energy Strategy Rev.*, 44, 100968, 2022.
5. Mohammad, H.M. and Ibrahim, R.H., Optimization of sustainable cutting conditions in turning carbon steel by CNC turning machine. *J. Mater. Sci. Eng.*, 6, 2, 1000319, 2017.
6. Flick, A. and Fliegel, G., Trends in the steel industry: innovative solutions to drive sustainability. *Trans. Indian Inst. Met.*, 66, 5–6, 535, 2013.
7. Cavaliere, P., Clean ironmaking and steelmaking processes: efficient technologies for greenhouse emissions abatement, in: *Clean Ironmaking and Steelmaking Processes*, pp. 1–37, Springer, Cham, Switzerland, 2019.
8. Sarkar, S. and Mazumder, D., Solid waste management in steel industry - challenges and opportunities. *Int. J. Soc. Behav. Educ. Econ. Bus. Ind. Eng.*, 9, 3, 984, 2015.
9. He, K. and Wang, L., A review of energy use and energy-efficient technologies for the iron and steel industry. *Renewable Sustainable Energy Rev.*, 70, 1022, 2017.
10. Wang., R., Energy saving technologies and optimisation of energy use for decarbonised iron and steel industry, PhD thesis, Durham University, UK, 2022, Available at Durham E-Theses Online: http://etheses.dur.ac.uk/14289/.
11. Larsson., M., Process integration in the steel industry - Possibilities to analyse energy use and environmental impacts for an integrated steel mill, PhD Thesis, 2004.
12. de Beer, J., Worrell, E., Blok, K., Future technologies for energy efficient iron and steel making. *Annu. Rev. Energy Environ.*, 23, 123, 1998.
13. Griffin, P.W. and Hammond, G.P., Industrial energy use and carbon emissions reduction in the iron and steel sector: A UK perspective. *Appl. Energy*, 249, 109, 2019.
14. Holappa., L., A general vision for reduction of energy consumption and CO_2 emissions from the steel industry. *Metals*, 10, 1117, 2020.
15. Shahabuddin, M., Brooks, Geoffrey, Rhamdhani, Muhammad Akbar, Decarbonisation and hydrogen integration of steel industries: Recent development, challenges and techno economic analysis. *J. Clean. Prod.*, 395, 136391, 2023.
16. Ren, L., Shi, H., Yang, Y., Liu, J., Ou, X., Carbon reduction cost of hydrogen steelmaking technology in China. *Energy*, 320, 135177, 2025.
17. dos Santos, G.T.L., Green hydrogen development: An analysis of the Brazilian innovation system, Master's Thesis, Lund University, Sweden, 2023.
18. Wiencke, J., Lavelaine, H., Panteix, PJ., Petitjean, C., Rapin, C., Electrolysis of iron in a molten oxide electrolyte. *J. Appl. Electrochem.*, 48, 115, 2018.

19. Norgate, T., Haque, N., Somerville, M., Jahanshahi, S., Biomass as a source of renewable carbon for iron and steelmaking. *ISIJ Int.*, 52, 8, 1472, 2012.
20. Abu Hanieh, A., Hasan, A., AbdElall, S., Krajnik, P., The role of resource efficiency in engineering education. *11th Global Conference on Sustainable Manufacturing*, pp. 90–95, Berlin, Germany, 23rd-25th September, 2013.
21. Raabe, D., Tasan, C.C., Olivetti., E.A., Strategies for improving the sustainability of structural metals. *Nature*, 575, 64, 2019.
22. Na, H., Sun, J., Qiu, Z., Yuan, Y., Du, T., Optimization of energy efficiency, energy consumption and CO_2 emission in typical iron and steel manufacturing process. *Energy*, 257, 124822, 2022.
23. Paul, S., Kanesalingam, B., Ma, Y., Villanova, J., Requena, G., Akpu, S.C., Raabe, D., Battiato, I., D-Marais, L., A percolating path to green iron. *ChemRxiv*, 2024, https://doi.org/10.26434/chemrxiv-2024-xxh1j.
24. D'Antimo, M., Optimised steel structures for a low carbon future, in: *Life-Cycle of Structures and Infrastructure Systems*, F. Biondini and D.M. Frangopol (Eds.), pp. 2421–2428, CRC Press, London, 2023.
25. Tibolt, M. and Ochojski, W., Steel solutions for a sustainable bridge infrastructure of today and tomorrow. *CE/Papers Proc. Civ. Eng.*, 7, 3–4, 27–32, 2024.
26. Pei, M., Petäjäniemi, M., Regnell, A., Wijk, O., Toward a fossil free future with HYBRIT: Development of iron and steelmaking technology in Sweden and Finland. *Metals*, 10, 972, 2020.
27. A consensus study report on health risk considerations for the use of unencapsulated steel slag, National Academies Press, Washington, DC, pp. 1–147, 2023.
28. Mishra, P.P., Sravan, C., Shivali, R., Circular economy in the Indian extractive industry: A preliminary analytical assessment of the major mining companies. *Clean. Waste Syst.*, 11, 100287, 2025.
29. Ariyama, T., Perspectives on the promising pathways to zero carbon emissions in the steel industry toward 2050. *ISIJ Int.*, 65, 2, 165, 2025.
30. Li, H., Zhao, Y., Guo, C., Li, J., Analysis of technological pathways and development suggestions for blast furnace low-carbon ironmaking. *Metals*, 14, 1276, 2024.
31. Kumar, S., Verma, A.K., Mirza, A., Digitalisation, artificial intelligence, IoT, and industry 4.0 and digital society, in: *Digital Transformation, Artificial Intelligence and Society*, Frontiers of Artificial Intelligence, Ethics and Multidisciplinary Applications, pp. 35–57, Springer, Singapore, 2024.
32. Zong, Z. and Guan, Y., AI-driven intelligent data analytics and predictive analysis in industry 4.0: transforming knowledge, innovation, and efficiency. *J. Knowl. Econ.*, 16, 864, 2025.
33. Mrida, M.S.H., Rahman, M.A., Alam, M.S., AI-driven data analytics and automation: a systematic literature review of industry applications. *Strateg. Data Manag. Innov.*, 02, 1, 21, 2025.

7

Minimizing Cr Loss during EAF Stainless Steel Making for Achieving Sustainability

Subham Ranjan[1,2], Rinku Singroha[2], Sayantan Mukherjee[3] and Sudipta Patra[3]*

[1]*Department of Materials Science and Engineering, Indian Institute of Technology, Kanpur, U.P., India*
[2]*R&D Center, Jindal Stainless Limited, Hisar, Haryana, India*
[3]*Department of Metallurgical Engineering, Indian Institute of Technology (BHU), Varanasi, India*

Abstract
EAF Stainless steel melting slag containing high percentage of Cr_2O_3 content is not only leading to Cr loss but also saving environmental problem as Cr_2O_3 convert to Cr^{6+} which damage the human life. The study designs and develops a pneumatic injection technique which can blow Al dross fines into the slag and reduce the Cr_2O_3 content to less than 4wt% in slag. The drastic reduction of Cr_2O_3 content about 60% from conventional practice can be explained by the slag condition, reduction efficiency achieved by Al and higher surface area induced by fine particle size. The thermodynamic analysis has been done to understand the Cr_2O_3 solubility, and a predictive equation has been formed. On the other hand, the Al dross which is a waste product of aluminum industry, can actually solve the double problem of both aluminum and stainless-steel industry in terms of sustainability.

Keywords: Cr recovery, EAF, stainless steel making, reduction, Al dross

7.1 Introduction

Stainless steel is among the highest growing material year-on-year basis due to its excellent combination of corrosion and mechanical properties along with good manufacturability, sustainability and low cost. Stainless

**Corresponding author*: sudipta.met@iitbhu.ac.in

Gurudas Mandal, Swarup Kumar Ghosh, Snehanshu Pal and Sandip Kunar (eds.) Sustainable Manufacturing of Advanced Materials, (137–150)

steel containing minimum ~11% Cr is generally manufactured through Electric Arc Furnace (EAF)-Argon Oxygen Decarburizer (AOD)-Ladle refining furnace (LRF) route. The raw materials firstly melted to close chemistry in EAF and then decarburized, desulfurized and further refined in AOD. Later on, the LRF is being used for temperature control, trimming addition, deoxidation and inclusion control before lifting continuous casting [1, 2].

Chromium is the most important alloying element that imparts corrosion resistance by producing passive films in surface layer of stainless steel [3, 4]. However, the problem arises in the electric arc furnace melting operation, large amount of chromium converted to chromium oxide and significant loss of Cr through the EAF slag occurs. Another problem is that the Cr_2O_3 (Cr^{3+}) in the EAF slag can be further oxidized to CrO_3(Cr^{6+}) at room temperatures under certain environmental condition. Cr^{6+} can be very dangerous for human as it can damage liver, kidney and also can cause cancer. Therefore, the EAF slag having high amount of Cr_2O_3 can't be used for land filling, construction etc. without treatment and elimination of Cr_2O_3 of the slag. Treatment and elimination of the EAF slag is very difficult and costly affair and thus there is always effort from the stainless steel producer and researcher to reduce the Cr_2O_3 in the slag during EAF operation itself [5–8].

There are a number of research paper and patent which suggested the methodology of chromium reduction in EAF slag and are listed below.

a. To control Cr_2O_3 content, the charge and silicon-rich FeSi alloy are used to ensure approximately 0.3% Si in the melt after processing [9–12].
b. Efficient Cr_2O_3 reduction is achieved by injecting carbon into the slag while simultaneously blowing oxygen [11–13].
c. Favorable reduction of Cr_2O_3 requires high slag basicity, high carbon activity ($a_c = 1$), low chromium activity in the metal, low partial pressure of oxygen, and high operating temperatures [11, 12].
d. Technology for the reduction with silicon and aluminum and the blowing in of carbon was also explored [11, 12].
e. Calcium carbide injection is another reported method for enhancing chromium oxide reduction [11, 12, 14].
f. Japanese Laid-Open Patent Application No. 2001-316712 describes a method for reducing chromium oxide by blowing reducing agents (such as aluminum, aluminum dross, or

carbon) with inert gas through a hollow electrode in an electric arc furnace [15, 16].

g. Similarly, Korean Laid-Open Patent Application No. 2000-0021329 describes a process where carbon powder is blown into an electric arc furnace to recover valuable metals from slag. However, this process suffers from slow chromium reduction rates due to relatively low temperatures [16, 17].

There has been reported problem of Si reduction. Accordingly, the amount of silicon used for chromium reduction does not reach 50% of amount of silicon added. Also, when a large amount of silicon is added to increase the amount of chromium recovered from the slag, a large amount of silicon oxide (SiO) is generated. Accordingly, the basicity (CaO/SiO) of the slag deteriorates so that the fluidity of slag is diminished. This, in turn, lowers the working efficiency of the process and represents a disadvantageous condition in the course of reducing chromium oxide in the slag [16].

The chromium oxide content in the slag generated as a by-product in the process for manufacturing the stainless steel using the electrical furnace is relatively high, e.g., on the order of 5% to 30%. Therefore, to reduce manufacturing costs and more effectively utilize resources, after the ferro-alloy and the scrap are melted, a reducing agent such as iron-silicon (Fe-Si) or aluminum is commonly added to the EAF used to raise the temperature of such molten steel. Addition of such a reducing agent reduces the chromium oxide in the slag and increases the content of metallic chromium found in the molten steel.

The current study studies the role of aluminum dross having Al content 18-30% as reducing agent in electric arc furnace stainless steel melting to reduce Cr_2O_3 content in the slag according to the equation below. Thus, the sustainability in both Cr recovery, Cr^{6+} removal from slag and also the waste of Al industry can be minimized.

7.2 Description of the Pneumatic Injection

Figure 7.1 shows the pneumatic injection machine, which was manufactured indigenously at Jindal stainless Limited, Hisar. The system can blow Al dross with nitrogen as carrier gas. The powdered aluminum dross reducing agent can be blown into the slag layer using a nitrogen gas to bring the powdered aluminum dross particles into contact with the chromium oxides within the slag. Nitrogen gas pressure was maintained in real heats so that it can effectively penetrate the slag layer and about 2bar pressure

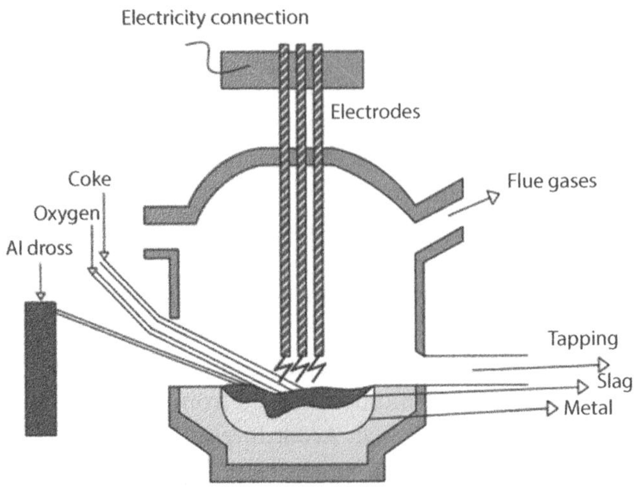

Figure 7.1 Schematic diagram of EAF with pneumatic injection machine in JSL.

was sufficient. A steel tube related to the initial flexible pipe to blow in the powdered aluminum dross using the nitrogen carrier gas to transport the particles. Gas and particles can be blown through this mild steel tube and into the furnace from the door where oxygen blowing is done.

18% Al dross was tried in the initial trial heats, but it was noticed that the required amount of the dross is more, and it increased the slag volume significantly; therefore, the losses could not be managed. About 1000 kg of Al dross was added to the slag layer considering the chromium oxide in a 40ton furnace.

To promote a relatively fast time of reaction between the metal oxides in the slag and the added aluminum dross reducing agent, the slag should be maintained in a liquid phase, and the slag should also have sufficient fluidity (e.g., sufficiently low viscosity) to permit the desired reduction of the metal oxides therein. Fluidity of slag becomes optimal when the slag basicity (defined herein as the CaO/SiO_2 weight ratio) ranges from about 1.1 to 1.7 and the alumina content of the slag is 10 wt % or more. Maintenance of appropriate slag fluidity increases the reduction reaction speed and also promotes the absorption of the reduced valuable metals from the slag into the molten steel according to the earlier studies [16].

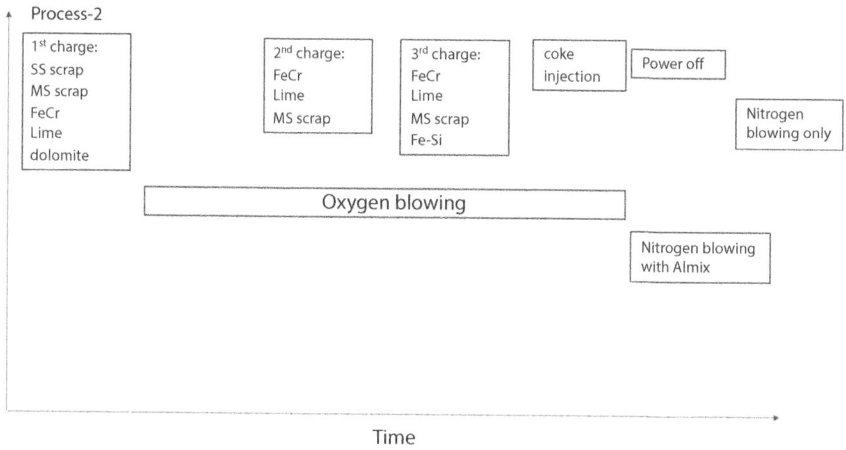

Figure 7.2 EAF process route.

7.3 Process

The schematic diagram of the process route is given in Figure 7.2 for a typical heat of 304 grade. The first charge mix contains SS scrap, MS scrap, FeCr, lime and dolomite. After the charging, the power continues to increase and after 5minutes, oxygen blowing from the slag door continues for about 15minutes, and then the second charge has been taken and oxygen blowing continued for another 15minutes, and 3rd charge has been taken and the O_2 blowing continues for another 5minutes. Then the coke injection was carried out along with O_2 for 10min and the complete melting of raw material with temperature about 1550°C was achieved. Then Al dross was started along with the nitrogen in the slag metal interface for 5min and only nitrogen blowing continued for 2 min. The tapping of slag and then metal was carried out.

7.4 Results

Table 7.1 shows the composition of the slags of the industrial trial heats where Cr_2O_3 content reached about ~3% with the above-mentioned method. The CaO content and SiO_2 content reaches 30%, FeO content decreased to 0.5%, Al_2O_3 content increased to 20%, MnO content about 2.5% and Cr_2O_3 content varies between 3-4% range.

Table 7.1 Industrial trial heats for Cr recovery in EAF.

CaO	SiO_2	MgO	FeO	Al_2O_3	MnO	Cr_2O_3
31.60	32.40	9.55	0.55	18.40	2.48	3.13
28.55	29.20	8.26	0.60	24.00	2.40	3.35
33.30	29.60	9.00	0.80	18.80	2.66	3.38
35.70	29.00	8.66	0.66	17.80	3.13	3.44
35.44	30.50	8.88	0.80	17.20	1.73	3.25
32.20	30.40	8.00	0.80	20.20	2.56	3.56
33.50	30.00	8.20	1.20	19.00	2.00	3.90
29.60	28.20	8.66	1.10	23.00	2.80	3.67
34.70	30.50	9.40	0.80	18.10	2.00	2.90
34.00	31.30	9.00	0.46	18.00	2.17	3.13
34.30	29.80	8.99	0.45	19.00	2.25	3.00

Table 7.2 shows non trial heats where CaO and SiO2 reach ~30%, FeO content increased to 1.4-2%, Al_2O_3 content reduced to 9-14%, MnO increases to 3.9-5.7% and Cr2O3 content increased to 10-12%.

Figure 7.3 plotted the Al_2O_3 vs Cr_2O_3 which shows clear trend that as the Al_2O_3 increases Cr_2O_3 content in the slag decreases which clearly tells that the Al in the dross reduced the Cr_2O_3 to Cr and through the reaction.

To understand the amount of Cr_2O_3 solubility in the slag phase under the temperature of 1450°C to 1600°C, thermodynamic calculation has been carried out and given in Figure 7.4; as the Cr_2O_3 content increases the solubility of chromium in slag phase decreases and in the spinel phase increases.

Table 7.2 Non trial heats.

CaO	SiO_2	MgO	FeO	Al_2O_3	MnO	Cr_2O_3
33.10	30.60	8.23	1.45	9.80	3.90	11.05
30.00	30.30	7.00	1.20	8.90	5.68	12.74
29.60	29.10	7.30	1.64	9.00	7.70	10.66
28.50	29.30	6.56	2.02	7.68	10.20	10.80

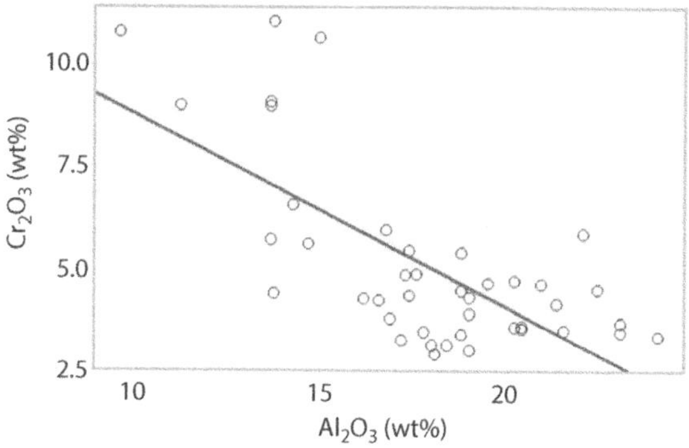

Figure 7.3 Effect of aluminum on chromium reduction.

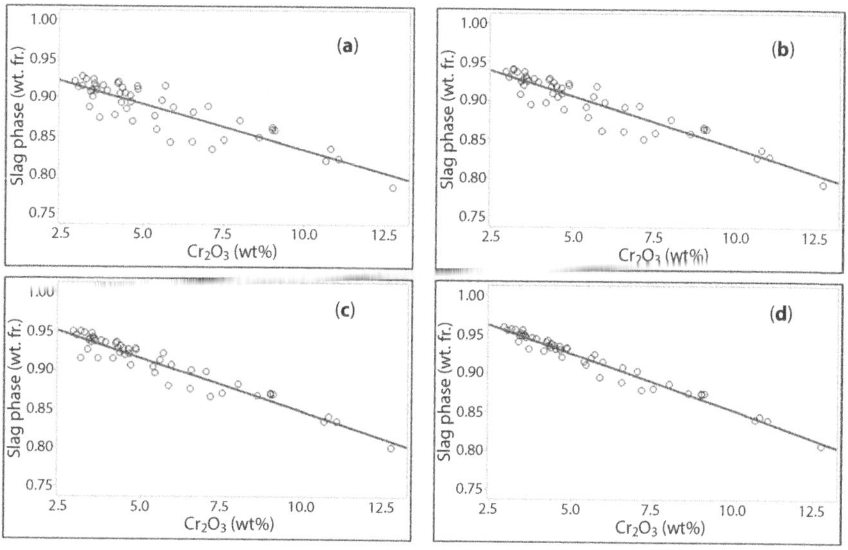

Figure 7.4 Effect of slag phase on chromium oxide formation in (a) 1450°C (b) 1500°C (c)1550°C (d) 1600°C.

Also, with increasing the temperature, the solubility of Cr_2O_3 in slag phase increases and in the spinel, phase decreases, while keeping the Cr_2O_3 content the same.

Based on the available data a predictive equation can be formed to understand the role of temperature and Cr_2O_3 on the amount of slag phase. Equation 7.1 can correlate it very well as observed in Figure 7.5.

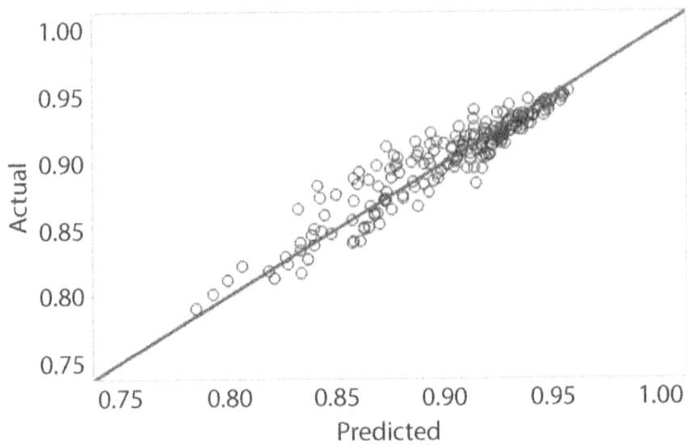

Figure 7.5 Actual vs predicted slag phase with varying chromium oxide and temperature.

$$Slag\ Phase\ (wt.\ fr.) = 0.599 - 0.0133 \times wt\%\ Cr_2O_3 + 2.1 \times 10^{-4} \times T \tag{7.1}$$

7.5 Discussion

7.5.1 The Role of Temperature and Slag Condition on the Reducibility

To reduce the Cr_2O_3, the activity of Cr_2O_3 should be understood as it has to take part in the reaction. The literature suggests that Cr_2O_3 activity in stainless steel refining slag increases with increasing the basicity and optimized slag composition saturated with $CaCr_2O_4$ and Cr_2O_3 was observed in the moderately reducing condition. MgO increase the activity of the slag when added until 10%; but then, it decreases the activity and Al_2O_3 slightly decreases the activity of Cr_2O_3. To minimize chromium loss, 44CaO-39SiO_2-11Al_2O_3 and 6MgO has been considered optimal at 1600°C. The optimum basicity has been suggested about 1.4-1.8. The higher basicity lowers the activity of SiO_2 and Al_2O_3 and increases the activity of Cr_2O_3 and better reduction can be achieved. In our study, it has been observed the basicity of the slag with respect to (CaO+MgO)/(SiO_2+Al_2O_3) in the final slag is about 1. As the reduction took place the quantity of Al_2O_3 increased which decreases the basicity from 1.3 to 1 after reduction [18].

The reduction behavior in aluminum has been observed in a stainless-steel melt and it was observed that within 5minutes, 90% MnO can be reduced, FeO and Cr_2O_3 was 80%, and 55% SiO_2. The reduction is controlled by diffusion in the slag and the mass transfer rate following sequences MnO>Cr_2O_3>FeO>SiO_2. Kinetically the reduction rate constant increases linearly with the product of diffusion coefficient and oxide activity. In our study the Al blowing time was about 4-5 minutes and 2-3 minute was the N_2 stirring time and the reduction of Cr_2O_3 has been completed. The MnO reduction and FeO reduction was also completed within that period of time [19].

Another option is to minimize the chrome loss during oxidation, which guides that the reducing condition should be maintained by adjusting the carbon blowing and oxygen blowing together, and oxygen should be blown in more than 1600°C where carbon efficiently oxidize, minimizing the chromium loss. It is very difficult to maintain the oxygen blowing as the difficult to melt pieces of FeCr and other ferroalloys need to be cut into small pieces by oxygen blowing. The carbon blowing was continued in our study along with oxygen blowing to save the silicon oxidation; but still, about 20%Cr_2O_3 of the slag can be observed before reduction [11].

Ferrosilicon plus carbon powder has been tried to promote CO gas bubble and foaming slag. Calcium carbide is also helpful even at lower temperature to promote the formation of CO and CaO. The practices were adopted but the Cr loss could not be guaranteed less than 10% in the slag It was also mentioned that the good steel-slag mixing during tapping enhances Cr_2O_3 reduction by silicon. But in practice, the silicon could not efficient reduce the Cr_2O_3 in the slag [11].

To account for unavoidable MgO dissolution from refractories and the beneficial effect of Al_2O_3 in lowering slag liquidus temperature, the effects of MgO and Al_2O_3 on chromium oxide activity were studied. With slag composition 41% CaO, 41% SiO_2, 10% Al_2O_3, 8% CrO_x, the Cr^{3+}/Cr^{2+} ratio decreased with the increasing MgO due to changes in CrO and $CrO_{1.5}$ activity coefficients. CrO_x activity coefficient increased initially with 2.5% MgO, then increased up to 10%. Beyond 10%, a decrease was attributed to precipitation of $Mg(Cr,Al)_2O_4$, confirmed by XRD. A slag containing 2.5–10% MgO is optimal for minimizing chromium loss in stainless steel refining. In our study, MgO was maintained at 8-10% continuously to effectively reduce the Cr_2O_3 loss; and it can also protect the refractory from erosion [20].

7.5.2 The Role of Al Dross Compared to Si on the Reducibility of EAF Slag

Silicon (as ferrosilicon) is effective for reducing chromium oxides (CrO_x) in slag. Adding about 7 kg/tonne of silicon decreases CrO_x to 30% of its original level. However, only ~36% of this silicon participates in the reduction reaction; the remaining ~64% dissolves into the steel, increasing its silicon content to ~0.5 wt-%. Excess silicon in steel causes several issues: it consumes more oxygen during subsequent converter refining, increases slag volume (reducing oxygen transfer efficiency), and raises bath temperature due to its highly exothermic oxidation, which can damage refractories. As a result, silicon acts more like a sacrificial reductant, preferentially oxidizing before chromium. Literature indicates that silicon "protects" chromium by being oxidized first. There is minimal benefit beyond ~0.4 wt-% silicon (~6 kg/tonne), and a practical upper limit of 0.3 wt-% silicon is recommended. Additionally, decreasing silica activity in the slag lowers chromium losses, supporting the use of a highly basic slag. Our findings suggest that silicon cannot fully reduce Cr_2O_3 content in the slag under these conditions [21].

Aluminum on the other hand is an effective method for removing chromium oxide (Cr_2O_3) from stainless steel slag. This study investigated aluminum reduction in synthetic $CaO–SiO_2–Al_2O_3–MgO–Fe_2O_3–Cr_2O_3$ slags under various conditions to optimize reduction efficiency and metal–slag separation. Higher temperature improves agglomeration of metal droplet and separation between metal and slag. Increased slag basicity enhances the reduction of SiO_2, Fe_2O_3, and Cr_2O_3. Addition of CaF_2 improves metal droplet agglomeration and metal–slag separation. The kinetics of the reaction can be written in the equations given below [22].

The reduction rate of each oxide MO (MO represents Cr_2O_3, MnO, FeO in the slag can be described by Equation 7.2 based on the two-film theory for interphase mass transfer [19].

$$-\frac{d(\%MO)}{dt} = \frac{A}{V_s} k_s \left[(\%MO)^b - (\%MO)^e\right] \tag{7.2}$$

where A, V_s, k_s are the slag/metal interfacial area [m^2], the volume of slag [m^3], and the slag phase mass transfer coefficient [m/s], respectively. Also, the superscript 'b' and 'e' refer to bulk and equilibrium concentration of MO in the slag.

A diffusional mass transfer coefficient in slag phase is defined as follow [19]

$$k_s = \frac{D_s}{\delta} \tag{7.3}$$

where D_S and δ are the diffusion coefficient in slag and the thickness of the concentration boundary layer, respectively.

Reduction rate constant of each oxide, K_O^{MO} could be in proportion to the product of diffusion coefficient and the activity of each oxide in molten slag as follows [19]:

$$K_O^{MO} \propto k_S^{MO}\ a_{MO} \propto D_M\ a_{MO} \tag{7.4}$$

D_M is the diffusion coefficient of element M constituting oxide MO in slag phase a_{MO} is the activity of MO in the slag.

In our study, the size of the aluminum dross was very fine (less than 1mm), which increases the surface area A for the reaction according to Equation 7.2. As the surface area increases, it also influences the distance to travel for the reaction. The mixing with N_2 blowing can lead to extensive mixing of Al dross with the slag, which also decreases the resistance to diffusion which is directly proportional to the kinetics. Thus, Al dross with very fine size, along with N_2 blowing, can minimize the Cr loss in the EAF slag.

7.6 Conclusion

Cr loss during EAF stainless steel making has been studied and the following conclusion can be drawn.

1. Cr_2O_3 can be decreased to less than 4% in the slag which is about 60% from the conventional practice by blowing Al dross fines from aluminum industry waste.
2. The reason behind the significant improvement of Cr recovery has been explained by the slag condition, reaction kinetics and benefit of aluminum as a reductant.
3. A pneumatic injection machine was developed for the blowing of Al dross with N_2 as carrier gas.
4. The thermodynamic analysis suggests that Cr_2O_3 solubility in slag varies with temperature and can be predicted well.
5. The process can provide two-fold sustainability as it can solve the loss of both stainless steel and aluminum industry, and can also solve waste problem.

Acknowledgement

The Authors would like to acknowledge the COO, JSL, Hisar plant for encouragement and guidance in the research work.

AI disclosure statement: The author acknowledges the use of AI-based tools for minor language editing or grammatical correction. However, all intellectual contributions, analyses, and conclusions presented in this chapter are entirely the author's own.

References

1. Patil, B.V., Chan, A.H., Choulet, R.J., Refining of Stainless Steels, pp. 715–741, 1998.
2. Ma, G., The characteristics, formation mechanisms and treatment processes of Cr (VI)- containing pyrometallurgical wastes: a review, 2006, http://www.springer.com/978-0-387-35416-3.
3. Miller-lhli, N.J., Chromium, in: *Hazard. Met. Environ.*, Elsevier Science Publishers B.V, pp. 373–404, 1992, https://doi.org/10.1016/S0167-9244(08)70112-4.
4. Sperling, M., Chromium ☆, in: *Mol. Sci. Chem. Eng.*, 2014, https://doi.org/10.1016/B978-0-12-409547-2.11125-4.
5. Kimbrough, D.E., Cohen, Y., Winer, A.M., Creelman, L., Mabuni, C., A critical assessment of chromium in the environment. *Crit. Rev. Environ. Sci. Technol.*, 29, 1–46, 1999, https://doi.org/10.1080/10643389991259164.
6. Prasad, S., Yadav, K.K., Kumar, S., Gupta, N., Cabral-Pinto, M.M.S., Rezania, S., Radwan, N., Alam, J., Chromium contamination and effect on environmental health and its remediation: A sustainable approaches. *J. Environ. Manage.*, 285, 112174, 2021, https://doi.org/10.1016/j.jenvman.2021.112174.
7. Singh, R., Kumar, M., Bishnoi, N.R., Development of biomaterial for chromium(VI) detoxification using Aspergillus flavus system supported with iron. *Ecol. Eng.*, 91, 31–40, 2016, https://doi.org/10.1016/j.ecoleng.2016.01.060.
8. Yu, Z., Xu, S.F., Zhao, J.L., Zhao, L., Zhang, A.Z., Li, M.Y., Toxic effects of hexavalent chromium (Cr6+) on bioaccumulation, apoptosis, oxidative damage and inflammatory response in Channa asiatica. *Environ. Toxicol. Pharmacol.*, 87, 103725, 2021, https://doi.org/10.1016/j.etap.2021.103725.
9. Kurka, V., Jonšta, P., Carbol, Z., Machovčák, P., Chromium reduction from high chromium slag with reducing agents of C and Si in an atmospheric induction melting furnace using an oxygen/fuel burner. *Met. 2016 - 25th Anniv. Int. Conf. Metall. Mater. Conf. Proc.*, vol. 84, pp. 84–90, 2016.
10. Yang, Q.X., Xu, A.J., Xue, P., He, D.F., Li, J.L., Björkman, B., Briquette Smelting in Electric Arc Furnace to Recycle Wastes from Stainless Steel

Production. *J. Iron. Steel Res. Int.*, 22, 10–16, 2015, https://doi.org/10.1016/S1006-706X(15)30131-X.
11. Arh, B. and Tehovnik, F., The oxidation and reduction of chromium during the elaboration of stainless steels in an electric arc furnace. *Mater. Tehnol.*, 41, 203–211, 2007.
12. Arh, B. and Tehovnik, F., The oxidation and reduction of chromium of stainless steels in an eletric arc furnace. *Metalurgija*, 50, 179–182, 2011.
13. Hilty, D.C. and Kaveney, T.F., Stainless Steel Melting. *Electr. Furn. Steelmak.*, pp. 143–160.
14. Björkvall, J., Ångström, S., Kallin, L., Reduction of chromium oxide containing slags using CaC2. *VII Int. Conf. Molten Slags Fluxes Salts*, pp. 663–670, 2004.
15. Kobayashi, JPH 2001, 316712, 2001.
16. Jung, S.-Y., Kim, Y.H., Byun, S.-M., Lee, S.-B., Chun, H.-C., Method for reducing to metallic chromium the chromumoxde in slag from stanless steel processing US 2008/0156144 A1, 2008.
17. Jae-yeol, K.Y.S., Recovery method of transition metal in stainless steel slag KR20000021329A, 1998.
18. Morita, K., Tsukiashi, K., Kimura, M., Sano, N., Activity of chromium oxide in CaO-SiO2 based slags at 1873 K. *Steel Res. Int.*, 76, 279–283, 2005, https://doi.org/10.1002/srin.200506009.
19. Park, J.H., Song, H.S., Min, D.J., Reduction behavior of EAF slags containing Cr2O3 using aluminum at 1 793 K. *ISIJ Int.*, 44, 790–794, 2004, https://doi.org/10.2355/isijinternational.44.790.
20. Kho, T.S., Swinbourne, D.R., Blanpain, B., Arnout, S., Langberg, D., Understanding stainless steelmaking through computational thermodynamics part 1: Electric arc furnace melting. *Trans. Inst. Min. Metall. Sect. C Miner. Process. Extr. Metall.*, 119, 1–8, 2010, https://doi.org/10.1179/174328509X431454.
21. Guo, M., Durinck, D., Jones, P.T., Heylen, G., Hendrickx, R., Baeten, R., Blanpain, B., Wollants, P., EAF Stainless Steel Refining - Part I Observational Study on Chromium Recovery. *Steel Res. Int.*, 78, 117–124, 2007, https://doi.org/10.1002/srin.200705869.
22. Lin, Y., Chou, K., Scheller, P., Shu, Q., Reduction of synthetic stainless steel slags by aluminium. *Ironmaking Steelmaking*, 46, 81–88, 2019, https://doi.org/10.1080/03019233.2017.1342411.

8

Bio-Inspired Steel Coatings for Corrosion Resistance and Sustainability

Sneha Roy[1] and Swarup Kumar Ghosh[2]*

[1]JSW Bhushan Power & Steel Limited, Sambalpur, Odisha, India
[2]Department of Metallurgy and Materials Engineering, Indian Institute of Engineering Science and Technology, Shibpur, Howrah, West Bengal, India

Abstract

Corrosion is a problem that is omnipresent, and it presents severe challenges that result in the degradation of steel infrastructure all over the world. Not only does this omnipresent problem present serious safety concerns, but it also results in economic losses amounting to billions of dollars annually. Considering the severity of this problem, there has been recent advancement in the field of biomimetics, which is an interdisciplinary field that draws inspiration from the intricate systems of nature. This ground-breaking research has resulted in the creation of novel coatings specifically tailored for corrosion applications. These ground-breaking coatings precisely replicate the structural and functional characteristics of numerous natural surfaces, such as the incredible lotus leaves, the special skin of sharks, and the tough shells of molluscs. Consequently, the surfaces treated with these coatings possess desirable characteristics, such as water repellence, self-cleaning, and resistance to fouling. This study elaborates on the different mechanisms, materials, fabrication techniques, and sustainability aspects pertaining to bio-inspired coatings designed specifically for steel and highlights pertinent case studies showcasing real-world applications and address future directions, identifying emerging trends and opportunities in this new and exciting area.

***Keywords*:** Biomimicry, bio-inspired coating, steel coating, corrosion resistance, sustainability, coating performance

**Corresponding author*: skghosh@metal.iiests.ac.in

Gurudas Mandal, Swarup Kumar Ghosh, Snehanshu Pal and Sandip Kunar (eds.) Sustainable Manufacturing of Advanced Materials, (151–178) © 2026 Scrivener Publishing LLC

8.1 Introduction

8.1.1 Overview of Steel Corrosion

Steel, especially carbon steel, is widely used in engineering due to its exceptional strength-to-cost ratio, durability, versatility, sustainability and cost-effectiveness. Steel, a critical structural material used across industries such as construction, automotive, energy, manufacturing, transportation and consumer goods (Figure 8.1), is inherently susceptible to corrosion.

Corrosion is an electrochemical process that occurs when steel is exposed to environmental agents like moisture, oxygen, chlorides, and pollutants, leading to its gradual degradation. This phenomenon compromises the structural integrity and load-bearing capacity of steel components, often resulting in catastrophic failures, costly repairs, and maintenance delays. Globally, the financial burden imposed by steel corrosion is staggering, which can be estimated to consume 3% to 4% of a nation's GDP in terms of direct and indirect costs [1]. Metal corrosion has significant economic implications, particularly in important sectors such as construction, automotive, energy, manufacturing and transportation (Figure 8.2), which can affect the national development of a country.

In industrial environments, corrosion manifests in multiple forms, such as uniform corrosion, pitting, crevice corrosion, galvanic corrosion, and stress-corrosion cracking, each of which is dictated by the nature of exposure and material design. Corrosion is influenced by multiple factors, including metal purity, corrosion product characteristics, temperature, pH levels, film formation, and environmental conditions (Figure 8.3). For instance, in marine structures, aggressive chloride ions accelerate localized corrosion, while atmospheric exposure leads to oxide layer formation that

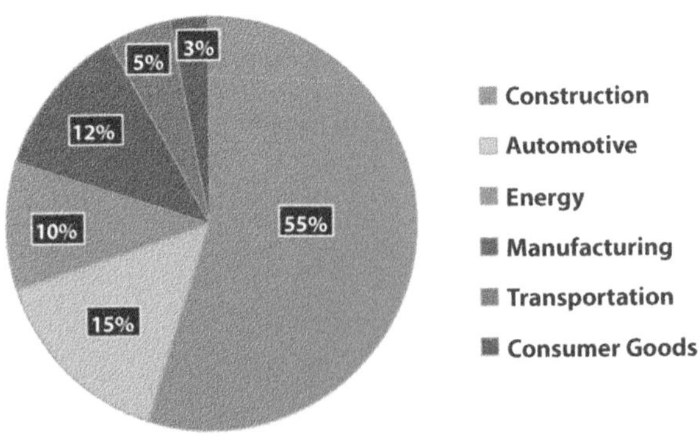

Figure 8.1 Application of steel across different sectors.

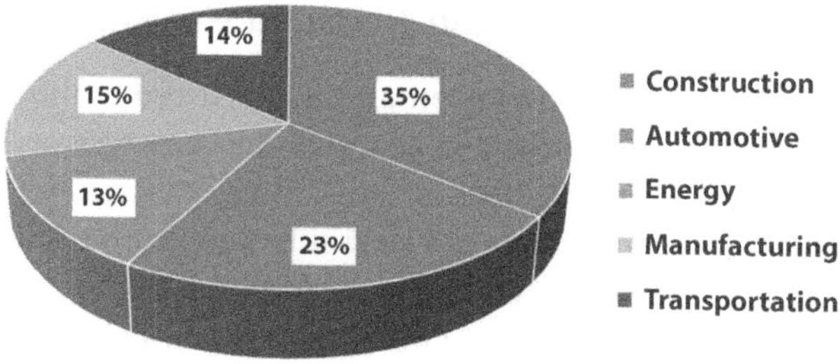

Figure 8.2 Economic losses from corrosion in key industries.

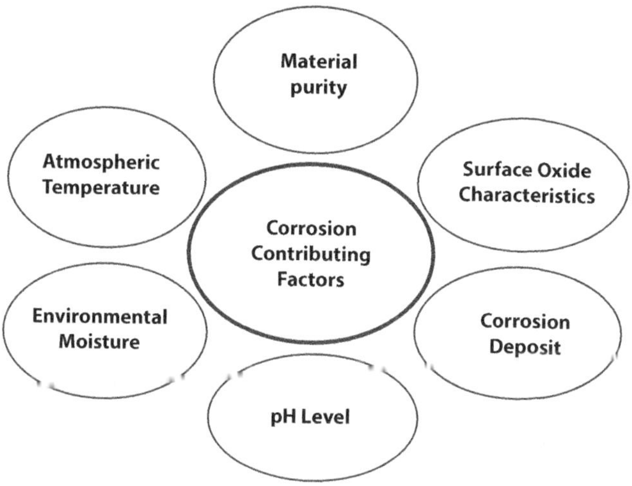

Figure 8.3 Schematic representation of corrosion influencing factors.

often lacks protective passivation [2]. The problem becomes more acute in ageing infrastructure and offshore energy systems, where accessibility for inspection and repair is limited.

Due to its poor corrosion resistance, steel requires protection when exposed to corrosive conditions. Depending on the environment and protection needs, metal, polymer, and organic-inorganic hybrid coatings offer diverse anti-corrosion solutions with varying performance levels [3]. In critical sectors like oil and gas, corrosion-induced failures have even led to environmental disasters and human fatalities. This has made corrosion control a central pillar in sustainable engineering design, where materials must not only be strong and functional but also resilient to environmental degradation over extended lifespans.

8.1.2 Importance of Corrosion-Resistant Coatings

Given the omnipresence of corrosion and its high economic impact, the development of corrosion-resistant coatings has emerged as a primary defense mechanism to protect steel surfaces. These coatings act as physical and chemical barriers that isolate the steel substrate from corrosive agents such as water, oxygen, acids, and salts. Traditionally, coatings based on epoxy resins, zinc-rich primers, and polyurethanes have been used to provide long-term protection under diverse service conditions [4]. However, many conventional coatings face limitations such as environmental toxicity, poor long-term adhesion, and susceptibility to mechanical damage. Their performance often diminishes in aggressive environments like coastal or industrial zones where fluctuating humidity, temperature, and chemical exposure are prevalent. In recent years, environmental regulations such as the European Union's Registration, Evaluation, Authorisation and Restriction of Chemicals (REACH) directive have restricted the use of hazardous materials in coatings, pushing researchers to seek more sustainable and environmentally benign alternatives [5].

Corrosion-resistant coatings are not merely protective layers but functional materials engineered to endure long-term service with minimal maintenance. Their performance parameters now include not only corrosion inhibition but also hydrophobicity, mechanical durability, thermal stability, UV resistance, and even self-healing capabilities [6]. The modern vision for steel protection thus converges toward multifunctional, eco-friendly, and intelligent coating systems.

8.1.3 Bio-Inspired Approach to Corrosion Protection

Nature, through millions of years of evolution, has perfected surface structures and chemical compositions that exhibit extraordinary properties such as self-cleaning, anti-fouling, drag reduction, water repellence, and mechanical robustness. The field of biomimetics, or bio-inspiration, seeks to replicate these natural systems in synthetic materials to achieve similar functionalities. Applying this philosophy to corrosion-resistant coatings involves designing surface topographies and chemistries inspired by organisms such as lotus leaves, shark skin, butterfly wings, and mollusc shells [7].

One of the most well-known examples is the "lotus effect," derived from the lotus leaf (*Nelumbo nucifera*), which exhibits superhydrophobicity due to its dual-scale roughness and waxy surface composition.

Coatings that emulate this surface texture and chemistry can drastically reduce the wettability of steel, thereby minimizing the residence time of moisture and corrosive ions [8]. When water beads roll off such surfaces, they also carry away contaminants, providing a self-cleaning function beneficial in polluted or marine environments. Similarly, shark skin is equipped with microscopic ridges called dermal denticles that reduce hydrodynamic drag and prevent the attachment of marine organisms. Coatings engineered with analogous micro-patterns can prevent biofouling, a major contributor to under-deposit corrosion in submerged steel components such as ship hulls and offshore platforms [9]. In addition to physical mimicry, chemical inspiration has led to the development of coatings using natural polymers like chitosan, cellulose, and gelatine, which offer intrinsic antimicrobial and corrosion-inhibiting properties. These bio-based materials not only reduce environmental impact but also add functionality, such as biodegradability and non-toxicity, to corrosion-resistant systems [10, 11].

The integration of bio-inspired design principles with modern fabrication techniques such as sol-gel processing, electrochemical deposition, and nanoimprinting has made it possible to realize coatings that are structurally complex, multifunctional, and environmentally sustainable. As a result, bio-inspired coatings are increasingly viewed as the next frontier in steel protection technologies by merging biology with materials science to deliver smarter, greener, and more effective solutions.

8.2 Bio-Inspired Coating Materials

Bio-inspired surfaces are gaining traction in corrosion protection due to their innovative designs and functional benefits [12]. The quest for high-performance, multifunctional, and sustainable corrosion protection systems has led to the exploration of bio-inspired coating materials where a class of coatings mimic the morphological and functional properties found in natural organisms. Nature offers a rich library of surface designs that demonstrate superior performance in water repellence, anti-fouling, and chemical resistance. These surfaces typically achieve their properties through hierarchical surface architecture and specialized surface chemistries. Mimicking these designs on steel substrates has opened new avenues in protective coatings technology. This section explores three major categories of bio-inspired coatings: lotus-leaf-inspired coatings, shark-skin-inspired coatings, and biomimetic polymer coatings.

8.2.1 Lotus-Leaf Inspired Coatings

The lotus leaf (*Nelumbo nucifera*) is globally renowned for its superhydrophobic and self-cleaning characteristics, a phenomenon often termed the "*lotus effect*." The unique surface of the lotus leaf comprises micro-scale papillae (5–10 μm in diameter), overlaid with nano-scale wax tubules (100–200 nm), which collectively reduce the contact area between water and the leaf surface. As a result, water droplets form nearly spherical beads with contact angles exceeding 150°, easily rolling off and collecting dust and contaminants in the process [8].

In engineered systems, replicating this behavior involves the creation of hierarchical surface structures combined with low-surface-energy materials. For steel applications, this has been achieved through various methods such as:

- **Sol-gel deposition** of silica or titania nanoparticles followed by fluorosilane treatment [13].
- **Etching and anodization** to create micro-roughened metal surfaces subsequently functionalized with hydrophobic agents [10].
- **Electrospinning and laser patterning** to impart multi-scale surface texture [14].

Such coatings not only reduce water wettability but also provide barrier protection against corrosion by minimizing the residence time of water and aggressive ions like Cl^- on the steel surface. Moreover, the self-cleaning nature of lotus-like coatings helps maintain their aesthetic and functional performance over time, reducing maintenance and cleaning costs.

Applications of lotus-inspired coatings have been demonstrated in automotive panels, outdoor infrastructure, energy towers, and even biomedical implants where both anti-wetting and anti-bacterial functions are desired [15]. In hostile environments like coastal regions, the significant reduction in corrosion current density (I_{corr}) observed in lotus-type surfaces has validated their potential for real-world deployment.

8.2.2 Shark-Skin Inspired Coatings

Shark skin has evolved to exhibit drag-reducing and anti-fouling properties, where a biological adaptation allows sharks to swim efficiently through water while preventing the colonization of algae and microorganisms. The secret lies in its microstructure: the skin is covered with overlapping, tooth-like scales called dermal denticles, aligned in flow direction, with riblet sizes ranging from 200 to 300 μm [16]. These structures generate vortices in the fluid boundary layer that reduce turbulent flow and inhibit the attachment of fouling organisms.

Translating this design to steel protection, researchers have developed micro-patterned coatings using:

- **Soft lithography and micromolding** can replicate shark-skin-like textures on polymeric films or metallic foils [10].
- **Laser interference lithography** allows large-area patterning of riblet structures with high fidelity [17].
- **3D printing and nanoimprint lithography** enable scalable production of drag-reducing topographies on complex surfaces.

In marine steel structures such as ship hulls, offshore platforms, and desalination pipes, shark-skin-inspired coatings have demonstrated a reduction in biofouling by up to 60% and drag reduction of 7%–10%, leading to substantial fuel savings and lower emissions [18]. Importantly, these coatings operate through passive anti-fouling mechanisms rather than toxic biocides, making them environmentally compliant under stringent maritime regulations. Furthermore, the durability and reusability of sharkskin-like polymer coatings add to their appeal for long-term industrial applications, especially where repainting or recoating is expensive and labor-intensive.

8.2.3 Biomimetic Polymer Coatings

While lotus and sharkskin coatings emphasize topological mimicry, biomimetic polymer coatings derive their inspiration from the chemical composition and functional performance of biological macromolecules. Natural polymers such as chitosan, cellulose, alginate, gelatine, and silk fibroin have garnered interest for use in corrosion-resistant coatings due to their abundance, biodegradability, non-toxicity, and functional versatility.

(i) Chitosan-Based Coatings

Chitosan, a deacetylated derivative of chitin obtained from crustacean shells, contains abundant amino and hydroxyl groups that can coordinate with metal ions. These groups can form protective complexes on steel surfaces, thereby inhibiting corrosion. Chitosan-based coatings have shown appreciable barrier performance in acidic and saline environments, particularly when cross-linked with glutaraldehyde or doped with nanoparticles [19].

(ii) Cellulose and Derivatives
Cellulose, the most abundant biopolymer on earth, provides a strong, water-insoluble matrix that can be functionalized with hydrophobic groups to enhance barrier properties. Nano-cellulose-based coatings have been reported to form dense, tortuous diffusion paths that delay the ingress of oxygen and water vapor, extending the service life of steel components [20].

(iii) Hybrid and Composite Biomimetic Coatings
Combining natural polymers with inorganic nanofillers such as graphene oxide, montmorillonite, or zinc phosphate can further improve their anti-corrosive efficiency. These hybrid coatings leverage the synergistic effects of polymer flexibility, chemical functionality, and nanofiller impermeability. For example, a chitosan–graphene oxide coating has shown over 90% reduction in corrosion current density compared to bare steel in 3.5 wt.% NaCl solution [21].

Biomimetic polymer coatings also offer the potential for stimuli-responsive behavior such as self-healing, pH sensitivity, or color change upon corrosion onset, which is valuable for smart coating development. In terms of sustainability, these coatings align with green chemistry principles by eliminating hazardous volatile organic compounds (VOCs) and reducing reliance on petrochemical-based polymers. Their end-of-life degradability further reduces the environmental burden, making them ideal for eco-sensitive applications in infrastructure and packaging.

8.3 Fabrication and Characterization of Bio-Inspired Coatings

The performance, functionality, and sustainability of bio-inspired coatings largely depend on the precision of fabrication techniques and the depth of material characterization. In replicating the sophisticated architectures found in nature, such as the lotus leaf's hierarchical surface or the riblet patterns of shark skin, fabrication methods must offer control at both micro- and nanoscales, ensure uniformity, and be compatible with the surface properties of steel. Equally important is the use of multi-modal characterization techniques to validate surface morphology, chemical composition, mechanical integrity, and corrosion resistance under simulated and real-world environments.

This section discusses two primary fabrication strategies, such as sol-gel processing and electrochemical deposition, followed by an extensive review of advanced characterization tools critical for the evaluation of bio-inspired coatings.

8.3.1 Sol-Gel Processing

Sol-gel processing is a bottom-up chemical synthesis technique that provides exceptional control over coating morphology, porosity, and composition. The method involves the transition of a system from a liquid "sol" (a colloidal suspension of particles) into a solid "gel" phase, typically through hydrolysis and polycondensation of metal alkoxide precursors [22]. The resulting gel can be deposited on metallic surfaces *via* methods like dip-coating, spin-coating, or spray-coating.

Chemical Mechanism:

- **Hydrolysis:**

$$M(OR)_x + xH_2O \longrightarrow M(OH)_x + xROH \tag{8.1}$$

- **Condensation:**

$$M(OH)_x + M(OH)_x \longrightarrow M\text{-}O\text{-}H + H_2O \tag{8.2}$$

After deposition, the gel undergoes drying and thermal curing (150–500°C), during which solvent evaporation and structural densification occur, resulting in a durable, adherent coating.

8.3.2 Electrochemical Deposition

Electrochemical deposition, or electrodeposition, is a faradaic process in which metal ions are reduced at the surface of a conductive substrate (steel) under an applied electrical potential. It offers excellent control over coating thickness, composition, and surface architecture, making it one of the most scalable techniques for industrial coating applications [23].

Working Principle:
In a typical three-electrode system:

- **Cathode (working electrode):** Steel substrate.
- **Anode (counter electrode):** Inert metal (e.g., Pt).
- **Electrolyte:** Aqueous solution containing metal salts and optional organic additives.

Through adjustments in current density, bath composition, pH, and temperature, one can tailor the microstructure of the coating, creating surfaces with unique anti-corrosive or anti-fouling properties.

Advantages

- **Localized deposition control:** Suitable for complex geometries and site-specific coating repair.
- **Template-assisted patterning:** Enables fabrication of sharkskin-like riblet microstructures by using patterned masks or soft lithography [9].
- **Co-deposition capability:** Incorporation of nanoparticles (TiO_2, Al_2O_3, graphene oxide) and biopolymers (e.g., chitosan) into the metallic matrix to form hybrid functional coatings [24].

Bio-Inspired Examples

- **Chitosan-metal composite coatings**, such as chitosan–Zn or chitosan–Ni, have been electrodeposited onto mild steel, showing reduced corrosion current density and improved barrier performance [25].
- **Shark-skin patterned nickel coatings** have been fabricated using pulse electrodeposition through micro-riblet masks, demonstrating antifouling and hydrodynamic benefits [18].

8.3.3 Characterization Techniques

The performance of bio-inspired coatings must be quantified through a multidisciplinary suite of techniques that assess both structural fidelity and corrosion resistance.

(i) Scanning Electron Microscopy (SEM)
SEM offers high-resolution imaging of surface topography and coating microstructure. It can verify:

- The presence of dual-scale roughness (microscale protrusions and nanoscale textures).
- Coating homogeneity, crack formation, or pinholes.
- The alignment and dimensions of shark-skin riblets or lotus-inspired architectures.

For instance, SEM images of fluorinated silica coatings have confirmed uniformly distributed nano-papillae structures, correlating directly to superhydrophobic behavior [13].

(ii) Atomic Force Microscopy (AFM)
AFM measures nanoscale roughness, mechanical stiffness, and adhesion forces. It provides 3D surface topography and root-mean-square (RMS) roughness data critical for understanding wettability. Bio-inspired coatings showing RMS roughness >100 nm often correspond to high contact angles (>150°), supporting the self-cleaning mechanism [26].

(iii) Electrochemical Impedance Spectroscopy (EIS)
EIS evaluates the electrochemical stability and corrosion protection efficiency of coatings in electrolytic environments (e.g., 3.5 wt.% NaCl). It is particularly sensitive to coating degradation over time.

Parameters derived from Nyquist and Bode plots:

- **R_{ct} (charge transfer resistance):** Higher R_{ct} = better corrosion protection.
- **C_{dl} (double-layer capacitance):** Lower C_{dl} = lower water and ion permeability.

Bio-inspired coatings with hierarchical structures often show R_{ct} values 10–100 times greater than bare steel, confirming their superior barrier performance [27].

(iv) Contact Angle Measurements
Static and dynamic contact angles are measured to assess hydrophobicity. A water contact angle above 150° and a roll-off angle below 10° indicates superhydrophobic behavior, essential for lotus-like coatings. Contact angle

hysteresis (difference between advancing and receding angles) is also critical; low hysteresis implies better self-cleaning and lower adhesion.

(v) Salt Spray (Fog) Testing (ASTM B117) [28]
This accelerated corrosion test involves exposing coated specimens to a continuous mist of 5% NaCl at 35°C for up to 1000 hours. It evaluates:

- The onset of rust or blistering.
- Edge creep or delamination.
- Resistance to coating defects (scratches or pinholes).

Well-designed bio-inspired coatings exhibit minimal corrosion products even after prolonged exposure.

(vi) Fourier Transform Infrared Spectroscopy (FTIR) & X-ray Photoelectron Spectroscopy (XPS):
These spectroscopic tools help to identify chemical bonding and surface functional groups, such as hydroxyl, amino, or fluorocarbon groups. They confirm the presence of hydrophobic agents, corrosion inhibitors, or polymeric binders within the coating.

8.4 Corrosion Resistance Mechanisms of Bio-Inspired Coatings

Steel corrosion is primarily driven by electrochemical reactions that involve the oxidation of iron in the presence of water and oxygen, often accelerated by chloride ions. Traditional protective coatings act as passive barriers; however, bio-inspired coatings enhance this function through multi-scale structural and chemical features that actively resist the ingress of corrosive species.

(i) Hydrophobic and Superhydrophobic Surface Structures
Bio-inspired coatings, particularly those mimicking lotus leaves, create superhydrophobic surfaces characterized by water contact angles greater than 150° and low roll-off angles (<10°), minimizing water retention on the surface [8]. The Cassie–Baxter state, wherein water droplets rest atop micro/nano air pockets, significantly lowers the contact between water and the steel surface, disrupting the electrochemical corrosion process. Studies have shown that such surfaces reduce corrosion current densities (I_{corr}) by several orders of magnitude. For instance, a silica-based superhydrophobic

coating reduced I_{corr} from 3.65×10^{-6} A/cm^2 (bare steel) to 1.8×10^{-8} A/cm^2 in 3.5% NaCl solution [13].

(ii) Antifouling and Drag-Reducing Textures

Inspired by shark skin, micro-riblet patterns on coatings can inhibit biofilm formation and reduce hydrodynamic drag, especially in marine or aqueous environments. These structures make it physically difficult for microbial organisms or salt deposits to adhere and colonize the surface, thereby mitigating microbial-induced corrosion (MIC) and under-deposit corrosion [9]. Reduced drag also enhances the efficiency of fluid-transport systems such as pipelines and ship hulls.

(iii) Ion-Blocking and Diffusion Barrier Layers

Biopolymer-based coatings, especially chitosan and cellulose composites, act as barrier layers by creating a tortuous diffusion path for corrosive agents like chloride ions and dissolved oxygen. These polymers can bind to the steel surface *via* functional groups such as amino and hydroxyl moieties, forming protective chelates that reduce active sites for oxidation [19]. In hybrid coatings, the inclusion of nanofillers (e.g., graphene oxide, montmorillonite) further enhances impermeability and mechanical integrity, making the coating resilient against mechanical scratches and cracking [21].

(iv) Self-Healing Capability

Some bio-inspired coatings incorporate microcapsules or nanocontainers loaded with corrosion inhibitors (e.g., cerium salts, benzotriazole), which are released upon mechanical damage or pH-triggered stimuli. This mimics the healing process in biological systems, where damage is autonomously repaired to restore functionality. Sol-gel coatings embedded with such self-healing agents have demonstrated the ability to repair micro-cracks and re-passivate exposed steel areas, extending the life of the coating substantially [29].

8.5 Environmental Sustainability of Bio-Inspired Coatings

In modern engineering, sustainability is as vital as performance. A coating that offers exceptional corrosion resistance but involves toxic precursors, high energy inputs, or poor recyclability cannot be considered viable in the long term. Bio-inspired coatings inherently align with sustainable design principles through their use of natural materials, benign synthesis processes, and long-term durability.

(i) Use of Renewable and Biodegradable Materials

Bio-inspired coatings often utilize natural polymers such as chitosan, cellulose, alginate, gelatine, or silk fibroin. These materials are:

- **Biodegradable**—they break down into environmentally safe components.
- **Renewable**—sourced from agricultural or marine waste streams.
- **Non-toxic**—safe for humans and ecological systems.

For example, chitosan extracted from shrimp shell waste has been successfully used to form corrosion-resistant coatings on mild steel, providing a sustainable alternative to petroleum-based polymers [20].

(ii) Low Energy and Eco-Friendly Synthesis

Techniques like sol-gel processing and electrochemical deposition are comparatively low-energy processes that avoid the use of high-temperature sintering or plasma treatments. Sol-gel chemistry, in particular, enables room-temperature synthesis with minimal solvent usage and allows *in-situ* doping with green inhibitors [30]. Moreover, the absence of VOCs in most bio-inspired systems eliminates air pollution risks and supports workplace safety standards.

(iii) Long-Term Durability and Reduced Maintenance

Bio-inspired coatings typically exhibit longer lifespans than conventional coatings due to their multifunctionality:

- Superhydrophobicity prevents water and contaminant accumulation.
- Self-healing reduces the need for recoating.
- Anti-fouling structures minimize microbial degradation.

This leads to reduced frequency of maintenance and material replacement, ultimately conserving resources and energy.

8.6 Life Cycle Assessment (LCA) and Circularity

Life Cycle Assessment (LCA) is a holistic methodology used to evaluate the environmental impacts of a product from raw material extraction to end-of-life disposal. When applied to coatings, LCA assesses factors like:

- **Embodied energy**
- **Greenhouse gas emissions**
- **Toxicological impact**
- **Waste generation**

Preliminary LCA comparisons indicate that bio-inspired coatings based on natural polymers produce a 30–60% reduction in CO_2 emissions and up to 90% less toxic waste compared to epoxy- or heavy-metal-based coatings [5, 31]. Furthermore, the emerging principles of circular economy are designing coatings that are repairable, upgradable, and recyclable, which can be naturally incorporated into bio-inspired systems due to their modular structure, renewable feedstock, and biodegradability.

8.7 Case Studies and Future Directions

Breakthroughs in smart coating research have resulted in more effective corrosion protection and extended service life. Advances in biomimetic materials have enabled the development of smart and multifunctional coatings [32]. This section showcases concrete examples of bio-inspired coatings, including lotus-leaf, shark-skin, and biopolymer-nanofiller systems, providing visual schematics, performance data, and critical discussions on future opportunities. These case studies demonstrate the potential of bio-inspired coatings to drive innovation and solve complex problems in various industries.

8.7.1 Lotus-Inspired Superhydrophobic Coatings

Lotus-inspired superhydrophobic coatings mimic the lotus leaf's unique properties, featuring water repellence and self-cleaning. The lotus leaf remains an archetype for superhydrophobic strategy: its surface comprises micron-scale papillae topped with nano-scale wax crystals (micro/nano hierarchy). This architecture traps air, enabling droplets to bead and roll off, ensuring self-cleaning (Cassie–Baxter wetting) [33].

Performance Case Examples

1. Silica-sol-gel with fluorosilane

- Water contact angle (WCA): > 160°
- Corrosion current density (I_{norr}): suppressed from 3.65×10^{-6} to 1.8×10^{-8} A/cm² in 3.5 wt. % NaCl [34]

2. Graphene and Polydimethylsiloxane (PDMS) spray-on stainless steel mesh

- WCA: ≈ 150°
- Maintained functionality post-ultrasonic testing; used for oil–water separation with efficiency ≈ 99.4%, flux = 27,400 L m^{-2} h^{-1}

Table 8.1 compares the performance of lotus-like superhydrophobic coatings based on key parameters of water contact angle (CA), corrosion medium, and protection outcome. The Silica sol-gel with fluorosilane system exhibits excellent superhydrophobicity with a water CA greater than 160°, and provides significant corrosion resistance in a 3.5 wt.% NaCl medium, reducing the corrosion current density (I_{corr}) from 3.65×10^{-6} A/cm² to 1.8×10^{-8} A/cm². This enhanced protection is attributed to the air

Table 8.1 A comparison of lotus-like coating performance.

Sl. no.	Coating system	Water CA (°)	Corrosion medium	Protection outcome
1.	Silica sol-gel + fluorosilane	>160	3.5 wt.% NaCl	Reduced I_{corr} from 3.65×10^{-6} A/cm² to 1.8×10^{-8} A/cm²
2.	Graphene@ PDMS (spray)	≈150	Ultrasonic + NaCl fog	High hydrophobicity sustained post-sonication
3.	Dual-scale SiO_2–PDMS coating	~155	Water, NaOH, HCl, abrasion	Durable hydrophobicity under severe wear

barrier created by the hierarchical micro-nanostructure and low surface energy of fluorosilanes, which minimizes electrolyte penetration [35, 36]. The Graphene@PDMS (polydimethylsiloxane) spray coating demonstrates a water CA of about 150° and maintains high hydrophobicity even after exposure to ultrasonic NaCl fog, indicating excellent chemical and mechanical stability. The graphene reinforcement not only contributes to surface roughness but also imparts electrical conductivity and corrosion shielding [37, 38]. Meanwhile, the dual-scale SiO_2–PDMS coating achieves a CA of ~155° and exhibits durable hydrophobicity under severe chemical medium (NaOH, HCl) and mechanical (abrasion) stress. It reveals its robust structural integrity and potential for real-world applications in harsh environments [39, 40]. These results collectively underscore the effectiveness of bio-inspired coating strategies for enhancing surface protection through a combination of surface roughness and low surface energy materials.

Superhydrophobic coatings are known for their remarkable water-repelling properties and enhanced corrosion protection. These coatings illustrate how micro/nano-structured topographies combined with hydrophobic chemistry significantly reduce wetting and electrochemical activity, which are considered the key mechanisms for mitigating steel corrosion [41, 42]. Nanostructured superhydrophobic coatings have shown great potential in corrosion prevention and are being explored for use in multiple industries. The synergy of low surface energy and high surface roughness in these coatings generates micro air pockets, which limit contact with corrosive agents and enhance corrosion resistance [43]. The creation of superhydrophobic coatings has the potential to reshape various industries, including aerospace and biomedical engineering, by enhancing material durability and performance.

8.7.2 Shark-Skin-Inspired Riblet Textures

Shark-skin-inspired riblet textures mimic the unique surface patterns found on shark skin. These textures have gathered significant attention in various fields, including engineering, materials science, and biomimicry. These textures are designed to mimic the unique surface patterns found on shark skin, which provide several benefits, such as

(i) **Drag reduction**: Shark skin's riblet patterns can reduce drag in fluid flows, making them useful for applications like aircraft, ships, and pipelines.

(ii) **Anti-fouling**: The texture can prevent the accumulation of microorganisms, plants, and animals on surfaces, reducing maintenance needs [44].

(iii) **Self-cleaning**: The riblet structure can facilitate easy removal of dirt and debris.

Table 8.2 shows the list of performance metrics of shark-skin-inspired riblet textures: Antifouling efficiency, drag reduction, and fabrication methods. Fabrication techniques such as soft lithography, nanoimprint lithography, and templated micro-molding have enabled scalable replication of these riblet patterns on polymers and metal substrates. For instance, Sharklet™ AF, a commercial shark-skin mimic made of polydimethylsiloxane (PDMS), demonstrated a 77% reduction in algal spore settlement of Ulva linza compared to a smooth control surface, with a reported

Table 8.2 Performance metrics of shark-skin-inspired riblet textures: Antifouling efficiency, drag reduction, and fabrication methods.

Sl. no.	Study/surface	Riblet dimensions	Fouling reduction	Drag reduction	Fabrication technique
1.	Sharklet™ AF (PDMS)	2 μm height, 2–5 μm spacing	77% reduction in *Ulva linza* spores	Not specified	Soft lithography
2.	Sharklet™ AF (PDMS)	Same as above	94% reduction in *S. aureus*, 86% in *P. aeruginosa*	Not specified	Micro-moulding
3.	Riblet film (Bechert *et al.*)	40–60 μm spacing	Not reported	Up to 10% in turbulent flow	Laser micromachining
4.	Riblet + Silane modification	2–8 μm height, ~4 μm spacing	65–80% bacterial reduction (marine biofilms)	5–8% (marine vessels)	Replica molding + chemical vapor
5.	Micropatterned epoxy resin	10 μm groove width	70% diatom reduction (*Navicula perminuta*)	Not specified	Microembossing
6.	Bioinspired superhydrophobic	<10 μm riblet spacing	90% reduction in *E. coli* adhesion	Not quantified	Electrospinning + spray coating

engineered roughness index of 9.5, showcasing effective protection for marine steel surfaces [18]. Similarly, experiments on bacterial adhesion showed a 94% reduction in Staphylococcus aureus and an 86% reduction in Pseudomonas aeruginosa colonization on Sharklet-patterned surfaces [45]. These findings suggest that the micro-topographical constraints significantly hinder the ability of microbes to establish adhesion due to size incompatibility and altered hydrodynamic microenvironments.

Further, the drag-reducing capabilities of riblet textures have also been quantified. Studies on flow over micro-riblet-coated surfaces reported up to a 10% reduction in turbulent drag in aqueous environments [46]. This not only contributes to energy savings in marine vessels but also reduces surface exposure time to fouling agents, thereby enhancing long-term material integrity. Combining riblet structures with superhydrophobic chemical treatments (e.g., fluorosilanes or perfluorinated coatings) further enhances anti-fouling properties by creating air pockets (Cassie–Baxter state) that reduce the effective contact area for contaminants. This synergy facilitates self-cleaning, bubble nucleation, and contaminant roll-off, thus enabling resistance against micromechanical fouling, microbial adhesion, and even macroscopic corrosion in submerged or high-humidity conditions [47, 48]. Shark-skin-inspired riblets are particularly valuable in protecting marine steel structures, ship hulls, offshore platforms, and biomedical implants where conventional coatings may degrade over time or pose environmental risks. Their biocompatibility, scalability, and energy-saving potential make them an attractive alternative in sustainable engineering. Additionally, a bioinspired superhydrophobic surface incorporating riblets with spacing below 10 μm was developed using a hybrid electrospinning and spray coating approach. This surface achieved a 90% reduction in *E. coli* adhesion, highlighting its significant antibacterial functionality. Although its drag reduction behavior was not quantified, the hierarchical structure and water-repellent nature suggest strong potential for dual performance in marine and biomedical settings [49].

8.7.3 Biomimetic Polymer–Nanofiller Coatings

Biomimetic polymer-nanofiller coatings combine biological inspiration with advanced materials to create innovative surfaces with enhanced properties. Natural polymers like chitosan, gum Arabic, and cellulose are biodegradable and receptive to functionalization. When augmented with nanofillers such as graphene oxide, CuO, ZnO, SiC, or Ag nanoparticles, these coatings exhibit robust corrosion inhibition, barrier properties, and stimuli-response.

Chitosan–Graphene Oxide (CHT-GO) Hybrids [50]

- In CO_2-saturated NaCl, chitosan–GO coatings:
  - Charge transfer resistance (R_t): increased from 61 to 361 $\Omega{\cdot}cm^2$ (25 ppm)
  - Corrosion current (I_{corr}): decreased from 855 to 88 $\mu A{\cdot}cm^{-2}$
- Inhibition efficiency (IE): increased from 0% to 89.7%

Oleic Acid–Chitosan–GO Coatings

- EIS & Tafel: ~100-fold improvement in resistance compared to pure chitosan coatings

Chitosan–CuO on X60 Steel in 5% HCl

Surface analysis indicates that nanocomposite adsorption leads to the formation of a protective barrier on carbon steel. Furthermore, the addition of KI improved inhibition efficiency (IE) as the temperature rose from 25°C to 60°C [51].

Table 8.3 summarizes the corrosion performance of X60 carbon steel in both uncoated (blank) and Chitosan–CuO (CHT–CuO) nanocomposite-coated conditions at varying concentrations (0.1%, 0.3%, and 0.5% for $CHT_{0.5}$–CuO, $CHT_{1.0}$–CuO, and $CHT_{2.0}$–CuO). The electrochemical behavior was assessed using potentiodynamic polarization (PDP) and linear polarization resistance (LPR) methods, yielding corrosion current density (i_{corr}), polarization resistance (R_p), and inhibition efficiency (IE%).The uncoated (blank) system exhibited the highest i_{corr} of 73.20 $\mu A/cm^2$ and the lowest R_p of 247.73 $\Omega{\cdot}cm^2$, indicative of significant corrosion activity in the acidic environment. Upon coating with $CHT_{0.5}$–CuO at 0.5%, the i_{corr} dropped to 11.73 $\mu A/cm^2$ with an R_p of 931.79 $\Omega{\cdot}cm^2$, corresponding to an inhibition efficiency of 83.98%. With 1.0% CHT–CuO, further improvements were noted. The sample coated with 0.5% $CHT_{1.0}$–CuO showed the lowest i_{corr} (9.75 $\mu A/cm^2$), the highest R_p (1174 $\Omega{\cdot}cm^2$), and maximum IE of 86.68%, signifying excellent corrosion resistance and strong barrier properties [52, 53]. However, at 2.0% CHT–CuO concentration, particularly at 0.5%, a rise in i_{corr} (11.80 $\mu A/cm^2$) and a marginally reduced IE (83.88%) was observed, potentially due to nanoparticle agglomeration and resulting micro-defect formation within the coating matrix [52, 54]. This decline in efficiency at higher concentrations is consistent with literature reports, which highlight that excessive nanoparticle loading can hinder dispersion

Table 8.3 Electrochemical corrosion parameters of mild steel in the presence of chitosan–CuO nanocomposite coatings at varying concentrations.

System/ Conc.	–Ecorr (mV)	icorr (μA/ cm²)	βa (mV/ dec)	–βc (mV/ dec)	IE (%) (PDP)	Rp (Ω·cm²)	IE (%) (LPR)
Blank	443.9	73.20	83.48	80.25	—	247.73	—
$CHT_{0.5}$–CuO 0.1%	418.0	14.91	56.54	86.09	79.63	880.38	71.86
0.3%	415.1	14.56	54.81	89.79	80.11	899.36	72.45
0.5%	409.1	11.73	49.41	90.86	83.98	931.79	73.41
$CHT_{1.0}$–CuO 0.1%	418.4	13.19	55.95	81.75	81.98	915.88	73.05
0.3%	409.6	11.24	50.61	86.34	84.64	955.61	74.07
0.5%	400.1	9.75	48.88	101.61	86.68	1174.00	78.89
$CHT_{2.0}$–CuO 0.1%	419.2	14.99	55.20	90.41	79.52	795.92	68.87
0.3%	412.3	14.25	51.90	93.58	80.53	865.96	71.39
0.5%	409.1	11.80	49.87	92.55	83.88	934.43	73.49

and uniformity, thereby compromising protective integrity [55]. These findings highlight the synergistic corrosion inhibition behavior of the biopolymeric chitosan and nano-CuO. Chitosan contributes film-forming and metal-chelating properties, while CuO nanoparticles offer a nano-barrier effect and surface reactivity that collectively enhance corrosion resistance [52, 55].

Chitosan–TiO_2/Fe_3O_4 Nanocomposites

- Efficiencies: ~97% (TiO_2) and ~95% (Fe_3O_4) for mild steel in 15% HCl [56]

Chitosan–Gum Arabic (Ch/GA) Nanocomposites

- In saline environments, enhanced polarization resistance (R_p) and reduced double layer capacitance (C_{dl}), indicating superior surface barrier performance [41].

8.8 Emerging Trends and Future Outlook

(i) Smart/Self-Healing Coatings
Integration with microcapsules, slippery liquid-infused porous surface (SLIPS) designs, and Ce^{+3} dopants offers autonomous repair and replenishment upon damage, which is analogous to regenerative systems in nature observed in plant-inspired sol-gels. Integration of self-healing mechanisms mimics biological systems to autonomously repair damaged coatings and extend lifespan. These technologies can be applied to various fields, including self-healing coatings for aerospace, automotive, and construction, as well as medical devices and implants [57].

(ii) Hybrid Topologies
Combining lotus leaf (self-cleaning) and shark-skin (anti-fouling) design principles in one surface aims to deliver dual functionality which is ideal for marine applications such as reducing drag, fuel consumption, and maintenance costs for ship hulls, minimizing fouling, corrosion, and inspection needs for offshore structures, enhancing performance, efficiency, and lifespan for marine equipment etc.

(iii) Multi-Scale Fabrication Techniques
Innovations in nanoimprinting for precise patterning at the nanoscale, 3D printing for complex geometries and structures, laser ablation for high-resolution surface modification, and electrospinning for fabrication of nanofibers now allow reproducible patterning across scales for manufacturing scalability. These advancements drive innovation, enabling the development of complex materials and surfaces with enhanced properties.

(iv) Sustainable Materials Initiatives
The development of coatings using recycled agricultural waste, low-VOC processes, and biodegradable systems aligns with circular economy objectives. Aligning with circular economy principles, these coatings promote resource efficiency, sustainable production and closed-loop systems.

(v) Artificial Intelligence (AI) and Machine Learning (ML)
The application of AI and ML is required to accelerate the discovery, optimization of bio-inspired coatings and identify new materials. AI and ML can predict the performance of bio-inspired coatings, reducing the need for extensive experimental testing. The discovery can lead to novel bio-inspired coatings for various industries, including aerospace, energy, and healthcare.

(vi) Scalability and Commercialization
Efforts are essential to scale up bio-inspired coating technologies for industrial applications through bridging the gap between laboratory research and commercial viability. Establishing standards can ensure consistency and reliability, and these coatings can offer sustainable solutions by reducing environmental impact.

8.9 Conclusions

This chapter investigated the evolving situation of bio-inspired steel corrosion protection coatings comprehensively, highlighting their outstanding performance, environmental friendliness, and potential for sustainability. The key findings of this study are listed below:

1. Bio-inspired coatings, mimicking natural systems like lotus leaves, shark skin, and mollusc shells, bring multifunctionalities like super hydrophobicity, drag reduction, self-cleaning, and anti-fouling. All these features significantly minimize moisture retention, biofilm growth, and chloride penetration, leading to steel structure corrosion.
2. Lotus-inspired surfaces also exhibited water contact angles >150° and reduced corrosion current densities by a few orders of magnitude (e.g., 3.65×10^{-6} to 1.8×10^{-8} A/cm^2). Shark-skin-inspired riblets reduced microbial adhesion by up to 94% and drag by 10% in fluidic environments, proving their dual functionality in aquatic environments.
3. Chitosan nanocomposites, when blended with metal oxide nanoparticles such as CuO or TiO_2, exhibited improved corrosion protection through their dual action mechanisms of film-forming along with chelating action of chitosan, and reactive, impermeable barriers of nanoparticles. The best performance was at 1.0% chitosan–CuO loading, where corrosion current density fell and polarization resistance

reached its peak, with inhibition efficiencies of more than 86%.

4. The performance of chitosan nanocomposites is reduced with the maximum loading of nanoparticles (e.g., 2.0% chitosan–CuO), which results in decreased efficiency due to defects caused by agglomeration and weak dispersion. Thus, optimal filler content and dispersion are still critical design parameters.
5. The use of biodegradable, non-toxic, and renewable material sources (e.g., chitosan, cellulose, gum Arabic) in coating formulations is conducive to sustainable production. These systems also minimize volatile organic compound emissions and facilitate circular economy principles by being lower in embodied energy and better in end-of-life degradability.
6. Bio-inspired steel coatings are a pioneering field that provides technically robust, multi-functional and environmentally benign solutions. With growing inter-disciplinary research and industrial upscaling, these coatings can potentially revolutionize corrosion protection for infrastructure, marine, biomedical, and transport sectors.

AI disclosure statement: The author confirms that no artificial intelligence (AI)-generated content or AI-assisted writing tools were used in the preparation of this chapter (including summaries and the abstract). All text, analysis, and conclusions are the result of the author's original work.

References

1. Koch, G.H., Brongers, M.P.H., Thompson, N.G., Virmani, Y.P., Payer, J.H., Corrosion cost and preventive strategies in the United States. Federal Highway Administration (FHWA-RD-01-156), 2002.
2. Revie, R.W. and Uhlig, H.H., *Corrosion and Corrosion Control: An Introduction to Corrosion Science and Engineering*, 4th Ed., John Wiley & Sons, New Jersey, USA, 2008.
3. Wang, L., Wang, X., Liu, T., Sun, F., Li, S., Geng, Y., Yao, B., Xu, J., Fu, J., Bio-inspired self-healing and anti-corrosion waterborne polyurethane coatings based on highly oriented graphene oxide. *npj Mater. Degrad.*, 7, 96, 2023.
4. Shreir, L.L., Jarman, R.A., Burstein, G.T., *Corrosion: Metal/Environment Reactions*, 3rd Ed., Elsevier, Oxford, UK, 2010.

5. Williams, E.S., Panko, J., Paustenbach, D.J., The European Union's REACH regulation: a review of its history and requirements. *Crit. Rev. Toxicol.*, 39, 7, 553, 2009.
6. Wei, Y., Yu, Y., Li, B., Hu, S., Niu, H., Qiu, R., Ouyang, Y., A bio-inspired solid–liquid compositing fluid-infused surface for prohibiting abiotic and microbiologically induced corrosion. *J. Mater. Sci.*, 57, 10100, 2022.
7. Bhushan, B., Biomimetics: lessons from nature – an overview. *Philos. Trans. R. Soc. A*, 367, 1445, 2009.
8. Barthlott, W. and Neinhuis, C., Purity of the sacred lotus, or escape from contamination in biological surfaces. *Planta*, 202, 1, 1, 1997.
9. Dean, B. and Bhushan, B., Shark-skin surfaces for fluid-drag reduction in turbulent flow: a review. *Philos. Trans. R. Soc. A*, 368, 4775, 2010.
10. Guo, Z., Liu, W., Su, B.-L., Superhydrophobic surfaces: from natural to biomimetic to functional. *J. Colloid Interface Sci.*, 353, 2, 335, 2011.
11. Hu, Y., Hong, T., Zhou, S., Wang, Y., Ye, Q., Sheng, S., Wang, S., He, C., He, H., Xu, M., Bibliometric analysis of coating protection from 2015 to 2025. *Coatings*, 15, 686, 2025.
12. Ma, Q., Yang, Q., Zhang, J., Ren, F., Xia, C., Chen, F., Anti-corrosion properties of bio-inspired surfaces: a systematic review of recent research developments. *Mater. Adv.*, 5, 2689, 2024.
13. Xiong, Z., Huang, J., Wu, Y., Gong, X., Robust multifunctional fluorine-free superhydrophobic fabrics for high-efficiency oil–water separation with ultrahigh flux. *Nanoscale*, 14, 5840, 2022.
14. Xu, W., Liu, H., Lu, S., Xi, J., Wang, Y., Fabrication of superhydrophobic surfaces with hierarchical structure through a solution-immersion process on copper and galvanized iron substrates. *Langmuir*, 24, 10895, 2008.
15. Collins, C.M. and Safiuddin, M., Lotus-Leaf-Inspired biomimetic coatings: different types, key properties, and applications in infrastructures. *Infrastructures*, 7, 46, 2022.
16. Bechert, D.W., Bruse, M., Hage, W., Meyer, R., Fluid mechanics of biological surfaces and their technological application. *Naturwissenschaften*, 87, 4, 157, 2000.
17. Carman, M.L., Estes, T.G., Feinberg, A.W., Schumacher, J.F., Wilkerson, W., Wilson, L.H., Callow, M.E., Callow, J.A., Brennan, A.B., Engineered antifouling microtopographies – correlating wettability with cell attachment. *Biofouling*, 22, 1, 11, 2006.
18. Schumacher, J.F., Carman, M.L., Estes, T.G., Feinberg, A.W., Wilson, L.H., Callow, M.E., Callow, J.A., Finlay, J.A., Brennan, A.B., Engineered antifouling microtopographies – effect of feature size, geometry, and roughness on settlement of zoospores of the green alga *Ulva*. *Biofouling*, 23, 1, 55, 2007.
19. Srivastava, M., Srivastava, S.K., Nikhil, Ji, G., Prakash, R., Chitosan-based new nanocomposites for corrosion protection of mild steel in aggressive chloride media. *Int. J. Biol. Macromol.*, 140, 177, 2019.

20. Badgar, K., Abdalla, N., El-Ramady, H., Prokisch, J., Sustainable applications of nanofibers in agriculture and water treatment: A review. *Sustainability*, 14, 464, 2022.
21. Amin, H.M.A. and Galal, A. (Eds.), *Corrosion Protection of Metals and Alloys Using Graphene and Biopolymer Based Nanocomposites*, CRC Press, Taylor & Francis Group, Boca Raton, FL, USA, 2021.
22. Brinker, C.J. and Scherer, G.W., *Sol-Gel Science: The Physics and Chemistry of Sol-Gel Processing*, Academic Press, San Diego, CA, USA, 1990.
23. Paunovic, M. and Schlesinger, M., *Fundamentals of Electrochemical Deposition*, 2nd Ed., Wiley-Interscience Pub., New Jersey, USA, 2006.
24. Narayanan, T.S.N.S., Park, I.S., Lee, M.H., Strategies to improve the corrosion resistance of microarc oxidation (MAO) coated magnesium alloys for degradable implants: Prospects and challenges. *Prog. Mater. Sci.*, 60, 1, 2014.
25. Francis, A.A., Chitosan-Based Conversion Coatings, in: *Conversion Coatings for Magnesium and Its Alloys*, V.S. Saji, T.S.N. Sankara Narayanan, X. Chen (Eds.), pp. 377–394, Springer, Cham, 2022.
26. Lima, A.C. and Mano, J.F., Micro-/Nano-Structured superhydrophobic surfaces in the biomedical field: Part I: Basic concepts and biomimetic approaches. *Nanomedicine*, 10, 1, 103, 2015.
27. Macdonald, D.D., Reflections on the history of electrochemical impedance spectroscopy. *Electrochim. Acta*, 51, 8–9, 1376, 2006.
28. ASTM B117-18: Standard Practice for Operating Salt Spray (Fog) Apparatus, ASTM International, West Conshohocken, PA, USA, 2018.
29. Kumar, N. and Subasri, R., Corrosion protection and self-healing by Sol-Gel process: A review of recent patent literature. *Recent Pat. Corros. Sci.*, 2, 2, 148, 2012.
30. Livage, J., Henry, M., Sanchez, C., Sol-gel chemistry of transition metal oxides. *Prog. Solid State Chem.*, 18, 4, 259, 1988.
31. ISO 14040:2006. Environmental management – Life cycle assessment – Principles and framework, International Organisation for Standardisation (ISO), Geneva, 2006.
32. George, J.S., P, P.V., Hoang, A.T., Kalarikkal, N., Nguyen-Tri, P., Thomas, S., Recent advances in bio-inspired multifunctional coatings for corrosion protection. *Prog. Org. Coat.*, 168, 106858, 2022.
33. Mohamed, A.M.A., Abdullah, A.M., Younan, N.A., Corrosion behaviour of superhydrophobic surfaces: A review. *Arabian J. Chem.*, 8, 6, 749, 2015.
34. Lyu, K., Liu, X., Liu, R., Yang, H., Qiao, Y., Shah, S.P., Evaluation of the anti-corrosion performance of CeO_2-modified graphene oxide nanocomposite epoxy coating subjected to simulated saline-alkali solution. *Polymers*, 14, 7, 1412, 2022.
35. Kim, D.C., Jeon, D.-Y., Chung, H.-J., Woo, YS., Shin, J.K., Seo, S., The structural and electrical evolution of graphene by oxygen plasma-induced disorder. *Nanotechnology*, 20, 375703, 2009.

36. Liu, K., Bin, R., Xiao, H., Long, Z., Hong, Z., Yang, H., Wu, S., Structural analysis of nitride layer formed on uranium metal by glow plasma surface nitriding. *Appl. Surf. Sci.*, 265, 389, 2013.
37. Zhang, L., Zhong, Y., Cha, D., Wang, P., A self-cleaning underwater superoleophobic mesh for oil-water separation. *Sci. Rep.*, 3, 2326, 2013.
38. Liu, M., Wang, S., Wei, Z., Song, Y., Jiang, L., Bioinspired design of a superoleophobic and low adhesive water/solid interface. *Adv. Mater.*, 21, 6, 665, 2009.
39. Wang, S. and Jiang, L., Definition of superhydrophobic states. *Adv. Mater.*, 19, 21, 3423, 2007.
40. Zhang, X., Shi, F., Niu, J., Jiang, Y., Wang, Z., Superhydrophobic surfaces: from structural control to functional application. *J. Mater. Chem.*, 18, 621, 2008.
41. Al Kiey, S.A., Hasanin, M.S., Heakal, F.E.-T., Green and sustainable chitosan–gum Arabic nanocomposites as efficient anticorrosive coatings for mild steel in saline media. *Sci. Rep.*, 12, 13209, 2022.
42. Mahalakshmi, P.V., Vanithakumari, S.C., Gopal, J., Kamachi Mudali, U., Raj, B., Enhancing corrosion and biofouling resistance through superhydrophobic surface modification. *Curr. Sci.*, 101, 10, 1328, 2011.
43. Mamgain, H.P., Pati, P.R., Samanta, K.K., Brajpuriya, R., Gupta, R., Pandey, J.K., Giri, J., Sathish, T., Kanan, M., A review on bio-inspired corrosion resistant superhydrophobic coating on copper substrate: recent advances, mechanisms, constraints, and future prospects. *Results Eng.*, 25, 103868, 2025.
44. Yang, H. and Wang, S., A Review on the effect of microstructure surface on the adhesion of marine fouling organisms. *IOP Conf. Ser.: Mater. Sci. Eng.*, 782, 052033, 2020.
45. Reddy, S.T., Chung, K.K., McDaniel, C.J., Darouiche, R.O., Landman, J., Brennan, A.B., Micropatterned surfaces for reducing the risk of catheter-associated urinary tract infection: An *in vitro* study on the effect of Sharklet micropatterned surfaces to inhibit bacterial colonization and migration of uropathogenic Escherichia coli. *J. Endourol.*, 25, 9, 1547, 2011.
46. Bechert, D.W., Bruse, M., Hage, W., Van Der Hoeven, J.G.T., Hoppe, G., Experiments on drag-reducing surfaces and their optimization with an adjustable geometry. *J. Fluid Mech.*, 338, 59, 1997.
47. Scardino, A.J. and de Nys, R., Mini review: Biomimetic models and bioinspired surfaces for fouling control. *Biofouling*, 27, 1, 73, 2011.
48. Damodaran, V.B. and Murthy, N.S., Bio-inspired strategies for designing antifouling biomaterials. *Biomater. Res.*, 20, 18, 2016.
49. Liu, Y., He, X., Yuan, C., Cao, P., Bai, X., Antifouling applications and fabrications of biomimetic microstructured surfaces: A review. *J. Adv. Res.*, 59, 201, 2024.
50. Eduok, U., Ohaeri, E., Szpunar, J., Anticorrosion allyl sulfonate graft chitosan/graphene oxide nanocomposite material. *Mater. Adv.*, 2, 1621, 2021.

51. Umoren, P.S., Kavaz, D., Umoren, S.A., Corrosion inhibition evaluation of Chitosan–CuO nanocomposite for carbon steel in 5% HCl solution and effect of KI addition. *Sustainability*, 14, 13, 7981, 2022.
52. Xue, H., Hu, L., Xiong, Y., Zhu, X., Wei, C., Cao, F., Zhou, W., Sun, Y., Endo, Y., Liu, M., Liu, Y., Liu, J., Abududilibaier, A., Chen, L., Yan, C., Mi, B., Liu, G., Quaternized chitosan-Matrigel-polyacrylamide hydrogels as wound dressing for wound repair and regeneration. *Carbohydr. Polym.*, 226, 115302, 2019.
53. Abdolmohammad-Zadeh, H., Galeh-Assadi, M., Shabkhizan, S., Mousazadeh, H., Sol–gel processed pyridinium ionic liquid-modified silica as a new sorbent for separation and quantification of iron in water samples. *Arabian J. Chem.*, 9, S587, 2011.
54. Mekewi, M.A., Ramadan, A.M., ElDarse, F.M., Abdel Rehim, M.H., Mosa, N.A., Ibrahim, M.A., Preparation and characterization of polyurethane plasticizer for flexible packaging applications: Natural oils affirmed access. *Egypt. J. Pet.*, 26, 1, 9, 2017.
55. Rahman, M.M., Khan, K.H., Parvaz, M.M.H., Irizarry, N., Uddin, M.N., Polymer nanocomposites with optimized nanoparticle dispersion and enhanced functionalities for industrial applications. *Processes*, 13, 994, 2025.
56. Kesari, P., Udayabhanu, G., Roy, A., Pal, S., Chitosan based titanium and iron oxide hybrid bio-polymeric nanocomposites as potential corrosion inhibitor for mild steel in acidic medium. *Int. J. Biol. Macromol.*, 225, 1323, 2023.
57. Dallaev, R., Advances in materials with self-healing properties: A brief review. *Materials*, 17, 10, 2464, 2024.

9

Designing Advanced Materials Using Embedded Atom Method Interatomic Potential: A Review

Sayak Bal and Snehanshu Pal*

Department of Metallurgy and Materials Engineering, Indian Institute of Engineering Science and Technology, Shibpur, Howrah, West Bengal, India

Abstract

Interatomic potentials occupy an essential role in the modeling of materials at the mesoscale. The Embedded Atom Method Interatomic Potential is a type of interatomic potential which is mostly used in the modeling and design of metals and alloys using molecular dynamics. To accurately replicate mechanical responses and other properties such as elastic constants, defect energies, thermal properties and microstructural evolution for the design and sustainable manufacturing of advanced materials, accurate and standard embedded atom potentials are needed. This article covers a review of the historic development of the embedded atom method interatomic potential; the development methodologies and the recent works being done in this field.

Keywords: Embedded atom method, interatomic potentials, review, molecular dynamics, sustainable manufacturing

9.1 Introduction

Interatomic potentials, also known as force fields, describe the interaction between atoms in a system. These are primarily described through potential energy landscapes [1]. Interatomic potentials are mathematically formulated, mostly in an analytical style, to help describe the interaction energy between the constituents of a system, from metallic to covalent systems to

**Corresponding author*: snehanshu@metal.iiests.ac.in

Gurudas Mandal, Swarup Kumar Ghosh, Snehanshu Pal and Sandip Kunar (eds.) Sustainable Manufacturing of Advanced Materials, (179–198) © 2026 Scrivener Publishing LLC

even strongly correlated systems. The mathematics describes the force on an atom based on its position in the system. Interatomic potentials are very much necessary backbones for driving molecular dynamics simulations; and hence, the development of interatomic potentials are of utmost importance for the mesoscale study of most physical systems.

The first interatomic potentials that were developed were of the pair potential style [2]. The Lennard-Jones (LJ) potential was formulated to describe the interaction between two atoms [3]. While it was a significant milestone in our understanding of interatomic behavior, it did not generalize well to most physical systems such as alloys and other correlated systems. Some other types of pair potentials were also developed to mitigate the shortcomings of Lennard-Jones potential like the Morse potential formulation [4], but even then, the underlying assumption of neglecting the interactions between neighboring atoms make them very much unusable to model any system with more complexity than isolated pairs of atoms. The problems with pair-style potentials led to the development of more complicated potentials which take into account many-body interactions. Many-body potentials include Finnis-Sinclair potentials [5], Embedded Atom Method (EAM) potentials [6], Modified EAM (MEAM) potentials [7], Alloy-EAM potentials [8], Tersoff potentials [9], Abell potentials [10], Brenner potentials [11], Long range Carbon Bond Order potential (LCBOP) [12], Stillinger-Weber (SW) potential [13], Reactive force fields (ReaxFF) [14], Charge Optimized Many Body (COMB) potentials [15] and Angular Dependent potential (ADP) [16]. Even though many body potentials were very much effective to model the interactions between complicated many body systems, these still encompassed many assumptions which made them ineffective in the modeling of more complicated systems, especially strongly correlated systems. To model increasingly complicated systems, Machine Learned Interatomic Potentials (MLIPs) are being developed, which aim to use Machine Learning to predict the potential energies of atoms in many body systems [17–24]. Figure 9.1 summarizes the evolution of interatomic potentials, from pair potentials to the EAM potential and beyond.

While all these potentials have been successful in modeling many body systems to varying degrees, EAM potentials particularly stand out in the modeling of metallic systems. This review primarily concerns the development of EAM potentials starting from the first EAM potential developed by Daw and Baskes [1], for the study of hydrogen embrittlement in steels to recently developed modern EAM potentials for Barium-Palladium alloy system [25], Nickel-Molybdenum system [26], and plasticity-oriented

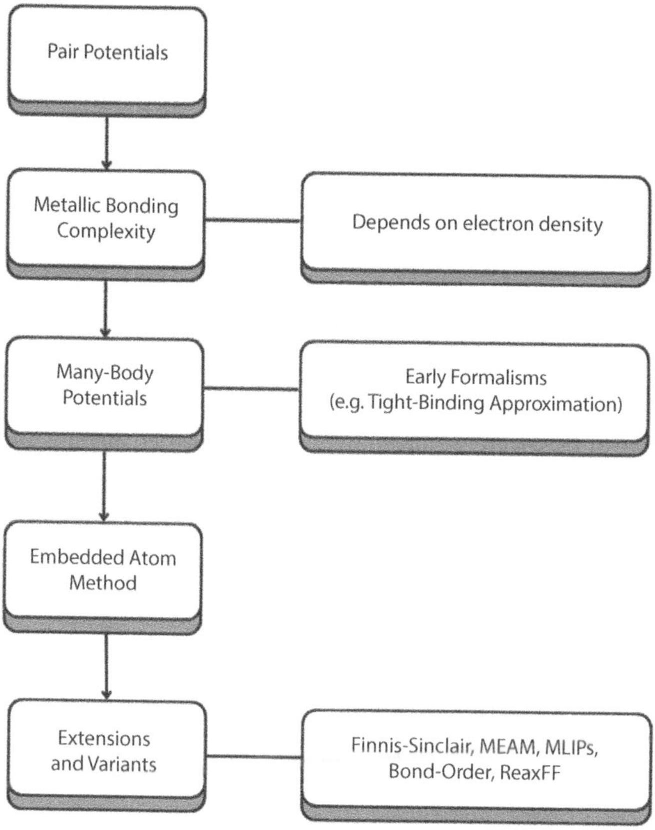

Figure 9.1 Evolution of interatomic potentials.

EAM potential for CrFeMnNi high-entropy alloys [27], and the applications of EAM potentials in sustainable manufacturing.

9.2 Theoretical Foundations of EAM Potentials

EAM potential is a semi-empirical interatomic potential. It was first developed by Daw and Baskes [1, 28] to simulate and consequently model hydrogen embrittlement in steels. The central idea behind EAM potential is the energy of a particular atom in a system is the sum of its interactions with its neighbors and its 'embedding energy' in the matrix of atoms to which it belongs. Mathematically, the potential energy (E) can be represented as follows in the EAM potential formalism:

$$E_i = F_\alpha(\sum_{j \neq i} \rho_\beta(r_{ij})) + \frac{1}{2}\sum_{j \neq i} \phi_{\alpha\beta}(r_{ij})$$

Where $\phi_{\alpha\beta}(r_{ij})$ represents the pair interaction energy energy term between the atom and its neighbors and is the embedding energy term, where r_{ij} is the cutoff radius for which the neighborhood of the atom is considered. The total energy of the system is consequently given by:

$$E_{tot} = \sum E_i$$

Thus, not only the pair interactions are considered but also the embedding energy of the atom is considered which approximates the many-body effects. The standard EAM energy expression thus becomes:

$$E_{tot} = \sum_{i=1}^{N}[F(\rho_i)] + \frac{1}{2}\sum_{i=1}^{N}\sum_{j \neq i} \phi(r_{ij})$$

where,

$$\rho_i = \sum_{j \neq i} \rho(r_{ij})$$

The $\rho(r_{ij})$ term is the host electron density and thus considers the electronic interactions of the system by modeling the embedding of the atom in a 'sea of electrons'. EAM potentials by their very nature model metallic systems very efficiently because the energy of each atom depends not just on its individual interactions with its neighboring atoms but also its contributing electron density, and therefore, modeling the cohesion of the metallic systems.

The data to fit EAM potential for any system is derived from Density Functional Theory (DFT) calculations, and therefore, incorporate electronic correlation and many body effects implicitly [28]. This property of EAM potentials make them much better suited to modeling weakly correlated many body systems, where traditional pair potentials fail.

EAM potentials are justified by the second-moment tight binding approximation, which can be attributed to the success of EAM potential in the modeling of many metallic properties such as cohesion, elasticity and vacancy formation [28].

The success of EAM potential in the modeling of metallic systems started from the classic Daw and Baskes paper and then was repeated extensibly for the modeling of many other systems such as development of EAM potentials of Face Centered Cubic (FCC) metals by Foiles, Baskes and Daw [29]. The following section shall highlight the development methodolgies usually used for the development of EAM potentials along with the tools typically used to perform the calculations for developing reference datasets, based on which the EAM potentials are fitted.

9.3 Development Methodologies

The EAM potential development methodology incorporates the fitting of many terms based on DFT calculations so that the fitted potential matches with experimental results or results from *ab-initio* molecular dynamics simulations. While fitting the pair interaction energy term is relatively easy, the challenge lies in fitting the cross-potential terms.

For a binary alloy $(\alpha - \beta)$, the total energy can be written as:

$$E_{tot} = \sum_{i} F_{\alpha}(\rho_i) + \frac{1}{2}\sum_{i \neq j} \phi_{\alpha\beta}(r_{ij})$$

where,

$$\rho_i = \sum_{j \neq i} \rho_{\alpha\beta}(r_{ij})$$

In this expression, three sets of cross-energy terms appear, the pair potential term, the electron density function and the embedding energy function. Now, while fitting EAM potentials, either all of the terms are fitted from scratch from DFT data or various mixing rules are applied to combine already existing terms for pure metals to approximate elemental pair cross energy terms. Two mixing models are the most prominent:

Johnson Alloy Model (JAM), in which $\phi_{\alpha\beta}$ is calculated by taking a weighted average of the elemental pair potentials [30].

$$\phi_{\alpha\beta} = \frac{w_{\alpha}\phi_{\alpha\alpha} + w_{\beta}\phi_{\beta\beta}}{w_{\alpha} + w_{\beta}}$$

Geometric mixing is a much simpler rule which is sometimes used as it preserves symmetry and easy to implement. The downside is, it can fail for strongly asymmetric alloys.

$$\phi_{\alpha\beta}(r) = \sqrt{\phi_{\alpha\alpha}(r)\phi_{\beta\beta}(r)},$$
$$\rho_{\alpha\beta}(r) = \sqrt{\rho_{\alpha\alpha}(r)\rho_{\beta\beta}(r)}$$

As discussed by Weiss *et al.* [31], concatenation of pure-element tables and then refitting only the cross ϕ (and optionally ρ) terms to DFT mixing enthalpies or alloy compound enthalpies. This retains the pure-element fidelity while capturing alloy behavior *via* minimal new fitting parameters. Various strategies have been used to fit the EAM potential energy terms. These have been highlighted below.

(i) Traditional property-based fitting

This is the most intuitive approach to EAM potential fitting. A range of target properties such as cohesive energies, elastic constants, equilibrium lattice constants, vacancy formation energies, surface energies and stacking fault energies are chosen and calculated using DFT calculations or experimental results and then fitted to the EAM potential terms using analytic optimization or spline functional forms. The cost function is usually a weighted least-squares sum of deviations:

$$\chi^2 = \sum_k w_k (P_k^{EAM} - P_k^{ref})^2$$

where each P_k a reference property from DFT calculations or experimental results and w_k its corresponding weight. Optimization techniques usually use local methods like Levenberg-Marquardt or global optimization techniques such as genetic algorithms. Some papers like Cu EAM by Mishin *et al.*, Ni-Al EAM by Purja-Pun and Mishin have used property-based fitting [29, 32–35, 41, 55].

(ii) Force-matching

The force-matching method is another method for the construction of EAM potentials which work by minimizing the difference between DFT predicted forces or experimental results and the forces predicted by the candidate potential. The typical error metric can be represented as:

$$\chi^2 = \sum_{i,\alpha}[F_{i,\alpha}^{EAM} - F_{i,\alpha}^{DFT}]^2 + \lambda \sum_{j}[E_j^{EAM} - E_j^{DFT}]^2$$

The force matching method was first pioneered by Ercolessi and Adams [41], to fit the Al EAM potential. Another modern example such as the $BaPd_2$ EAM potential developed by Pal and Mukhopadhyay has used the force-matching framework using the PotFit package [25, 35, 36].

(iii) Multi-objective and hybrid fitting

Multi-objective fitting through Pareto optimization has also been used to replace a single cost function for the optimization of the candidate potential to fit match many DFT calculated properties and experimental results. Hybrid methods combine property-based and force-matching terms into one composite cost or introduce regularization penalties to enforce physical constraints. Some examples include the hybrid Fe-C EAM potential constructed by Allera *et al.* [37], and the hybrid EAM potential blended with surrogate Machine Learning model constructed by Seko *et al.*, [36, 37].

(iv) High-throughput and active learning workflows

Modern EAM potential development is greatly accelerated by active learning loops which iteratively identify configurations where a candidate potential fails and adds them to a DFT based training set which then updates the candidate potential. This approach uses high-throughput frameworks to draw training data from large already established material databases and supplements the DFT based training datasets. Some examples of this framework include the RAMPAGE framework developed by Weiss *et al.* [31], and the KLIFF framework developed by Wang *et al.* [38–40].

All of these EAM potential fitting strategies make use of optimization of a candidate EAM potential function to match target values which are either generated through DFT calculation or are obtained experimentally for validation. Validation of the fitted EAM potentials are also done through *ab-initio* molecular dynamics calculations so that maximum number of possible configurations can be captured and predicted accurately.

Some of the tools which have been used to fit EAM potentials are highlighted in Table 9.1.

In addition to this, many tools are used to perform DFT calculations and validate the results. A famous tool is the Vienna *Ab-initio* Simulation Package (VASP) and Quantum Espresso, both of which are used to perform DFT and *ab-initio* molecular dynamics calculations. Some of the

Table 9.1 Tools used in the development of EAM potentials.

Tool	Primary method	References
PotFit	Force-Matching	[41]
Atomicrex	Hybrid and Multi-Objective	[42]
GULP	Property based fitting	[40]
EAPOTc	Multi-Objective and High Throughput	[43]
EAPOTs	Multi-Objective and High Throughput	[44]
KLIFF	Active Learning	[38]
RAMPAGE	Active Learning	[31]

Table 9.2 Validation and reference data generation software.

Tool	Application	References
Quantum Espresso	DFT, *Ab-initio* Molecular Dynamics	[45]
VASP	DFT, *Ab-initio* Molecular Dynamics	[46, 47]
LAMMPS	Molecular Dynamics	[48, 49]
ASE	DFT, Molecular Dynamics	[39]

tools used for validation and performing calculations are listed in Table 9.2 along with their specific applications in materials modeling.

The whole development methodology can be summarized with a flowchart given in Figure 9.2.

9.4 Some Existing EAM Potentials

Nearly every existing EAM potential was developed using one of the aforementioned methodologies. For example, Olsson *et al.*, have constructed EAM potentials for Fe, Ta, and W by fitting to cohesive energies, lattice constants and elastic constants [50]. As mentioned above, Williams *et al.*, have developed EAM potentials for Cu and Ag using force matching methodology which has accurately replicated lattice parameters, cohesive energies and elastic constants [51]. Ercolessi and Adams's classic Al potential

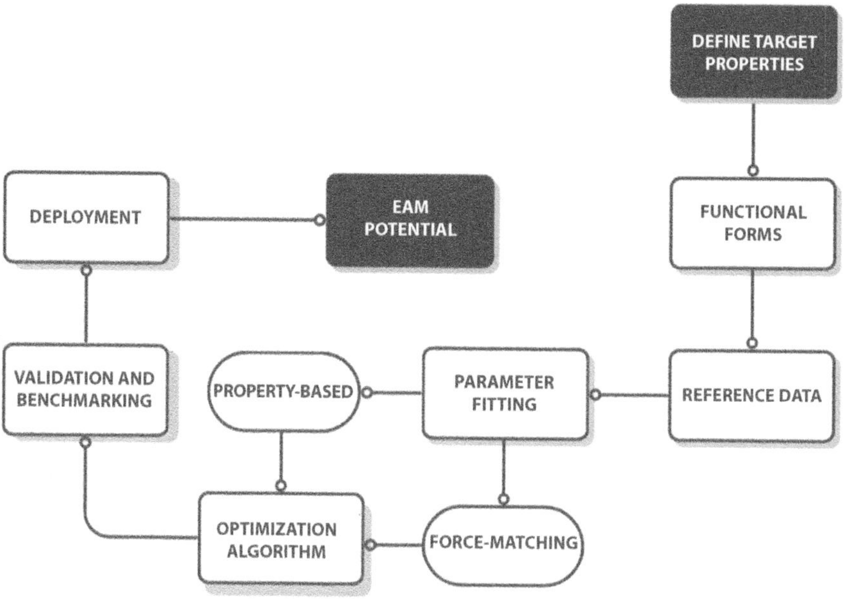

Figure 9.2 EAM potential development methodology.

used force-matching with cubic splines to reproduce FCC *Al* properties [41]. Another EAM potential which was produced by Purja-Pun *et al.*, is the EAM for NiAl which was produced by fitting to Ni and Al data, reproducing the NiAl lattice constant and cohesive energy [33].

In addition to single element EAM potentials, many multicomponent alloy EAM potentials have also been produced using the methodologies mentioned before. The iron-chromium EAM potential was created by fitting to mixing enthalpies and experimental phase data so that experimental results could be reproduced using the EAM potential [52]. Another example is the $BaPd_2$ alloy EAM potential developed by Pal and Mukhopadhyay which was fitted using the PotFit package implementing the force-matching methodology [25]. The AlMg alloy EAM potential was produced by Liu and Adams by fitting to *ab initio* forces and experimental data *via* force matching [53].

EAM potentials are extensively used for the design and sustainable manufacturing of advanced materials as any Molecular Dynamics simulation on any system runs on the basis of its EAM potential. Figure 9.3 illustrates how EAM potentials are used in the design of advanced materials. The National Institute of Standards and Technology and OpenKIM database

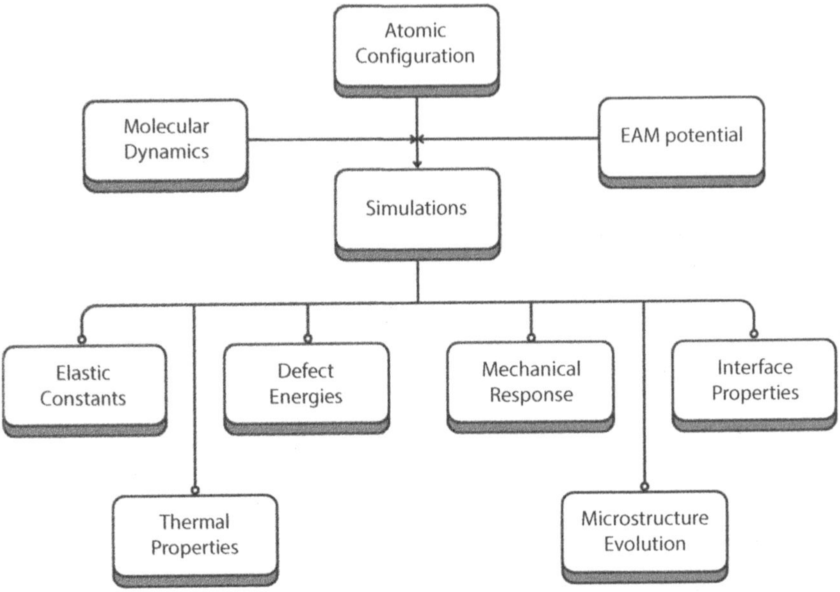

Figure 9.3 Use of EAM potential in design of materials.

provides a standard list of accepted EAM potentials currently available for different material systems.

9.5 Validation of EAM Potentials

The generated EAM potentials must be validated against many test conditions so that the robustness of the EAM potential can be ensured. Many studies have been done for the benchmarking and validation of EAM potentials. Rassoulinejad-Mousavi *et al.*, studied transferability of EAM potentials to various conditions by testing tens of EAM potentials for Cu, Ni, Al at 300K, finding that potentials fit only at 0K often fail at room temperatures [54, 55]. Some other benchmarking studies include the comparative study of Al EAM potentials by Mendelev *et al.* [56], and the study of Fe EAM potential by Starikov *et al.* [57].

Another very significant study which highlighted the effectiveness of well-fitted EAM potentials over complex Machine Learned Interatomic Potentials (MLIPs) was done by Jacobs *et al.*, where it was shown that traditional EAM models run orders of magnitude faster than MLIPs in simulations [58].

9.6 Recent Works and Future Directions in EAM Development

While EAM potential has proven to be a very successful framework in the field of materials modeling, there are still multiple research gaps left to be filled which hinder the development of more advanced materials. One particularly glaring research gap is the problem of transferability of EAM potentials across multiple simulation conditions. Because EAM potentials are trained from data on a very specific set of conditions, simulations using EAM potentials correspond to experimental results only in those conditions and generalize very badly when the conditions vary [55]. This problem is more amplified in case of multi-component systems like high and medium entropy alloys where the cross-interaction terms might have been calculated on the basis of simple mixing rules, and therefore, not capturing the more complex interactions. Although traditional EAM potentials have proven very useful for the modeling of metallic alloys, the development of newer and more sophisticated EAM potential models are being undertaken so that a much wider variety of properties can be modeled. One avenue of research is the development of Machine Learned corrections to EAM potentials which use a hybrid framework with machine learning, typically gaussian processes, to generate more accurate EAM potentials [59, 60].

Uncertainty Quantification remains a pressing research gap in EAM potential development. In the work by Artrith and Behler, Bayesian frameworks have been used to assign error bars to EAM-ML hybrid potential predictions [60]. Other studies have also explored ensemble fitting and dropout-based uncertainty quantification in EAM parameter estimation [61]. Much work still is left to be done in this direction. While frameworks like RAMPAGE and EAPOTc automate alloy EAM fitting, Talirz *et al.*'s Materials Cloud EAM suite demonstrate fully automated workflows [62].

Multi-fidelity approaches such as coupling EAM potential development with continuum mechanics along with quantum mechanical calculations will enhance transferability. New EAM+ potentials have been developed which integrate long range Coulombic [63], and London [64], dispersion force corrections and thus integrating long-range order in the EAM potential framework. This has helped in the accurate modeling of polar and ionic metals (e.g., MgO, Li and AlN).

Additional developments to the EAM potential framework also include the embedment of charge-dependent terms thus allowing the EAM potential to handle various oxidation states. Polarizable EAMs have emerged which merge EAM's embedding with Drude oscillators [63].

Descriptor based expansions such as moment tensor based EAM potentials have been introduced with adaptively fitted coefficients which replace fixed analytic or spline forms [65].

Work on reactive EAM variants that incorporate bond-order terms [66, 67] or link to ReaxFF-like reactive schemes [68], is being done so that EAM potentials can incorporate bond breaking and formation. As more complex EAM potential forms are developed, there has been an increase in need of more interpretable embedding functions linked to electronic descriptors [69].

There have also been efforts to integrate EAM potentials for real-time digital twin architectures for process control requiring adaptive EAM potential frameworks, such as in sustainable manufacturing processes [70].

Work has also been done on optimizing the EAM generation pipeline for calculations using Graphics Processing Units (GPUs) and exascale systems ensuring scalable and energy-efficient simulations. The development of more sustainable production systems for EAM potential development will be direly required in the future [70–76].

9.7 Applications of EAM in Advanced Materials

Some examples of EAM potentials being used for the development and design of advanced materials are shown below.

(i) High Entropy and Refractory High Entropy Alloys

EAM potentials for High Entropy Alloys (HEAs) and Refractory High Entropy Alloys (RHEAs) are very much needed for the real world design of these advanced materials. EAM potential for CrFeMnNi quaternary HEA was developed for the design of alloys with excellent strength-ductility tradeoff [29, 77]. This has enabled the study of how varying Mn and Ni content affects the stacking fault energy, directly influencing the alloy's mechanical properties and deformation behavior.

Another example is the development of EAM potential for FeNiCrCoCu quinary HEA system. Here the development of accurate EAM potential has enabled the study of the resistance of the HEA to radiation damage [78].

MLIPs were developed for the MoNbTaVW RHEA for the study of segregation and defect behavior in the alloy system at high-temperatures so that practical high-temperature components could be designed from it, especially in the Aerospace and Energy sectors [79].

(ii) Shape Memory Alloys

The EAM Potential for NiTi binary alloy was developed to predict the shape memory properties of the alloy system. MD simulations of the phase changes using the developed potential accurately predict the structural properties of this alloy. This has been very significant in the development of advanced functional materials [80].

EAM potential has also been developed for NiTiHf shape memory alloy system which helps in the design of components for high-temperature Aerospace and Automotive applications [81].

(iii) Lightweight Aerospace Alloys

EAM potentials have been developed for the TiAl alloy system for design and use in high-temperature conditions in Aerospace applications [82].

(iv) Superalloys for High Temperature Applications.

EAM potentials have been developed for NiAlRe alloy system which helps in the design of γ / γ' Ni based superalloys with enhanced creep resistance. These alloys are used in the design of turbine blades [83].

In addition to these, EAM potentials are continuously being developed for other classes of advanced materials too, such as, metal-matrix ceramic particle reinforced nanocomposites and Complex Concentrated Alloys for sustainable additive manufacturing [84]

9.8 Conclusion

The EAM potential framework has proven to be one of the most indispensable computational tools for the modeling of metals and alloy systems. Through fitting of EAM potentials on DFT calculated and experimental data, EAM potentials can be used to run reliable and efficient molecular dynamics simulations for the study of metal systems. This helps in the prediction of thermomechanical behavior of materials which is then used for the design of advanced materials. While a significant number of EAM potentials for metal elements have been developed, much work still remains to be done in the development of EAM potentials for advanced material systems such as high-entropy alloys, refractory high-entropy alloys, and other emerging classes of materials. The EAM potential framework thus remains one of the very essential tools for advanced material design, development and sustainable manufacturing processes.

AI disclosure statement: The author confirms that no artificial intelligence (AI)-generated content or AI-assisted writing tools were used in the preparation of this chapter (including summaries and the abstract). All text, analysis, and conclusions are the result of the author's original work.

References

1. Daw, M.S. and Baskes, M.I., Semiempirical, quantum mechanical calculation of hydrogen embrittlement in metals. *Phys. Rev. Lett.*, 50, 17, 1285–1288, 1983.
2. Schwerdtfeger, P. and Wales, D.J., 100 Years of the Lennard-Jones Potential. *J. Chem. Theory Comput.*, 20, 9, 3379–3405, 2024.
3. Jones, J.E., On the determination of molecular fields—I. From the variation of the viscosity of a gas with temperature. *Proc. R. Soc. Lond. A*, 106, 441–462, 1924.
4. Morse, P.M., Diatomic molecules according to the wave mechanics. II. Vibrational levels. *Phys. Rev.*, 34, 57–64, 1929.
5. Finnis, M.W. and Sinclair, J.E., A simple empirical N-body potential for transition metals. *Philos. Mag. A*, 50, 45–55, 1984.
6. Daw, M.S. and Baskes, M.I., Embedded-atom method: Derivation and application to impurities, surfaces, and other defects in metals. *Phys. Rev. B*, 29, 6443–6453, 1984.
7. Baskes, M.I., Modified embedded-atom potentials for cubic materials and impurities. *Phys. Rev. B*, 46, 2727–2742, 1992.
8. Zhou, X.W., Wadley, H.N.G., Johnson, R.A., Larson, D.J., Tabat, N., Cerezo, A., Kelly, T.F., Atomic scale structure of sputtered metal multilayers. *Acta Mater.*, 49, 4005–4015, 2001.
9. Tersoff, J., Empirical interatomic potential for silicon with improved elastic properties. *Phys. Rev. B*, 38, 9902–9905, 1988.
10. Abell, G.C., Empirical chemical pseudopotential theory of molecular and metallic bonding. *Phys. Rev. B*, 31, 6184–6196, 1985.
11. Brenner, D.W., Empirical potential for hydrocarbons for use in simulating the chemical vapor deposition of diamond films. *Phys. Rev. B*, 42, 9458–9471, 1990.
12. Los, J.H. and Fasolino, A., Intrinsic long-range bond-order potential for carbon: Performance in Monte Carlo simulations of graphitization. *Phys. Rev. B*, 68, 024107, 2003.
13. Stillinger, F.H. and Weber, T.A., Computer simulation of local order in condensed phases of silicon. *Phys. Rev. B*, 31, 5262–5271, 1985.
14. van Duin, A.C.T., Dasgupta, S., Lorant, F., Goddard, W.A., ReaxFF: A reactive force field for hydrocarbons. *J. Phys. Chem. A*, 105, 9396–9409, 2001.

15. Shan, T.R., Devine, B.D., Kemper, T.W., Sinnott, S.B., Phillpot, S.R., Charge-optimized many-body potential for the alumina system. *Phys. Rev. B*, 81, 125328, 2010.
16. Mrovec, M., Nguyen-Manh, D., Pettifor, D.G., Vitek, V., Bond-order potential for transition metals: Theory and application to modeling dislocation behavior in tungsten. *Phys. Rev. B*, 69, 094115, 2004.
17. Bartók, A.P., Payne, M.C., Kondor, R., Csányi, G., Gaussian Approximation Potentials: The Accuracy of Quantum Mechanics, without the Electrons. *Phys. Rev. Lett.*, 104, 136403, 2010.
18. Thompson, A.P., Swiler, L.P., Trott, C.R., Foiles, S.M., Tucker, G.J., Spectral neighbor analysis method for automated generation of quantum-accurate interatomic potentials. *J. Comput. Phys.*, 285, 316–330, 2015.
19. Zhang, L., Han, J., Wang, H., Car, R., E, W., Deep Potential Molecular Dynamics: A Scalable Model with the Accuracy of Quantum Mechanics. *Phys. Rev. Lett.*, 120, 143001, 2018.
20. Shapeev, A.V., Moment Tensor Potentials: A Class of Systematically Improvable Interatomic Potentials. *Multiscale Model. Simul.*, 14, 1153–1173, 2016.
21. Drautz, R., Atomic cluster expansion for accurate and transferable interatomic potentials. *Phys. Rev. B*, 99, 014104, 2019.
22. Batzner, S., Musaelian, A., Sun, L., Geiger, M., Mailoa, J.P., Kornbluth, M., Molinari, N., Smidt, T.E., Kozinsky, B., E(3)-equivariant graph neural networks for data-efficient and accurate interatomic potentials. *Nat. Commun.*, 13, 2453, 2022.
23. Smith, J.S., Isayev, O., Roitberg, A.E., ANI-1: an extensible neural network potential with DFT accuracy at force field computational cost. *Chem. Sci.*, 8, 3192–3203, 2017.
24. Behler, J., Perspective: Machine learning potentials for atomistic simulations. *J. Chem. Phys.*, 145, 170901, 2016.
25. Pal, S. and Mukhopadhyay, S., Development of embedded-atom method (EAM) potential for Palladium–Barium alloy. *Mol. Simul.*, 50, 14, 1–10, 2024.
26. Gupta, A., Dahale, C., Maiti, S., Srinivasan, S.G., Rai, B., Development of an embedded-atom method potential of Ni–Mo alloys for electrocatalysis and surface compositional studies [Preprint]. *arXiv*:2409.07320, 2024.
27. Daramola, A., Bonny, G., Adjanor, G., Domain, C., Monnet, G., Fraczkiewicz, A., Plasticity-oriented interatomic potential for CrFeMnNi high-entropy alloys [Preprint]. *arXiv*:2201.02126, 2022.
28. Daw, M.S., Embedded atom method: Many-atom description of metallic cohesion, in: *Atomistic Simulation of Materials*, V. Vitek and D.J. Srolovitz (Eds.), pp. 181–191, Springer US, Boston, MA, 1989.
29. Foiles, S.M., Baskes, M.I., Daw, M.S., Embedded-atom-method functions for the fcc metals Cu, Ag, Au, Ni, Pd, Pt, and their alloys. *Phys. Rev. B*, 33, 7983–7991, 1986.
30. Johnson, R.A., Analytic EAM alloy model. *J. Mater. Res.*, 4, 379–388, 1989.

31. Weiss, E.J., Ward, L., Oberdorfer, C., Withrow, T., Riegner, D.C., Agrawal, A., Windl, W., Rapid production of accurate embedded-atom method potentials for metal alloys (RAMPAGE) [Preprint]. *arXiv*:2208.02223, 2022.
32. Mishin, Y., Mehl, M.J., Papaconstantopoulos, D.A., Voter, A.F., Kress, J.D., Structural stability and lattice defects in copper: *Ab initio*, tight-binding, and embedded-atom calculations. *Phys. Rev. B*, 63, 224106, 2001.
33. Purja-Pun, G.P. and Mishin, Y., Development of an interatomic potential for the Ni–Al system. *Philos. Mag.*, 89, 3245–3267, 2009.
34. Cleri, F. and Rosato, V., Tight-binding potentials for transition metals. *Phys. Rev. B*, 48, 22–33, 1993.
35. Brommer, P. and Gähler, F., Potfit: Effective potentials from *ab initio* data. Modell. *Simul. Mater. Sci. Eng.*, 15, 295–304, 2007.
36. Seko, A., Hayashi, H., Nakayama, K., Takahashi, A., Tanaka, I., Representation of compounds for machine-learning prediction of physical properties. *Phys. Rev. B*, 95, 144110, 2017.
37. Allera, A., Ribeiro, F., Perez, M., Rodney, D., Carbon-induced strengthening of bcc iron at the atomic scale. *Phys. Rev. Mater.*, 6, 013608, 2022.
38. Wang, L., *et al.*, KLIFF: An active-learning framework for automated interatomic potential development. *Comput. Phys. Commun.*, 264, 107847, 2021.
39. Larsen, A.H., *et al.*, The atomic simulation environment—a Python library for working with atoms. *J. Phys.: Condens. Matter*, 29, 273002, 2017.
40. Gale, J.D. and Rohl, A.L., The general utility lattice program (GULP). *Mol. Simul.*, 29, 291–341, 2003.
41. Ercolessi, F. and Adams, J.B., Interatomic potentials from first-principles calculations: The force-matching method. *Europhys. Lett.*, 26, 583–588, 1994.
42. Stukowski, A., Fransson, E., Mock, M., Erhart, P., Atomicrex—a general purpose tool for the construction of atomic interaction models. *Modell. Simul. Mater. Sci. Eng.*, 25, 055003, 2017.
43. Yao, B., Liu, Z., Zhang, R., EAPOTc: An integrated empirical interatomic potential optimization platform for compound solids. *Comput. Mater. Sci.*, 211, 111551, 2022.
44. Yao, B.N., Liu, Z.-R., Zhang, R.F., EAPOTs: An integrated empirical interatomic potential optimization platform for single elemental solids. *Comput. Mater. Sci.*, 197, 110626, 2021.
45. Giannozzi, P., *et al.*, Quantum ESPRESSO: a modular and open-source software project for quantum simulations of materials. *J. Phys.: Condens. Matter*, 21, 395502, 2009.
46. Kresse, G. and Furthmüller, J., Efficiency of *ab-initio* total energy calculations for metals and semiconductors using a plane-wave basis set. *Comput. Mater. Sci.*, 6, 15–50, 1996.
47. Kresse, G. and Furthmüller, J., Efficient iterative schemes for *ab-initio* total-energy calculations using a plane-wave basis set. *Phys. Rev. B*, 54, 11169–11186, 1996.

48. Thompson, A.P., *et al.*, LAMMPS—a flexible simulation tool for particle-based materials modeling at the atomic, meso, and continuum scales. *Comput. Phys. Commun.*, 271, 108171, 2022.
49. Plimpton, S.J., Fast parallel algorithms for short-range molecular dynamics. *J. Comput. Phys.*, 117, 1–19, 1995.
50. Olsson, P.A.T., Semi-empirical atomistic study of point defect properties in BCC transition metals. *Comput. Mater. Sci.*, 47, 135–145, 2009.
51. Williams, P.L., Mishin, Y., Hamilton, J.C., An embedded-atom potential for the Cu–Ag system. *Modell. Simul. Mater. Sci. Eng.*, 14, 817–833, 2006.
52. Eich, S.M., Beinke, D., Schmitz, G., Embedded-atom potential for an accurate thermodynamic description of the iron–chromium system. *Comput. Mater. Sci.*, 104, 185–192, 2015.
53. Liu, X.Y. and Adams, J.B., Grain-boundary segregation in Al-10 at.% Mg alloys at hot working temperatures. *Acta Mater.*, 46, 3467–3476, 1998.
54. Rassoulinejad-Mousavi, S.M., Mao, Y., Zhang, Y., Evaluation of copper, aluminum, and nickel interatomic potentials on predicting the elastic properties. *J. Appl. Phys.*, 119, 244304, 2016.
55. Rassoulinejad-Mousavi, S.M. and Zhang, Y., Interatomic potentials transferability for molecular simulations: A comparative study for platinum, gold and silver. *Sci. Rep.*, 8, 2424, 2018.
56. Mendelev, M.I., Kramer, M.J., Becker, C.A., Asta, M., Analysis of semi-empirical interatomic potentials appropriate for simulation of crystalline and liquid Al and Cu. *Philos. Mag.*, 88, 1723–1750, 2008.
57. Starikov, S., Smirnova, D., Pradhan, T., Lysogorskiy, Y., Chapman, H., Mrovec, M., Drautz, R., Angular-dependent interatomic potential for large-scale atomistic simulation of iron: Development and comprehensive comparison with existing interatomic models. *Phys. Rev. Mater.*, 5, 063607, 2021.
58. Jacobs, R., Morgan, D., Attarian, S., Meng, J., Shen, C., Wu, Z., Wood, B.M., A practical guide to machine learning interatomic potentials—Status and future. *Curr. Opin. Solid State Mater. Sci.*, 35, 101214, 2025.
59. Dragoni, D., *et al.*, Achieving DFT accuracy with a machine-learning interatomic potential: Thermomechanics and defects in bcc ferromagnetic iron. *Phys. Rev. Mater.*, 2, 013808, 2018.
60. Artrith, N. and Behler, J., High-dimensional neural-network potentials for metal oxidation states. *Phys. Rev. B*, 85, 045439, 2012.
61. Köfinger, J. and Hummer, G., Empirical optimization of molecular simulation force fields by Bayesian inference. *Eur. Phys. J. B.*, 94, 245, 2021.
62. Talirz, L., *et al.*, Materials Cloud, a platform for open computational science. *Sci. Data*, 7, 299, 2020.
63. Lemkul, J.A., *et al.*, Implementation of extended Lagrangian dynamics in GROMACS for polarizable simulations using the classical Drude oscillator model. *J. Comput. Chem.*, 36, 1473–1479, 2015.
64. Pártay, L.B., Bartók, A.P., Csányi, G., Kernel EAM: Descriptor-based embedding. *Phys. Rev. B*, 92, 024108, 2015.

65. Yang, J., *et al.*, Graph-EAM: An Interpretable and Efficient Graph Neural Network Potential Framework. *J. Chem. Theory Comput.*, 19, 5910–5923, 2023.
66. Ling, S., *et al.*, High-throughput theoretical design of lithium battery materials. *Chin. Phys. B*, 25, 018208, 2016.
67. Stukowski, A., Sadigh, B., Erhart, P., Alfredo, C., Efficient implementation of the concentration-dependent embedded atom method for molecular-dynamics and Monte-Carlo simulations. *Modell. Simul. Mater. Sci. Eng.*, 17, 075005, 2009.
68. Christensen, A.S. and von Lilienfeld, O.A., Transfer learning for EAM potentials. *Nat. Commun.*, 11, 5123, 2020.
69. Yang, Y.G., Johnson, R.A., Wadley, H.N.G., Kinetic Monte Carlo simulation of heterometal epitaxial deposition. *Surf. Sci.*, 499, 141–151, 2002.
70. Yang, L., *et al.*, Implementation of metal-friendly EAM/FS-type semi-empirical potentials in HOOMD-blue: A GPU-accelerated molecular dynamics software. *J. Comput. Phys.*, 359, 352–360, 2018.
71. Scogland, T., *et al.*, A first look at integrated GPUs for green high performance computing. *ENAM-HPC10 Workshop*, 2009.
72. Schoonhoven, R., Veenboer, B., van Werkhoven, B., Batenburg, K.J., Going green: Optimizing GPUs for energy efficiency through model-steered auto-tuning [Preprint]. arXiv, 2022.
73. Glaser, J., Nguyen, T.D., Anderson, J.A., *et al.*, Strong scaling of general-purpose molecular dynamics simulations on GPUs. *Proceedings of ACM-SIGSIM PADS*, pp. 187–195, 2015.
74. Perez, D., *et al.*, EXAALT: Exascale atomistics for accuracy, length, and time. Exascale Computing Project, 2023, https://www.exascaleproject.org/research-project/exaalt/.
75. Anderson, J.A., Glotzer, S.C., *et al.*, The performance and energy efficiency of HOOMD-blue on GPU clusters. *J. Comput. Phys.*, 374, 31–44, 2019.
76. Villa, O., *et al.*, Scaling the power wall: A path to exascale. *SC14 Proceedings*, 2014.
77. Daramola, A., Bonny, G., Adjanor, G., Domain, C., Monnet, G., Fraczkiewicz, A., Development of a plasticity-oriented interatomic potential for CrFeMnNi high entropy alloys. *Comput. Mater. Sci.*, 203, 111165, 2022.
78. Deluigi, O.R., Pasianot, R.C., Valencia, F.J., Caro, A., Farkas, D., Bringa, E.M., Simulations of primary damage in a High Entropy Alloy: Probing enhanced radiation resistance. *Acta Mater.*, 213, 116951, 2021.
79. Byggmästar, J., Nordlund, K., Djurabekova, F., Modeling refractory highentropy alloys with efficient machinelearned interatomic potentials: Defects and segregation. *Phys. Rev. B*, 104, 10, 104101, 2021.
80. Ko, W.-S., Grabowski, B., Neugebauer, J., Development and application of a Ni-Ti interatomic potential with high predictive accuracy of the martensitic phase transition. *Phys. Rev. B*, 92, 13, 134107, 2015.

81. Ataollahi, S. and Mahtabi, M.J., An interatomic potential for ternary NiTiHf shape memory alloys based on modified embedded atom method. *Comput. Mater. Sci.*, 227, 112278, 2023.
82. Cheng, C., Ma, Y., Bao, Q., Wang, X., Sun, J., Zhou, G., Wang, H., Liu, Y., Xu, D., Development and application of EAM potentials for Ti, Al and Nb with enhanced planar fault energy of Ti. *Comput. Mater. Sci.*, 173, 109432, 2019.
83. Du, J.P., Wang, C.Y., Yu, T., Construction and application of multi-element EAM potential (Ni–Al–Re) in γ/γ′ Ni-based single crystal superalloys. *Modell. Simul. Mater. Sci. Eng.*, 21, 1, 015007, 2013.
84. Dongare, A.M., Zhigilei, L.V., Rajendran, A.M., LaMattina, B., Interatomic potentials for atomic scale modeling of metal–matrix ceramic particle reinforced nanocomposites. *Compos. Part B Eng.*, 40, 6, 461–467, 2009.

10

Green Processing of Particles Dispersed Composite Materials

Rudra Narayan Mohapatro[1], Sunita Routray[2*] and Ranjita Swain[1]

[1]Department of Chemical Engineering, C.V. Raman Global University, Bhubaneswar, Odisha, India
[2]Department of Mechanical Engineering, Centurion University of Technology and Management, Ramachandrapur, Jatni, Odisha, India

Abstract
The development of particle reinforced composites over green processing techniques has developed as a sustainable alternative to outdated composite fabrication methods. These eco-friendly approaches aim to decrease environmental effect by utilizing renewable resources, ecological matrices, recycled fillers, and energy-efficient methods. This study explores the synthesis, characterization, and routine evaluation of particle composites produced using green processing routes such as solvent-free partying, bio-based binders, low-temperature curing, and additive manufacturing. Particular attention is given to natural and industrial waste-derived particles such as rice husk ash, fly ash, nano-clays, and cellulose nanocrystals used as assistances in polymeric and ceramic atmospheres. The properties of the composites are analyzed to assess their suitability for applications in automotive, packaging, and construction areas. The results prove that green processed composites not only subsidize to environmental sustainability but also offer modest or improved functional properties. This study highlights the probable of integrating waste valorizations with green engineering to promote circular economy ideologies in material science.

Keywords: Sustainable, green processing, composite, waste, ecological

**Corresponding author*: sunitaroutray77@gmail.com

Gurudas Mandal, Swarup Kumar Ghosh, Snehanshu Pal and Sandip Kunar (eds.) Sustainable Manufacturing of Advanced Materials, (199–222) © 2026 Scrivener Publishing LLC

10.1 Introduction

The rising demand for sustainable development and ecologically friendly technologies has led to substantial interest in green processing techniques for composite materials. In particular, the fabrication of particles dispersed composite materials, which involve reinforcing a matrix with micro or nano-scale particles has seen major advancements through the adoption of green methods. These mixtures are widely used in aerospace, automotive, biomedical, and construction applications due to their enhanced properties [1].

Green processing refers to the use of biodegradable materials, energy-efficient systems, and minimal hazardous waste generation during composite manufacture. This includes techniques such as solvent-free synthesis, use of bio-based or recyclable matrices, low-temperature processing, and combination of natural or waste-derived supporting particles (like biochar, fly ash, or cellulose). Such approaches not only diminish ecological impact but also often lower manufacture costs and improve sustainability across the product lifespan [2].

This field links materials science, ecological engineering, and nanotechnology, aiming to achieve high-performance composites while obeying to principles of green harmony and circular economy. The development and optimization of green processing methods for particles dispersed composites are therefore crucial for advancing cleaner industrial practices and indorsing sustainable material revolution.

10.2 Background and Need for Green Processing

Traditional processing methods for fused materials often involve energy-intensive measures, toxic solvents, synthetic polymer mediums, and the use of non-renewable or non-degradable supports. These practices contribute suggestively to pollution, carbon emissions, and long-term ecological degradation. In light of universal challenges such as climate change, resource depletion, and firmer environmental regulations, there is a serious need to invent green processing methods that reduce waste, energy use, and environmental toxicity during the material's lifespan.

Green processing is allied with the values of green chemistry and engineering, which highlight the reduction or removal of hazardous substances, better energy efficiency, and the use of renewable or recyclable components.

When applied to particles dispersed fused materials, green processing seeks to maintain or enhance the properties while significantly reducing their environmental impact [3].

10.3 Understanding Particles Dispersed Composite Materials

Particles detached composite resources contain a medium (often a polymer, ceramic, or metal) rooted with separate particles such as ceramics, carbonous materials, or metallic powders at micro or nano scales. These elements assist to support the matrix, refining qualities like strength, stiffness, thermal stability, electrical conductivity, and resistance to wear or corrosion.

The performance depends on:

- Type and dimension of the dispersed elements.
- The dispersion uniformity within the matrix.
- Type of interface.
- The processing conditions used during fabrication.

The combination of natural particles (e.g., clay, cellulose nanocrystals, rice husk ash, biochar) and manufacturing waste particles (e.g., fly ash, slag, silica fume) is gaining approval due to their low cost, abundance, and ecological benefits. By utilizing such resources in green processing, we can report both performance and sustainability objectives [4].

10.4 Green Processing Approaches

A wide range of green dispensation methods are being explored and applied for manufacturing particles dispersed composites. These methods aim to decrease carbon emanations, limit solvent convention, and promote the use of sustainable raw materials. Common green processing approaches include:

- **Solvent-Free and Water-Based Processing:** Avoiding the use of volatile organic compounds (VOCs) by using water as a dispersal medium or retaining melt processing techniques can significantly reduce discharges and toxicity [5].

- **Low-Temperature Curing and Processing:** Using room temperature or low-temperature curative systems (e.g., UV-curable resins, bio-based thermosets) lowers energy consumption and helps in achieving energy-efficient manufacturing [6].
- **Use of Renewable or Used Materials:** Employing natural polymers (like polylactic acid, cellulose, or starch-based resins) or recycled plastics as mediums and strengthening them with agricultural or industrial waste particles supports circular economy goals.
- ***In-Situ* Particle Synthesis:** *In-situ* formation of consolidation elements inside the intermediate through green chemical paths (e.g., sol-gel processing, precipitation) offers healthier control of elemental diffusion and removes the need for separate particle production and management [7].
- **Additive Manufacturing (3D Printing):** When combined with green feedstock, 3D printing can harshly minimize substantial waste and energy usage, presenting a sustainable route to produce customized composites with disseminated particles [8].

10.5 Role of Bio-Based and Waste-Derived Particles

A crucial aspect of green composite growth is the choice of strengthening particles. Bio-based and waste-derived materials offer a dual advantage: they decrease dependence on virgin raw materials and help in waste vaporization. Some examples include:

- **Biochar:** Derived from the pyrolysis of biomass, biochar has outstanding thermal stability and strength properties. It helps as a sustainable underpinning agent in polymer matrices.
- **Fly Ash and Slag:** These manufacturing by-products are rich in aluminosilicates and can improve compressive strength and durability when dispersed in cement or polymer composites.
- **Nanocellulose:** Extracted from plant matter, nanocellulose exhibits high asset, biodegradability, and renewability. It can be used to strengthen biopolymers in packaging, medical, and structural submissions.

- **Rice Husk Ash, Bagasse Ash, and Coconut Shell Powder:** Agricultural residues like these offer silica-rich or carbonaceous structures that subsidize to the strength and thermal lining of composites.

The appropriate surface behavior or functionalization of such elements is often required to recover their compatibility and closeness with the host medium, which can be done using eco-friendly agents or plasma treatment techniques [9].

10.6 Benefits and Challenges

Benefits of green processing in particle-dispersed composites include:

- Reduced ecological impact through energy proficiency and waste reduction.
- Lower making costs by using nearby available normal or waste materials.
- Enhanced life cycle performance and possible for biodegradability or recyclability.
- Prospects for carbon credit and acquiescence with ecological regulations.

However, several challenges persist:

- Reaching uniform particle dispersion deprived of toxic wetter or high-energy mixing.
- Safeguarding reliable strength and thermal performance with bio-based inputs.
- Partial scalability and market receipt of new green materials.
- Essential for robust classification and long-term durability studies.

Overcoming these contests requires a multidisciplinary method concerning materials science, chemical engineering, environmental science, and industrial design [10].

10.7 Future Outlook and Applications

As sustainability develops a key design criterion in existing manufacturing, the upcoming of green processing in particle-dispersed composites appears promising. Developments in nanotechnology, biopolymers, and clean energy bases are likely to quicken the expansion of greener composites.

Emerging submissions include:

- Lightweight motor-powered parts using bio-filled thermoplastics to recover fuel efficiency.
- Eco-friendly structure panels with industrial ash fillers in cement matrices.
- Biodegradable wrapping materials strengthened with nanocellulose or starch.
- Functional fabrics and coverings including biochar or silver nanoparticles with antimicrobial possessions.
- Sustainable energy strategies like solar boards or battery electrodes made with green composites.

The integration of life cycle calculation (LCA) and techno-economic investigation (TEA) will be serious in evaluating and optimizing the sustainability performance of these materials across different sectors [11].

10.8 Particles Composite Processing in Green Processing of Particles Dispersed Composite Materials

Particle-dispersed composite resources consist of a matrix (polymer, metal, or ceramic) reinforced with fine elements that improve the material's strength, thermal, and barrier properties. These particles, which may be micro- or nano-sized, are disseminated consistently throughout the matrix to produce a synergistic upgrading in performance. The green processing of these composites highlights ecological sustainability by accepting eco-friendly materials, cleaner technologies, and energy-efficient fabrication techniques.

Green processing of particle-dispersed mixtures focuses on dipping ecological influences during both the substantial selection and manufacturing stages. In outdated composite processing, artificial matrices and risky chemicals are commonly used, foremost to ecological effluence and health hazards. In contrast, green processing uses bio-based or biodegradable polymers (e.g., polylactic acid, starch-based polymers), recycled thermoplastics, and renewable or waste-derived elements such as fly ash, rice husk ash, bagasse, coconut shell powder, silica, and biochar. These selections not only reduce confidence on virgin resources but also support the recycling and reuse of manufacturing or agricultural waste, aligning with circular economy values [12].

Several processing methods are working in green composite manufacture, including solution forming using water or ethanol-based thinners, melt amalgamation at low temperatures, *in-situ* polymerization with non-toxic substances, and power-driven approaches like compression molding and extrusion. Advanced dispersal techniques such as ultrasonication or strength emotive are used to allocate elements evenly within the matrix, safeguarding unvarying goods throughout the composite. These low-energy approaches significantly decrease carbon emissions associated to outdated high-temperature or chemical-intensive procedures.

The advantages of green dispensation include ecological sustainability, compact energy practice, lower substantial costs when using excess fillers, and the production of harmless and more biodegradable goods. It also delivers a stage for lightweight, high-strength materials appropriate for eco-friendly claims in automotive, construction, packaging, and electronics.

However, the green processing of particle mixtures is not without tests. Attaining proper compatibility between hydrophilic usual elements and hydrophobic polymer mediums often wants surface treatments or green compatibilizers, which may add difficulty. Additionally, the inconsistency in element size, shape, and composition specifically from natural or waste-derived causes can result in unpredictable composite properties. Ensuring unwavering dispersal without the use of conventional chemical dispersants ruins another difficulty, as poor dispersal can lead to weak spots in the material. Furthermore, some bio-based polymers may have lesser thermal or strength performance associated to their synthetic complements, limiting their use in high-demand claims unless hybrid strengthening approaches are functional [13].

Despite these confines, ongoing study and growth in the field of green materials discipline are creating talented results. Innovations in nanoparticle mixture from renewable sources, advanced dispersion methods, and biodegradable matrix resources are helping to overwhelmed current obstacles and increase the potential of green particle dispersed complexes.

Advantages of Green Processing of Particle Dispersed Composites

- ✓ Decreases reliance on fossil possessions
- ✓ Lowers carbon footmark
- ✓ Enables ecological or recyclable composites
- ✓ Improved asset, stiffness, or hardiness
- ✓ Better thermal and chemical resistance
- ✓ Skilful degradation rates
- ✓ Consumption of low-cost industrial/agricultural waste
- ✓ Decrease the processing costs *via* low-energy methods
- ✓ Reduces use of toxic substances and solvents [14].

Disadvantages of Green Processing of Particle Dispersed Composites

- ✓ Inconsistency in natural elemental properties (size, shape, purity)
- ✓ Possible compatibility questions between matrix and filler
- ✓ Imperfect processability of bio-based mediums
- ✓ Struggle in achieving unvarying particle dispersion without chemical dispersants
- ✓ Occasionally lesser strength or thermal performance associated to conventional composites
- ✓ Moisture compassion in natural particle systems
- ✓ High cost of some biopolymers or green solvents
- ✓ Deficiency of adjustment and industrial-scale accessibility [14].

10.9 Natural Fiber and Polymer Matrix Materials in Green Composites

Green complexes are a class of ecologically friendly resources made from usual fibers and biodegradable or recyclable polymer matrices. These mixtures are calculated to decrease the environmental influence accompanying with conventional synthetic fiber-reinforced plastics by endorsing sustainability, biodegradability, and the utilization of renewable resources.

10.9.1 Natural Fibers Used in Green Composites

Natural fibers are derivative from plants, animals, or minerals, but plant-based fibers are most common in green composites due to their abundance, renewability, and favorable strength properties.

Common Plant-Based Natural Fibers:

Fiber	Source	Key properties
Jute	Strong woody plant of jute plant	High strength, moral thermal insulation
Hemp	Strong woody plant of hemp plant	Outstanding tensile strength, rot-resistant
Flax	Strong woody plant of flax plant	High arduousness, biodegradable
Kenaf	Strong woody plant of kenaf plant	Lightweight, good strength act
Sisal	Greeneries of agave plant	Rigid, good influence resistance
Coir	Coconut husk	High durability, low price
Banana	Banana plant pseudo-stem	Biodegradable, reasonable strength
Bagasse	Sugarcane waste	Renewable, lightweight

These fibers are mainly calm of cellulose, hemicellulose, lignin, and pectin, with cellulose providing tensile strength and difficulty [15].

10.9.2 Polymer Matrix Materials in Green Composites

Polymer matrices in green composites can be [15]:

A. Biodegradable Polymers

Polymer	Origin	Properties
Polylactic Acid (PLA)	Agitated corn starch or sugarcane	Recyclable, transparent, rigid
Polyhydroxyalkanoates (PHA)	Bacterial fermentation	Ecological, thermoplastic
Polycaprolactone (PCL)	Synthetic biodegradable	Stretchy, low melting point
Starch-based Polymers	Natural stiffener (corn, potato)	Compostable, low strength
Cellulose Acetate	Derived from cellulose	Biodegradable, good strength

B. Recyclable/Green Synthetic Polymers

Polymer	Notes
Recycled Polypropylene (r-PP)	Used with accepted fibers in hybrid green mixtures
Recycled Polyethylene (r-PE)	Profitable, less biodegradable but eco-conscious
Polymers with Green Additives	Improved recyclability or degradability

10.9.3 Selection Criteria for Natural Fibers and Polymer Matrix

- **Fiber Aspect Ratio**: Lengthier fibers with advanced aspect ratios recover load transfer.
- **Moisture Absorption**: Usual fibers are hydrophilic; matrix collection must version for moisture sensitivity.
- **Thermal Stability**: Fibers degrade >200°C, so matrix dispensation temperature must be below that.
- **Strength Compatibility**: Match fiber stiffness and forte with matrix behavior.
- **Processing Method Compatibility**: Deliberate melt blending, compression molding, extrusion, or injection molding.
- **Biodegradability Goals**: Both fiber and medium should be biodegradable if full degradation is anticipated [16].

10.9.4 Compatibility Issues and Solutions

Natural fibers are hydrophilic, though most polymers are hydrophobic, instigating poor interfacial adhesion.

Solutions to Improve Compatibility:

- **Fiber Surface Treatments:**
  - Treatment with alkali (NaOH)
  - Silane coupling agents
  - Acetylation or bleach treatment
- **Use of Compatibilizers:**
  - Maleic anhydride grafted polymers (e.g., MAPP for PP matrix)
- **Coating Techniques:**
  - Applying biodegradable coupling agents to fibers

These methods enhance fiber–matrix interlocking, stress transfer, and overall composite concert.

The combination of natural fibers and eco-friendly polymers in green amalgams offers a sustainable pathway for modern material growth. By sensibly selecting compatible fiber-matrix pairs and smearing suitable treatment methods, high-performance green composites can be tailored for requests in automotive interiors, building materials, packaging, furniture, and consumer goods [17].

10.10 Chemical Treatment of Natural Fibers

Alternative to synthetic fibers, natural fibers are gaining increasing attention as sustainable alternatives in composite materials. The natural fibers are mainly jute, flax, hemp, coir, and sisal etc. They have many advantages which include biodegradability, low cost, relatively good strength-to-weight ratio and renewability. Their inherent incompatibility with hydrophobic polymer matrices put a major limitation in utilizing natural fibers in advanced composite applications. Cellulose, hemicellulose, lignin, pectin, and waxes are the primary constituents of nnatural fibers, which make it hydrophilic in nature. Poor interfacial bonding with most polymer matrices, especially thermoplastics and thermosets has resulted due to that [18, 19].

Chemical treatment is done to improve fiber-matrix adhesion by reducing its moisture absorption tendency, which is achieved through modification of the fiber surface. Alkali treatment is the most common treatment, where sodium hydroxide (NaOH) solution is used to soak fibers. This treatment removes impurities and more cellulose on the fiber surface are exposed to increasing roughness. strength interlocking with the matrix increases due to enhanced surface roughness. Hydrophilic groups reduce in number which enhance dimensional stability and bonding strength. Due to alkali treatment, fiber bundles break into smaller filaments increasing surface area and interaction with the matrix [20, 21].

Silane coupling, is another treatment, where silane agents act as molecular bridges between the fiber and the matrix. Covalent bonds are formed between silanes and hydroxyl groups on cellulose. Stronger chemical bonding forms at the fiber-matrix interface, improving strength properties, especially tensile and flexural strength [22].

Some esterification processes such as acetylation and benzoylation are also adopted where hydroxyl groups in the cellulose are replaced with acetyl or benzoyl groups, respectively. These treatments make the fibers more compatible with non-polar matrices such as polypropylene (PP) or polyethylene (PE) by reducing the hydrophilic nature of the fibers. Composites show lower moisture uptake, better fiber dispersion, and more uniform stress transfer due to these modifications. Improved strength properties, enhanced thermal stability, reduction in water absorption, resistance to biological degradation are the results of chemical treatments [23, 24].

The area of applications of chemically treated fibers are automotive sectors, construction, packaging, and consumer goods. Compatibility with different matrix materials can be enhanced with chemical treatment of natural fibers. The performance of composites is improved by modifying the fiber surface chemistry, improving roughness, and reducing moisture affinity. The fiber type, matrix material, and intended application decide the type of treatment.

10.11 Strength Properties and Performance

Matrix is generally composed of polymers, metals, or ceramics and discrete particles are incorporated into it to synthesize particle-dispersed composite materials. These composites have enhanced strength properties and are widely used in structural, automotive, aerospace, and civil engineering applications. The type, size, distribution, volume fraction, and interfacial bonding of the dispersed particles decide the performance of these materials. The strength properties analyzed in these materials, reflecting different modes of deformation and failure [25].

The uniform distribution of hard particles, such as silica, alumina, or carbon nanotubes improve the tensile strength in particle-dispersed composites. Dislocation motion and crack propagation under load are hindered by the reinforced particles. The interfacial adhesion between the matrix and the particles decide the quality of the composite.

Stress concentration zones are created due to poor bonding, which result in premature failure. Bonding strength can be enhanced by surface treatments which optimizes load transfer from the matrix to the particles.

The presence of particles also affect the flexural strength of a material, by which a material resists deformation under bending. When a material bends, a compressive force acts at the upper surface of the material and a tensile force acts at the lower surface. Flexural rigidity is increased due to particle inclusion by resisting these stresses. Nanoparticles such as clay or graphene oxide can improve stiffness and delay the onset of plastic deformation in polymer matrix composites.

A material's ability to absorb and dissipate energy during sudden loading or shock is termed as impact resistance. In protective gear or automotive components, impact resistance is an important parameter. Enhancement of impact resistance is dependent on the way the particles are distributed. If particles are not well distributed, it may act as stress risers, causing brittle failure. The energy dissipation mechanisms are greatly influenced by the matrix's ability to yield and deform plastically around the particles [26].

The particle-dispersed composite materials, when optimized with dispersion, interfacial bonding, and particle characteristics show superior strength properties than the unreinforced counterparts. With advancement in nanotechnology, the performance of composites is improving continuously, which shows enhancement in the tensile, flexural, and impact strength.

10.12 Sustainable Processing Techniques

The eco-friendly processing methods for particle-dispersed composite materials are in demand due to increasing global focus on environmental sustainability. Traditional techniques of manufacturing processes are energy-intensive and generally rely on non-renewable resources, which produce hazardous emissions and waste.

Whereas, utilizing green materials, minimizing energy consumption, and adopting cleaner technologies, sustainable processing techniques aim to reduce environmental impacts. Sustainable processing techniques reduces the carbon footprint of materials throughout their lifecycle.

The use of bio-based and recyclable matrices in sustainable composite processing is one of the great initiatives. Epoxy or polyester, which are derived from petroleum-based sources are used as matrices in conventional processes. In the recent trend, it is emphasized to use biodegradable or renewable alternatives like polylactic acid (PLA), bio-based epoxy resins, and thermoplastic starch. These approaches reduces dependence on fossil fuels and make the disposal process easier through composting or thermal recycling. Fully green composites suitable for low to medium-load applications can be developed, when combined with natural particles such as cellulose nanocrystals, lignin, or agricultural waste-derived powders. The Low-energy manufacturing methods such as solvent-free processing, resin transfer molding (RTM) at room temperature, and compression molding with pre-dried materials help to reduce the energy input and harmful volatile organic compounds. The additive manufacturing (3D printing) generates minimal waste with precise deposition of composite layers. Direct ink writing (DIW) and fused deposition modeling (FDM) allows efficient material use and reduces scrap generation by adopting customized designs and efficient material use [27].

To improve matrix-particle adhesion, researchers are exploring environmentally benign treatments like plasma modification, UV-ozone cleaning, or bio-based coupling agents such as tannins and plant extracts rather than harsh chemicals or surfactants. This kind of treatment is called green surface treatment of particles. By doing so, the environmental integrity is maintained. For fabrication of particle-dispersed composites, water-based processing techniques are found to be very effective. The need for organic solvents can be eliminated by using the waterborne resins and dispersions for mixing particles [28].

Product life cycle assessment tools are also employed to assess the environmental benefits of the processes. These tools analyses the product from raw material sourcing to end-of-life by evaluating the energy use, emissions, and recyclability of composite materials from raw material sourcing to end-of-life. Combination of green materials, low-impact manufacturing, and innovative treatment technologies produces high-performance composites that meet both functional and environmental objectives.

10.13 Applications and Future Directions

The green composites are gaining demand in different sectors due to their lightweight nature, reduced environmental impact, and sufficient strength. These composites are in high demand in automotive sectors, the interior panels, seat backs, door trims, and dashboards are manufactured using these composites which help to reduce vehicle weight and enhance recyclability. In the aerospace sector also, researchers are trying to use green composites in non-critical aerospace structures, where sustainability and weight savings are beneficial. In critical aircraft parts, the use of green composites are restricted due to safety standards and performance requirements. Construction industries also use green composites for making partition panels, roofing, door frames, insulation materials, and flooring components. Apart from these, green composites are also used for packaging, consumer goods, and sports equipment. These are the eco-friendly alternatives to traditional plastic-based components [29].

Future research studies focus on enhancement of bonding between natural fibers and matrices through physical, chemical, and enzymatic surface treatments. To get balance between sustainability and performance, hybrid green composites are also developed. Use of nano particles for particle dispersion gives superior strength, thermal, and barrier properties.

Design and performance prediction of green composite can be predicted by integrating AI and machine learning. It helps with material selection, property optimization, and structural integrity analysis. The adoption of green composites across diverse sectors is accelerated to reduce carbon emissions, consumer demand for sustainable products, and rapid technological advancements.

10.14 Feasibility Studies on Investigating the Use of Industrial Waste Materials Reinforced Hybrid Composites

In recent years, the industrial and agricultural waste has been used as the raw materials for the development of sustainable composite materials. Fly ash is one of the byproducts can be used as another raw material for preparing a polymer in addition of other organic waste from industries and agriculture. The natural fiber is mixed with fly ash to form the composite polymer which can be used in many areas in the daily life of the human being.

The coconut fibers are used as one of the raw material mixes with the fly ash to form a composite polymer, that enhances the packing efficiency due to increase in the composite density. The coconut fibers having the low density as it is a lignocellulosic material which reduces the overall density while using in higher proportion.

Thus, optimizing the ratio of fly ash to coconut fibers allows for tailoring composite density for specific applications. The dimension stability of the composites is crucial while exposed to humid or wet environments as the composites thickness swell due to the water absorption. Coconut fiber is hydrophilic in nature that increases the swelling using water that can give rise to negative impact in a long run. The fly ash addition to the coconut fibers reduces the porosity and form's structure which increases the density, prevents the moisture entrance. The balanced composition of fly ash and coconut fibers forms a hybrid composite which improves the properties of the composite without any effect on the strength parameters. This type of composites can be used in the construction and automotive sectors. The physical properties enhanced by optimizing the compatibility of the matrix and treatment of the fibers [30].

This investigations of the pineapple leaf fiber (PALF) incorporate with the fly ash filler materials to form the epoxy reinforced composites that aims to enhance the strength and water resistance characterisation for the biomedical applications. PALF is one of the potential agricultural waste fibers used for the formation of synthetic reinforcements. The epoxy matrix composites were fabricated by adding PALF in a weight percentage of 10%, 20%, and 30% by weight with FA at 3%, 6%, and 9% wt. There are strength tests such as tensile, flexural, impact shows the optimum addition of 6% fly ash to enhance the tensile strength up to 65.3% as the flexural strength increases by 31.9%. The improvement in impact resistance increases to 74.2% by increasing the amount of fly ash by 9 wt%. The stiffness and load transfer efficiency by uniform dispersion of fly ash particles *via* SEM. By using Taguchi design and response surface methodology, (RSM) is applied to optimize the composition of PALF 20 wt% and FA 6 wt%. It shows that the higher FA content and more water immersion period reduces water uptake to improve resistance to moisture content (ASTM D 570) [31]. The investigation influences the fly ash filler to enhance the strength properties and water absorption behavior.

The research conducted by Santhosh Nagaraja *et al.*, investigates the influence of fly ash filler mix with the pineapple leaf fiber to form epoxy polymer composites reinforced to enhance the strength properties and water absorption, that focuses on the biomedical application where the material performance is serious. The low water absorption quality improves the durability and longevity of materials in biological environments. The use of pineapple leaf fiber as a natural reinforcement not only contributes to the sustainability of the composites but also enhances their strength performance. This combination of natural fibers with fly ash presents a promising approach to developing eco-friendly materials [32].

This study explores the hybrid composite reinforced made up of oil pulp empty fruit bunches and fly ash which focused on the thermal and physicochemical properties. There are different filler loading that enhance the quality of the composite to determine the strength, thermal stability, and overall material properties. Adding fly ash to the oil pulp enhances the composite thermal stability which enhances the material that can withstand at higher temperature. It is also observed that the filling of fly ash particles improves the tensile strength and flexural properties of the composites, making them more suitable for various applications [33].

The study reveals that the reinforced polymer composites was made by using banana fiber residue with fly ash. These materials properties changes by using different percentage of filling material with the waste residue, hence optimum conditions have been chosen for the best results. The fly ash content helps the drilling properties of the composite, that fly ash results in reduced drilling forces and improved surface quality of the drilled holes, indicating better machinability. The drilling performance varied with the varying composition of fly ash content. highlights the optimal fly ash content that balances strength strength and drilling performance, suggesting that a careful selection of filler can affect the overall characteristics of the composites. The addition of banana fiber, a natural and renewable resource, added with industrial waste fly ash not only contributes to the sustainability of the materials but also enhances their performance. The composite fabrication promotes the improvement in the functionality of the polymer composite for various industrial applications that is eco-friendly. This composite is an alternative material to develop high-performance composites that meet the growing demands of the manufacturing sector [34].

There is a demand for materials, which have high performance and are sustainable in day-to-day life. The sisal fiber reinforced hybrid composites utilizing bio-epoxy resin and fly ash as a filler. The combination of the natural fibers with bio-based resins and industrial waste creates eco-friendly composites. The sisal fibers increase the strength performance of the composites, providing improved strength and durability. Addition of the fly ash as a filler material subsidize the overall strength properties but also reduces the weight of the composites, making them suitable for various applications in industries such as automotive and construction. The material strength and performance maximized by using the optimum sisal fibers and filler material fly ash content maintain the sustainability of the material. Bio-epoxy resin, that is resulting from renewable resources, combined with natural fibers and waste materials, encourage a circular economy. This development of the material is eco- friendly and strongly robust which addresses the growing need for sustainable material solutions [35].

The investigation on the production of the hybrid polypropylene composites is reinforced from Himalayan agave fiber and graphene oxide (GO) coated fly ash. The polypropylene composites performance increases by adding the natural fibers and waste materials for various applications. The result indicates that the assimilation of Himalayan agave fiber significantly progresses the strength properties of the composites, including tensile strength, flexural strength, and impact resistance. GO-coated fly ash was added to fibers for further enhancement of the strength properties by providing better inter facial bonding and dispersion within the polymer matrix. The thermal stability exhibits the material application in different applications that is suitable for applications that require resistance to high temperatures. The morphological studies using scanning electron microscopy (SEM) demonstrate the effective distribution of agave fibers and

GO-coated fly ash within the polypropylene matrix, contributing to the overall performance of the composites. The findings underscore the potential of Himalayan agave fiber and GO-coated fly ash as effective reinforcements in enhancing the strength and thermal properties of polypropylene composites [36].

The hybrid composites reinforced formed using bamboo fiber and filled with fly ash, focusing on their potential applications in various industries. The manufacturing of the high-performance materials, the natural fibers and industrial waste combined in a proper proportion to enhance the strength properties. The importance of the inter facial bonding between the bamboo fibers and the polymer matrix, plays a crucial role in determining the overall performance of the composites. Polymer matrix, plays a crucial role in determining the overall performance of the composites. The potential of bamboo fiber and fly ash as effective reinforcements in enhancing the strength properties of hybrid composites, paves the way for their application in various sectors, including construction, automotive, and consumer goods [37]. The potential of plant fiber and industrial waste used to for a reinforced polymer composite is one of the viable substitutes for traditional wood materials. Construction and manufacturing industries demand increases day-to day life, so the plant fibers, such as jute, hemp, or bamboo, significantly enhance the strength properties of the composites, including tensile strength, flexural strength, and impact resistance. The use of industrial waste, such as fly ash or recycled plastics, not only improves the overall performance of the composites but also contributes to waste reduction and environmental sustainability. Optimization of the fiber and waste content is used to achieve the best balance of performance and sustainability. Thermal stability shows the material suitable for high temperature thermal application in the industries. It emphasizes the potential of a high-performance composites to replace conventional wood products, thereby reducing deforestation and promoting the use of renewable resources [38].

The strength behavior of woven bidirectional jute/glass hybrid nanocomposites filled with nanometer and micrometer-sized fly ash particles were evaluated for the industrial application. The experimental result indicates that the addition of both nanometer and micrometer fly ash particles helps to enhances the strength properties of the jute/glass hybrid composites, with nano-fly ash showing superior performance due to its larger surface area and better dispersion within the polymer matrix. Thermal analysis reveals that the addition of fly ash contributes to improved thermal stability, making these composites suitable for applications requiring resistance to heat and environmental factors [39].

The study conducted on the surface response methodology of Taguchi method was applied with various parameters of different components of epoxy composites reinforced with Prosopis Juli flora fiber and filled with silicon carbide (SiC) and fly ash under dry sliding wear conditions for the optimization. The main aim is to increase the wear resistance and role of the composite, which is used in the various industrial applications. On the wear composite characteristics, different filler materials have different properties. The results show that the incorporation of SiC and fly ash produces remarkable changes on the wear resistance of the Prosopis juliflora fiber-reinforced epoxy composites, to maintain the optimal conditions to increase the surface integrity to reduce the wear rate. Different parameters like filler content, sliding speed, and load are used to analyze the most effective conditions for minimizing wear through Taguchi method. For suitable applications required in high wear resistance the results prove that the synergistic effect of SiC and fly ash was increasing the strength properties and durability of composite [40].

The pineapple leaf fibre is used with the filler material fly ash to make an epoxy polymer composite reinforced which impacts on the strength properties and water absorption behavior that can be used for the biomedical applications. The biomedical field requires the biocompatible and sustainable materials. In view of this, the potential of utilizing natural fibers and industrial waste helps to enhance the performance of epoxy composites in biomedical sectors. It is also noticed that the incorporation of pineapple leaf fiber improves the strength properties of the composites, including tensile strength, flexural strength, and impact resistance, whereas the fly ash enhances these strength properties and plays a crucial role in reducing water absorption, that plays a vital role for maintaining the integrity and longevity of biomedical materials. It is also seen that the interfacial bonding between the fiber, filler, and epoxy matrix is very important, which significantly effects the overall performance of the composites [32]. The viability of using fly ash as a filler in banana fiber-reinforced hybrid composites, fixing on the potential benefits of combining natural fibers with industrial by-products to produce a sustainable materials. It indicates that the integration of banana fiber pointedly improves the strength properties of the composites, containing tensile strength, flexural strength, and impact resistance. The fly ash as a filler not only progresses these properties but also subsidizes to the reduction of the overall weight of the composites, making them suitable for lightweight applications [41].

10.15 Conclusion

In conclusion, the green dispensation of particle discrete composite materials signifies a vital step toward sustainable materials engineering. By incorporating waste utilization, renewable possessions, and energy efficient processing methods, this approach offers a viable substitute to outdated composites conducive positively to both ecological protection and industrial innovation.

The research highlights the potential of FA as a costeffective, ecofriendly filler that significantly enhances the strength integrity and water resistance of natural fiber composites. The composite's property profile—strength, durability, water stability—makes it a promising candidate for certain biomedical applications, where both strength performance and moisture resistance are essential. The findings suggest that fly ash fillers can be effectively utilized to improve the strength and water absorption properties of epoxy composites reinforced with natural fibers, opening avenues for the development of advanced materials suitable for various biomedical applications and emphasizing the importance of sustainable practices in material science.

The tensile strength shows an enhancement of up to 6.3%; flexural strength experiences an increase of 68%, impact strength proves a boost of 28%, and scratch stiffness elevates by 17% which were achieved at 20% fiber and 5% fly ash weight fraction. It provides valuable insights into the development of hybrid composites that leverage the advantages of renewable resources and industrial waste, paving the way for their application in diverse industries. The findings underscore the potential of banana fiber and fly ash as effective reinforcements in creating high-performance, eco-friendly composites that meet the growing demand for sustainable material solutions.

AI disclosure statement: The author acknowledges the use of AI-based tools for minor language editing or grammatical correction. However, all intellectual contributions, analyses, and conclusions presented in this chapter are entirely the author's own.

References

1. Khalfallah, A. and Leitao, C., Sustainable Manufacturing and Green Processing Methods. *Machines*, 12, 11, 815, 2024, https://doi.org/10.3390/machines12110815.

2. Mohamed, H.H., Chapter 1 - Introduction to green processing for sustainable materials, in: *Sustainable Materials and Green Processing for Energy Conversion*, pp. 1–42, Elsevier, 2022, https://doi.org/10.1016/B978-0-12-822838-8.00004-1.
3. Meher, A.K. and Zarouri, A., Green Analytical Chemistry—Recent Innovations. *Analytica*, 6, 1, 10, 2025, Pages 1–32. https://doi.org/10.3390/analytica6010010.
4. Lavallee, J.M., Soong, J.L., Francesca Cotrufo, M., Conceptualizing soil organic matter into particulate and mineral-associated forms to address global change in the 21st century. *Global Change Biol.*, 26, 261–273, 2020, https://doi.org/10.1111/gcb.14859.
5. Yao, A., Li, H., Luo, H., Zhao, J., Xiang, J., Chen, Y., Fan, H., A novel strategy for preparation of solvent-free high-solid content water-based polyurethane. *J. Polym. Res.*, 30, 258, 2023, https://doi.org/10.1007/s10965-023-03628-3.
6. Nath, S.S., Dineshkumar, R., Prasad, K.D.V., Gupta, D., Jangir, P., Kiruba, S., Computational approaches to optimize polymerization and curing in low-temperature thermosetting composites for enhanced additive manufacturing efficiency. *Therm. Sci. Eng. Prog.*, 58, 103223, 2025, https://doi.org/10.1016/j.tsep.2025.103223.
7. Sun, S., Jin, Z., Liu, X., Han, Z., Wang, Y., *In situ* consolidation process-based fabrication and interlaminar modification mechanism associated with CF/PEEK multiscale nanocomposites characterized by interlaminar doping of CNTs. *Compos. Sci. Technol.*, 222, 109356, 2022, https://doi.org/10.1016/j.compscitech.2022.109356.
8. Pongwisuthiruchte, A. and Potiyaraj, P., Challenges and innovations in sustainable 3D printing. *Mater. Today Sustainability*, 31, 101134, 2025, https://doi.org/10.1016/j.mtsust.2025.101134.
9. Khedulkar, A.P., Bobade, R.G., Doong, R., Pandit, B., Ky, N.M., Ambare, R.C., Hoang, T.-D., Kumar, K.J., Bio-based nanomaterials as effective, friendly solutions and their applications for protecting water, soil, and air. *Mater. Today Chem.*, 46, 102688, 2025, https://doi.org/10.1016/j.mtchem.2025.102688.
10. Javed, M.N., Sajid, A., Bangash, K., Abbas, M., Ahmed, S., Kaplan, A., Iqbal, S., Khan, M.N., Adnan, M., Ali, A., Zaman, F., Wahab, S., Potential and Challenges in Green Synthesis of Nanoparticles: A Review. *J. Xi'an Shiyou Univ., Nat. Sci. Ed.*, 19, 2, 1155–1165, 2023, http://xisdxjxsu.asia.
11. Ceschin, F. and Gaziulusoy, I., Evolution of design for sustainability: From product design to design for system innovations and transitions. *Des. Stud.*, 47, 118–163, 2016, https://doi.org/10.1016/j.destud.2016.09.002.
12. Tatami, J. and Nakano, H., Chapter 10 - Green Processing of Particle Dispersed Composite Materials, in: *Green and Sustainable Manufacturing of Advanced Material*, pp. 245–265, Elsevier, 2016, https://doi.org/10.1016/B978-0-12-411497-5.00010-2.
13. Paul, P., Green Nanoparticle Synthesis: Towards Sustainable Nanotechnology. *Int. J. Creat. Res. Thoughts*, 11, 7, 315–320, 2023.

14. Parveen, K., Banse, V., Ledwani, L., Green synthesis of nanoparticles: Their advantages and disadvantages. *Am. Inst. Phys.*, 020048, 2016, http://dx.doi.org/10.1063/1.4945168.
15. Mohammed, L., Ansari, M.N.M., Pua, G., Jawaid, M., Saiful Islam, M., A Review on Natural Fiber Reinforced Polymer Composite and Its Applications. *Int. J. Polym. Sci.*, 15, 243947, 2015, http://dx.doi.org/10.1155/2015/243947.
16. Mahesh, V., Joladarashi, S., Kulkarni, S.M., A comprehensive review on material selection for polymer matrix composites subjected to impact load. *Def. Technol.*, 17, 1, 257–277, 2021, https://doi.org/10.1016/j.dt.2020.04.002.
17. Islam, T., Chaion, M.H., Abdul Jalil, M., Rafi, A.S., Mushtari, F., Dhar, A.K., Hossain, S., Advancements and challenges in natural fiber-reinforced hybrid composites: A comprehensive review. *SPE Polym.*, 5, 481–506, 2024, https://doi.org/10.1002/pls2.10145.
18. Elfaleh, I., Abbassi, F., Habibi, M., Ahmad, F., Guedri, M., Nasri, M., Garnier, C., A comprehensive review of natural fibers and their composites: An eco-friendly alternative to conventional materials. *Results Eng.*, 19, 101271, 2023, ISSN 2590-1230, https://doi.org/10.1016/j.rineng.2023.101271.
19. Karimah, A., Ridho, M.R., Munawar, S.S., Adi, D.S., Ismadi, Damayanti, R., SubiyantoFatriasari, B.W., Fudholi, A., A review on natural fibers for development of eco-friendly bio-composite: characteristics, and utilizations. *J. Mater. Res. Technol.*, 13, 2442–2458, 2021, ISSN 2238-7854, https://doi.org/10.1016/j.jmrt.2021.06.014.
20. Kamaruddin, Z.H., Jumaidin, R., Ilyas, R.A., Selamat, M.Z., Alamjuri, R.H., Yusof, F.A.M., Influence of Alkali Treatment on the Mechanical, Thermal, Water Absorption, and Biodegradation Properties of Cymbopogan citratus Fiber-Reinforced, Thermoplastic Cassava Starch–Palm Wax Composites. *Polymers*, 14, 14, 2769, 2022, https://doi.org/10.3390/polym14142769.
21. Geremew, A., De Winne, P., Demissie, T.A., De Backer, H., Surface modification of bamboo fibers through alkaline treatment: Morphological and physical characterization for composite reinforcement. *J. Eng. Fibers Fabr.*, 19, 2024, doi:10.1177/15589250241248764.
22. Yang, M., Su, J., Zheng, Y., Fang, C., Lei, W., Li, L., Effect of Different Silane Coupling Agents on Properties of Waste Corrugated Paper Fiber/Polylactic Acid Composites. *Polymers*, 15, 17, 3525, 2023, https://doi.org/10.3390/polym15173525.
23. Varma, M. and Chandran, S., Surface treatment of natural fibers for enhancing interfacial adhesion and mechanical properties in biocomposites - A comprehensive review. *Compos. Interfaces*, 1–37, 2025, https://doi.org/10.1080/09276440.2025.2498795.
24. Liao, Z., Hu, Y., Shen, Y., Chen, K., Qiu, C., Yang, J., Yang, L., Investigation into the Reinforcement Modification of Natural Plant Fibers and the Sustainable Development of Thermoplastic Natural Plant Fiber Composites. *Polymers*, 16, 24, 3568, 2024, https://doi.org/10.3390/polym16243568.

25. Phiri, R., Rangappa, S.M., Siengchin, S., Oladijo, O.P., Ozbakkaloglu, T., Advances in lightweight composite structures and manufacturing technologies: A comprehensive review. *Heliyon*, 10, 21, e39661, 2024, ISSN 2405-8440, https://doi.org/10.1016/j.heliyon.2024.e39661.
26. Lazarus, B.S., Velasco-Hogan, A., Gómez-del Río, T., Meyers, M.A., Jasiuk, I., A review of impact resistant biological and bioinspired materials and structures. *J. Mater. Res. Technol.*, 9, 6, 15705–15738, 2020, ISSN 2238-7854, https://doi.org/10.1016/j.jmrt.2020.10.062.
27. Jefferson Andrew, J. and Dhakal, H.N., Sustainable biobased composites for advanced applications: recent trends and future opportunities – A critical review. *Composites, Part C: Open Access*, 7, 100220, 2022, ISSN 2666-6820, https://doi.org/10.1016/j.jcomc.2021.100220.
28. Nyabadza, A., McCarthy, E., Makhesana, M., Heidarinassab, S., Plouze, A., Vazquez, M., Brabazon, D., A review of physical, chemical and biological synthesis methods of bimetallic nanoparticles and applications in sensing, water treatment, biomedicine, catalysis and hydrogen storage. *Adv. Colloid Interface Sci.*, 321, 103010, 2023, ISSN 0001-8686, https://doi.org/10.1016/j.cis.2023.103010.
29. Khan, F., Hossain, N., Mim, J.J., Maksudur Rahman, S.M., Iqbal, M.J., Billah, M., Chowdhury, M.A., Advances of composite materials in automobile applications – A review. *J. Eng. Res.*, 13, 2, 1001–1023, 2025, ISSN 2307-1877, https://doi.org/10.1016/j.jer.2024.02.017.
30. Mulana, F., Prayogie Aulia, M., Aprilia, S., Fly Ash/Coconut fiber reinforced polymer Composites: Effect on physical properties (Density, water Absorption, and thickness Swelling). *Mater. Today Proc.*, 87, Part 2, 180–186, 2023, https://doi.org/10.1016/j.matpr.2023.02.394.
31. Das, O., Babu, K., Shanmugam, V., Sykam, K., Tebyetekerwa, M., Neisiany, R.E., Försth, M., Sas, G., Gonzalez-Libreros, J., Capezza, A.J., Hedenqvist, M.S., Berto, F., Ramakrishna, S., Natural and industrial wastes for sustainable and renewable polymer composites. *Renewable Sustainable Energy Rev.*, 158, 112054, 2022, https://doi.org/10.1016/j.rser.2021.112054.
32. Nagaraja, S., Anand, P.B., H. D., S., Ammarullah, M.I., Influence of fly ash filler on the mechanical properties and water absorption behaviour of epoxy polymer composites reinforced with pineapple leaf fibre for biomedical applications. *RSC Adv.*, 14, 14680–14696, 2024, DOI: https://doi.org/10.1039/d4ra00529e.
33. Zuhra, Z., Duan, C.M., Zuwanna, I., Rahmah, K., Mulana, F., Rosnelly, C.M., Meilina, H., Amin, A., Aprilia, S., Effect of filler loading on hybrid composites reinforced with oil pulp empty fruit bunches/fly ash: physicochemical properties and thermal analysis. *Rasayan J. Chem.*, 17, 4, 1517–1527, 2024, http://doi.org/10.31788/RJC.2024.1748822.

34. Kannan, G. and Thangaraju, R., Effect of industrial waste fly ash on the drilling characteristics of banana fiber residue reinforced polymer composites. *J. Ind. Text.*, 51, 2S, 2665S–2687S, 2022, DOI: https://doi.org/10.1177/15280837221102641.
35. Mulenga, T.K., Ude, A.U., Chinnasamy, V., Mechanical analysis of sisal fiber/bio-epoxy/fly-ash reinforced hybrid composites. *J. Compos. Mater.*, 57, 28, 2023, https://doi.org/10.1177/00219983231209881.
36. Kumar, P., Gupta, H.S., Singh, M., Chaudhari, A.S., Maurya, A.K., Manik, G., Mechanical, Thermal, and Morphological Analysis of Himalayan Agave Fiber/GO Coated Fly Ash Hybrid Polypropylene Composites. *Chem. - Eur. J.*, 31, 2, 2024, https://doi.org/10.1002/chem.202402393.
37. Venkateswara Rao, T., Venkatarao, K., Lakshmi Kanth, C., Mechanical Properties of Bamboo Fibre filled with Fly Ash filler Reinforced Hybrid Composites. *Int. J. Eng. Res. Technol. (IJERT)*, 3, 9, 2014, ISSN: 2278-0181, www.ijert.org.
38. Saxena, M., Morchhale, R.K., Prasad, B.K., Plant Fiber — Industrial Waste Reinforced Polymer Composites as a Potential Wood Substitute Material. *J. Compos. Mater.*, 42, 4, 2008, https://doi.org/10.1177/0021998307087014.
39. Raghavendra, G., Ojha, S., Acharya, S.K., Pal, S.K., Ramu, I., Evaluation of mechanical behaviour of nanometer and micrometer fly ash particle-filled woven bidirectional jute/glass hybrid nanocomposites. *J. Ind. Text.*, 45, 6, 1268–1287, 2016, DOI: https://doi.org/10.1177/1528083714557058.
40. Thirupathi, S., Solai, A., Pitchumani, S.V., Gopalan, V., Mallichetty, E., Evaluation of surface response of SiC and fly ash-filled Prosopis juliflora fibre-reinforced epoxy composites under dry sliding wear conditions using Taguchi method. *J. Eng. Appl. Sci.*, 71, 157, 2024, DOI: https://doi.org/10.1186/s44147-024-00485-9.
41. Venkateshwaran, N., Santhanam, V., Alavudeen, A., Feasibility Study of Fly Ash as Filler in Banana Fiber-Reinforced Hybrid Composites, in: *Processing of Green Composites*, pp. 31–47, Springer, 2019, DOI: https://doi.org/10.1007/978-981-13-6019-0_3.

11

Advanced Multifunctional Ceramics: Foundations, Properties, Characterization, Processing and Emerging Trends

Jagadeesha T.[1*] **and Sandip Kunar**[2]

[1]*Dept. of Mechanical Engineering, National Institute of Technology, Calicut, Kerala, India*
[2]*Department of Mechanical Engineering, Aditya University, Surampalem, India*

Abstract

Functional ceramics belongs to advanced materials class and responds well to metaphysical fields such as electrical, magnetic, optical, stress, temperature and pressure. They exhibit exceptionally good properties such as ferroelectricity, pyroelectricity, piezoelectricity, ionic conductivity and coupled fields. In this chapter, classification, properties, fabrication, characterization techniques and advanced applications are discussed in great details. Classification is explained in terms of mechanisms like piezo effect, pyro effect, electro-magnetic coupling mechanisms in functional ceramics. Detailed discussion on properties of functional ceramics is presented with mathematical models and equations—behaviors such as polarization, insulation, thermal diffusivity, electromagnetic coupling, electromechanical coupling. Understanding of properties and effect on microstructure is important from application point of view. Advanced manufacturing method like additive manufacturing such as direct ink writing, stereolithography, binder jetting, plasma and microwave-based fabrication systems are discussed in this chapter. Furthermore, characterization of functional ceramics using microscopy, X ray diffraction, thermal, electrical, mechanical, methods are presented in this chapter.

Keywords: Functional ceramics, electromechanical properties, additive manufacturing, characterization techniques, Piezo effect

**Corresponding author*: jagdishsg@nitc.ac.in

Gurudas Mandal, Swarup Kumar Ghosh, Snehanshu Pal and Sandip Kunar (eds.) Sustainable Manufacturing of Advanced Materials, (223–244) © 2026 Scrivener Publishing LLC

11.1 Introduction

Functional ceramics belongs to class of inorganic metals, and it is non-metallic material. Functional ceramics combines the strength of many materials like mechanical, electrical, chemical, optical, thermal magnetic properties. The main objective is to combine the strength of the different materials to fabricate a functional material. For example, aluminium is light weight material and non-corrosive, but its mechanical properties and wear resistance are poor. Silicon carbide is a ceramic material with high hot hardness and good wear resistance. When it is combined with aluminium, composite materials combines both good properties of aluminum and silicon carbide. Some materials are designed to respond to external stimuli.

Functional materials are those in which composition is controlled very precisely along with proper treatment to get correct crystal structure with correct control of defects, alloying elements and treatment to improve sub surface and surface properties, sometimes bulk properties as well.

Functional ceramics finds its use in diverse field such as electronic device fabrication, medical application, biotechnology-based devices, sensors, bio medical devices. It is truly multi-disciplinary domain which included material science, physics, chemistry, biology, engineering of various streams. Functional ceramics basically can adopt atomic and crystallographic structures as in perovskite, spinel, garnet and fluorites, which are quite flexible in terms of ionic substitutions and precise control of defects in materials [1]. Precise composition and defects control will improve permittivity, conductivity and polarizability when it is applied with electric field coupled with magnetic field. Yttria stabilized zirconia is one such good example.

Precise control of defects in the material is mainly responsible factor in functional ceramics for their improved properties. Presence of defects such as vacancies, interstitials and dopants strongly affect the movement of free electrons, ions and other charged particles, hence the property like magnetics field strength, permittivity, conductivity and other electro-magnetic properties greatly affect the functional ceramics. If ionic structures are considered, then the presence of Frenkel and Schottky defect changes the movement of charged particles and influence greatly on hysteresis and fatigue behavior. Enthalpy requirement for defects and interstitials are needed to get correct material properties [2]. In crystal structures, defects are mapped and analyzed using advanced techniques called atom probe tomography, position annihilation spectroscopy, electron paramagnetic resonance etc. Some of these advanced techniques are able to present 3 dimensional images of the defects, which will be helpful in detecting the size of the defects and by controlling the size of the defects, many crystallographic related properties can be controlled, manipulated and changed for better material behavior.

The fabrication of functional ceramics is complex process since good phase purity is needed with proper control of crystallographic structure as well. First step is to prepare fine powder which are very accurately stoichiometrically correct. Most commonly used process for powder preparations is hydrothermal route, sol-gel route and co precipitation route. Second step is to compact these powders to get green molds. Common process used are tape casting, cold isostatic pressing, uniaxial pressing. Third step is sintering. Controlled sintering is needed to give correct grain size, phase and density [3]. The common processes are microwave sintering, spark plasma sintering and additive manufacturing. Proper control of density or porosity is important, and it depends on pressure and temperature control.

Post processing is needed for sintering process of ceramics. Chemical and electrical properties change depending on the grain boundaries as it may act as discontinuity for charge flow. Grain boundaries act as site for corrosion by acting as runway for motion of ions, depending upon segregation of impurities at grain boundary. Considerable dielectric loss which decreases overall performance of functional ceramics is possible with the presence of grain boundaries. Focused ion beam and transmission electron microscopy images can help to change the grain boundaries and get the optimized structures with good functional abilities of ceramics. Also, electrical characteristics such as impedance may vary drastically along the grain boundary and also in the subsurface and identification is important to control the properties of functional ceramics. Impedance spectrometer is used for optimizing the grain boundaries at various phases. Although ceramics can withstand high temperature, but oxidation occurs which can alter the state from para to ferro, which in turn affects various properties such as coefficient of thermal expansion, polarization, permittivity, dielectric constant [4]. These properties are extremely important in electronic devices as these sharp discontinuities affect the electrical, electronics, superconductivity, oxygen stoichiometry and distortion of lattice.

Polarization ability of the functional ceramics depends on various factors such as polarization due to interaction between dipoles, interfaces, charges particles, ions etc. These interactions are dependent on frequency, and hence some properties such as dielectric constant, loss tangent modulus, permeability, coercivity etc. will change with the polarization mechanism. These are not only frequency dependent but also nonlinear and temperature depended in nature which makes it difficult to control [5].

Complex ionic and electronics interactions are different in ferro ceramics and Para ceramics, due to difference in their spin order and metal cations exchange.

In devices like batteries, sensors and electrochemical devices, ionic conductivity is important properties which depends on doping element, dosage, location, defect density etc. ion conductivity is high if barrier is low. Low barriers can be designed by controlling grain boundaries, structure and compositions. Higher activation energy, which can be evaluated by Nernst Einstein equation and Arrhenius equation gives low ionic conductivity and vice versa. Transport mechanism of ionic particles depends on grain boundaries, strains at the structure interfaces and multilayers, stoichiometry etc.

Consistent properties of functional ceramics are equally important. Although the material is designed and fabricated to give required functional properties, its values and behavior may change. For example: ionic properties, dielectric constant changes with moisture, exposure to harsh gas environment, oxidation environments, chemical, marine conditions etc. Different conditions cause different failures [6]. For example, if the electrical on off cycling is more and rapid, then electrical fatigue will occur, and which directly influence and reduce polarization. in some ceramics such as superconducting ceramics, conditions at grain boundaries and segregation in surface and sub surfaces affect the charge flow and transport of charges.

Mitigation strategies include dopant modification, surface coating, and encapsulation, along with multilayer device design to isolate sensitive regions. Thermal analysis techniques, such as TGA and DSC, are used to monitor phase stability and thermal degradation, and accelerated aging tests simulate long-term exposure.

11.2 Classification of Functional Ceramics

The main factors considered for classification of functional ceramics depends on how well the functional ceramics responds to stimuli such as temperature, current fields, mechanical stress field, chemical field and magnetic. Also, as discussed earlier, environmental factors such as relative humidity, temperature effect are also to be considered so that they are able to perform specific functional properties for which they are designed and fabricated. Classification also depends on defect distribution within crystallographic structure, electron and ion configuration, dynamic movement of charged particles such as electrons, ions and symmetry.

They are classified as dielectric ceramics, ferroelectric ceramics, piezoelectric ceramics, pyroelectric ceramics, magnetic ceramics, superconducting ceramics, and solid electrolytes or ionic conductors. The details are given in the subsequent sections.

11.2.1 Dielectric Ceramics

These are basically non-conducting materials whose insulating properties can be changed depending upon the polarization under the influence of externally applied electric field. Polarization could be ionic, electronics, space or even dipolar. Generally, these band width of these materials is high with low charge concentration and transport and has symmetric structure of lattice which makes it insulating under applied electrical field stimuli [7]. Their characterization is measured by three important parameters such as dielectric loss, dielectric strength and dielectric constant. If the material has high dielectric constant, then lattice arrangement affects strongly affect polarization under external fields in some materials while it exhibits low permittivity in other material. They are temperature and frequency dependent as well.

11.2.2 Ferroelectric Ceramics

These are basically sub-classification of dielectric ceramics whose insulating properties are due to reversal of spontaneous polarization under the influence of electric field. Unlike dielectric ceramics, they have non-symmetric crystal structure at curie temperature. If the temperature is below curie temperature, there will be polar displacement of ions which gives electrical dipole, which is responsible for hysterias behavior of the ferroelectric ceramics. This hysterias loop is due to orientation and reorientation of dipoles under the influence of external stimuli. Ferro ceramics are characterized by high piezoelectric coefficient, good permittivity and improved optical properties [8]. Their fatigue and other dynamic behavior depend on defect pinning, movement and arrest, as well as domain wall motion. These types of ceramics are temperature sensitive and can change the properties under the influence of frequency, temperature, humidity and other external factors, in spite of good thermo-mechanical coupling.

11.2.3 Piezoelectric Ceramics

These materials can produce electric charges when mechanical load is applied; and when electric field is applied, they can produce mechanical displacement. They do not have crystal symmetry, and therefore, there is no dipole effect under loading by mechanical means. There are highly anisotropic, and material properties are different in different directions and their directional dependent properties are expressed by piezoelectric constants, quality factor and electro-mechanical coupling factors.

11.2.4 Pyroelectric Ceramics

Pyro means temperature. When temperature is changed, it produces electric charges. These are temperature dependent sensitive materials which has spontaneous polarization which depends on temperature field. Temperature changes the vibration pattern of the crystal structure and lattice position. Change of lattice shift or position changes the ionic location within crystal which affect thermodynamic properties such as entropy, temperature and polarization [9]. They are characterized by pyroelectric coefficient, which depends on lattice vibration, thermal expansion of bonds, movement of charges, ions and their centers.

11.2.5 Magnetic Ceramics

The main composition of these ceramics are oxides of iron and metal oxides which gives spinel or garnet structure. Magnetic property is obtained due to interplay between cations and anions in tetra or octahedral space. The types of the ceramic obtained depends on alignment of spin. If alignment is antiparallel then ferrimagnetic ceramics are obtained; if the alignment is parallel, then ferromagnetic is obtained. In ferrimagnetic ceramics, net magnetic moment is because of unequal spin, because of nonalignment of spin. Properties like saturation magnetic ability, coercivity and permeability represents the magnetic ceramics. Soft magnetic ceramics has low coercivity and high permeability and hard magnetic ceramics has high remanence [10]. Hard magnetic ceramics are more stable compared to soft magnetic ceramics. Table 11.1 shows the summary of all functional ceramics.

11.2.6 Superconducting Ceramics

Superconducting ceramics are those ceramics whose resistivity becomes equal to zero when critical temperature is reached. They are produced by complex formation of oxides and has very strong spin fluctuations and strong interaction between photon and electron. The superconductivity

Table 11.1 Summary of all functional ceramics.

Type	Main mechanism	Structural feature	Properties	Examples
Dielectric Ceramics	Poor conductivity and polarization take place due to electrical field	Symmetric lattice or very low polarizable lattice	Very high dielectric constant and poor stability	Barium titanate, Alumina
Ferroelectric Ceramics	Spontaneous polarization with good reversibility	Non-symmetric lattice	Hysteresis is high	PZT
Piezo electric ceramics	When mechanically loaded charge is produced	Non-symmetric lattice	Highly anisotropic response	PZT, ZnO
Pyro electric ceramics	When there is change in temperature there is generation of charge	Uniaxially polarized lattice	Highly temperature sensitive	$LiTaO_3$, $PbTiO_3$.
Magnetic Ceramics	Cation exchange occurs due to magnetic field	Spinel or Garnet structure	High romance and permeability	$NiFe_2O_4$, $MnFe_2O_4$, $Y_3Fe_5O_{12}$
Superconducting ceramics	No conductivity	Layered structure	Highly anisotropy with good current density	YBCO
Ionic conductor ceramics	Good ion transport	Ions transfer through vacancy rich route	High ionic conductivity	YSZ and LLZO

in ceramics is not fully understood but believed to occur due to non-traditional pairing mechanism or charge stripes. They are characterized by curie temperature or critical temperature, critical magnetic field strength and current density. They are found to be very sensitive to grain boundary misalignment, defects and its density and also stoichiometry of oxygen chemistry [11]. Although they are very brittle in nature, still they are extensively used in electrical and magnetic circuit and device fabrication.

11.2.7 Ceramic Ionic Conductors

Ceramic ionic conductors are stable insulating material allowing only movement of specific type of ions only. Many ions like oxygen, lithium, sodium and hydrogen move both crystalline and amorphous phases due to gradient generated by electro chemical interactions. They have very high ionic conductivity due to good interaction between defects like vacancy and dopant. Dopants with very low activation energy is selected so that high ionic conductivity is possible [12]. Amount of ionic conduction depends on temperature and also lattice structure within the material. They are chemically and thermally stable functional ceramic material.

11.3 Properties of Functional Ceramics

Functional ceramics possesses unique properties compared to conventional ceramics in terms of crystal lattice and structure, defect density and map, coupling, ionic and electronic configuration etc. Their behavior is different under mechanical stress field, thermal field, magnetic field, electric field. Each property is represented by unique parameters which is characterized using advanced techniques [13].

Electrical Properties: They represent the dielectric properties of functional ceramics and dielectric properties is given by

$$\delta = \xi_0 E + P$$

δ is the displacement of electric field, E is the electric field applied, and P is polarization, ξ_0 is the permittivity of free space. If linear isotropy dielectric is assumed, then the amount of polarization is directly proportional to electric field.

$$P = \xi_0 \psi_0 E$$

ψ_0 is the electrical susceptibility. Then permittivity ξ of the ceramics is represented by

$$\xi = \xi_0 (1 + \psi_0)$$

The relative permittivity or dielectric constant is given by

$$\kappa = \frac{\xi}{\xi_0}$$

κ is the dielectric constant. Dielectric constant is very important property in design of capacitors and memory devices using functional ceramics. When the material is subjected to fatigue electrical field then there is energy dissipation due to alternatively changing electric field. Due to this energy dissipation, there is a dielectric loss. It is represented by

$$\tan\eta = \frac{\xi''}{\xi'}$$

ξ'' is imaginary part of the complex permittivity and ξ' is the real part of complex permittivity which is given by

$$\xi^c = \xi' + j\xi''$$

Complex permittivity is also frequency dependent and use Debye relaxation model and frequency dependent complex permittivity is given by.

$$\xi^C = \xi_\infty + \frac{\xi_s - \xi_\infty}{1 + j\omega\zeta}$$

ξ_s is the static permittivity, ξ_∞ is permittivity at infinite frequency, ω is the angular velocity and ζ is the relaxation time.

Spontaneous polarization is possible in ferroelectric functional ceramics, which occurs even when there is no presence of external electric field.

Coercive field and remnant polarization are important parameters to explain the behavior of ferroelectric ceramics [14].

$$\varpi = \oint EdP$$

ϖ is the energy density stored and lost during fatigue electrical switching. Landau-Ginzburg-Devonshire equation is widely used to express the phase transition in ferroelectric functional ceramics.

$$F(P) = F_0 + \frac{1}{2}\alpha P^2 + \frac{1}{4}\beta P^4 + \frac{1}{6}\gamma P^6 - EP$$

$\alpha, \beta \text{ and } \gamma$ are known as phenomenological coefficients which are temperature dependent, and E is the applied electric field and P is the degree of polarization. F is the free energy. P must be selected such that F is made minimum. It is a nonlinear equation involving second, fourth and sixth power of polarization.

In **piezoelectric ceramics** mechanical stress is applied to get electrical charges and both fields are linearly coupled with piezoelectric coupling factor. Load is related to stress and stress and strain are related by Hooks law. Stress is caused mechanically, and strain is produced by electric field.

$$\delta_i = d_{ijk}\sigma_{ij}$$
$$S_{ij} = d_{kij}E_k$$

δ_i is displacement due to electric field. S_{ij} is the strain tensor matrix, E_k is the electric field and d_{ijk} is the piezoelectric coefficient tensor. Simple notation is given as

$$\delta = d\sigma$$
$$S = d^T E$$

In pyroelectric ceramics, temperature changes the polarization, and it is expressed in terms of pyroelectric coefficient p_r

$$p_r = \left(\frac{dP_s}{dT}\right)_E$$

P_s is the spontaneous polarization which is a strong function of temperature T. The current generated in pyroelectric ceramics due to change in temperature is given by

$$i = p_r A\left(\frac{dT}{dt}\right)$$

A is the area. i is the current generated due to pyroelectric effect.

Spin configuration and magnetic behavior determines the properties of magnetic ceramics. M amount of magnetization and H is the magnetic field and χ_m is magnetic susceptibility

$$M = \chi_m H$$

Now the magnetic permeability η is given by

$$\eta = \eta_0(1+\chi_m)$$

η_0 is the permeability of free space. Based on the coercivity and permeability, magnetic ceramics can be soft or hard ceramics. Hard magnetic functional ceramics has high coercivity, and therefore, finds their use in permanent magnets. On the other hand, soft magnetic functional ceramics has low coercivity [15].

Superconducting functional ceramics has zero electrical resistance below curie temperature, and therefore, they have high conductivity. London equation is used to characterize superconducting behavior in ceramics.

$$\frac{\partial J}{\partial t} = \frac{\rho_s e^2}{m} E$$

J is the current density, ρ_s is the density of carriers, m is the effective mass of carriers and E is the applied electrical field which can also be expressed in terms of magnetic field. London equation clearly shows that rate of change of current density is upon the density of available carriers and also their charges and applied field.

Transport of ions in crystal lattice is important in ionic conducting functional ceramics. The ionic conductivity can be evaluated using Nernst Einstein Equation

$$\sigma_i = \frac{nq^2 D}{k_B T}$$

σ_i is ionic conductivity, q is charge of ions, D is the diffusion coefficient which depends on concentration and temperature. T is the absolute temperature. Transport of ions is strongly dependent on temperature and can be expressed using Arrhenius equation

$$\sigma_i = \sigma_o exp\left(-\frac{E_a}{k_B T}\right)$$

E_a is activation energy required for transport of ions in the crystal lattice.

Thermal properties like conductivity, diffusivity, coefficient of thermal expansion is important in functional ceramics. Heat conduction in functional ceramics is governed by Fourier heat conduction law

$$\frac{\partial}{\partial x}\left(k\frac{\partial T}{\partial x}\right)+\frac{\partial}{\partial x}\left(k\frac{\partial T}{\partial x}\right)+\frac{\partial}{\partial x}\left(k\frac{\partial T}{\partial x}\right)+\dot{q}=\rho c_p\left(\frac{\partial T}{\partial t}\right)$$

Where k is the thermal conductivity, T is the temperature, t is the time, $\dot{q}$ is the heat generation rate ρ is the density and c_p is the specific heat capacity. There will be change of length due to temperature and coefficient of thermal expansion and is given by

$$\Delta L = L_0 \alpha \Delta T$$

Mechanical properties of functional ceramics are equally important and should be evaluated using advanced techniques and measuring instruments. Mechanical properties include strength, stuffiness, hardness, toughness, ductility, malleability, brittleness.

$$E = \frac{\sigma}{\varepsilon}$$
$$G = \frac{E}{2(1+\nu)}$$
$$K = \frac{E}{3(1-2\nu)}$$

σ is the applied stress, ε is the corresponding strain, E is the young's modulus or modulus of elasticity, G is the modulus of shear or rigidity modulus, K is bulk modulus, ν is the poisons ratio.

11.4 Characterization of Functional Ceramics

Properties of functional ceramics are designed and tailor made for specific application, and thus, optimization of the function ceramics in terms of their behavior, microstructure and properties are very important. The properties are to be characterized using advanced characterization techniques to evaluate dielectric, piezoelectric, magnetic, super conducting, ionic, ferromagnetic properties. In addition, microstructural and secondary operations are needed to arrive at optimized functional ceramics [16]. Several techniques are used to find chemical, physical, mechanical, thermomechanical, electrical, thermo electrical, magnetic, thermomagnetic, chemical, structural properties.

Structural characterization is the first step in characterization of functional ceramics. X-ray diffraction is used to find crystal structure, purity of phase, lattice constants and parameters, orientation of crystal lattice. In X-ray diffraction, monochromatic beam is incident on the sample and diffracted beams are detected as a function of scattering angles. W.L. Bragg's law is used to find out the diffraction conditions. Figure 11.1 shows diffraction phenomenon.

From the Figure 11.1,

$$\begin{aligned}
&\mathrm{AB} = \mathrm{d}\sin(\theta) = \mathrm{BC} \\
&\mathrm{ABC} = 2\mathrm{d}\sin(\theta) \text{ by symmetry} \\
&\mathrm{A'B'} = 2\mathrm{d}\sin(\theta) = \mathrm{B'C'} \\
&\mathrm{A'B'C'} = 2\mathrm{ABC} = 4\mathrm{d}\sin(\theta) \text{ by symmetry}
\end{aligned}$$

In general, $n(\lambda) = 2d\sin\theta$

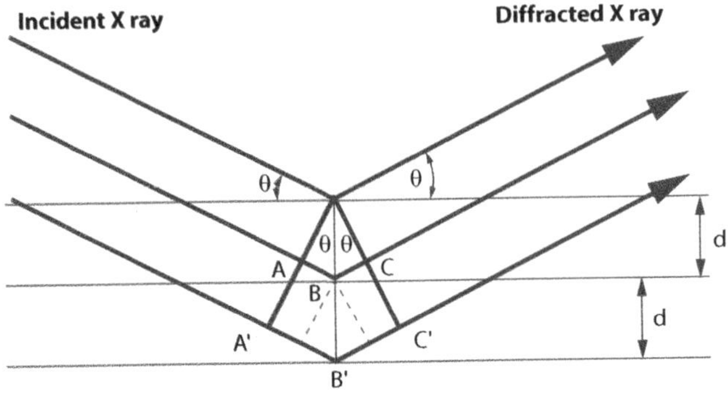

Figure 11.1 Diffraction phenomenon.

n is an integer in Bragg's equation and λ is the wavelength and d are the interatomic planar distance—is the angle of incidence of x ray which is incident on sample. Detectors are used for phase analysis and measurement of residual stress. Crystalline strain and size can also be found using the Braggs diffraction measuring instrument.

Transmission electron microscopy is used when high resolution images of internal crystal lattice structure, defects are needed to be found out. Higher atomic resolution is helpful in detection of dislocation, vacancies, interstitials grain boundaries, intergranular cracks, trans granular cracks etc. Many versions of TEM like scanning transmission electron microscopy and high-angle annular dark-field imaging are used to find atomic level composition [17]. Table 11.2 shows the summary of all types of TEMs used for functional ceramics.

Scanning electron microscopy gives high resolutions compared to conventional microscopy and they can provide distribution of grain size and chemical composition if they are equipped with x-ray spectroscopy detectors. If detailed chemical analysis is needed, then wavelength dispersive spectroscopy is used [18]. X ray photoelectron spectroscopy (XPS) and Auger electron spectroscopy (AES) are useful for studying the chemical bonding locally. Valency stages, binding energies, local chemistry, inner shell ionization, surface elemental analysis is possible with XPS and AES instruments.

Fourier transform infrared (FTIR), and Raman spectroscopy are used for characterizing the functional ceramics. Principle of inelastic scattering is used to get information such as vibration of lattice phonons, light elements detection, symmetry of crystal, phase transitions and state of stress [19]. Confocal setup gives spatial images which are used for vibrational modal analysis.

Table 11.2 Summary of TEMs used in characterisation of ceramics.

Feature	TEM	HAADF high-angle annular dark-field imaging	STEM scanning transmission electron microscopy
Basic Principle	Electron beam is passed through thin specimen. Image is obtained using transmitted electrons	Electrons are scattered at high angles and collected.	Electron probe scans workpiece to get image
Type of beam	Parallel electron beam	Scattered beam	Focused beam
Type of detector	Direct electron detectors	HAADF detectors	EELS detectors
Contrast mechanism	Mass thickness contrast	Z contrast	Depends on type of detectors employed
Resolution	High – sub angstrom	Atomic Resolution	Atomic resolution
Composition analysis	EDS mapping	Atomic mapping	Chemical Mapping
Image Modes	Bright and dark field	Z sensitive images	Bright and dark field, EELS mapping
Information derived	Defects, grain boundaries, lattice details, phase detection	Atomic no, strain and defect mapping	Atomic structure
Complexity	High, need high thin samples	Require abrasion corrected STEM	Very high due to focused probes

Thermal analysis techniques are used to characterize the stability of functional ceramics under thermal loading. Differential scanning calorimeter measures the difference in rate of heat flow into and out of ceramics samples. This will help in detecting the phase changes in functional ceramics like ferro and para electric ceramics [20]. TGA is used to find the rate of decomposition of material known as charr as a function of temperature. Loss of mass is a function of temperature, number of oxidation stages, environmental humidity etc. Dilatometry measures change in dimensions as a function of temperature, amount of yielding due to temperature effect and behavior of material under thermal loading.

Electrical characterization is used to measure dielectric constant and also loss of dielectric with respect to change of frequencies and wide range of temperature. Inductance-capacitance-resistance (LCR) meters are used to measure the impedance. Grain boundary resistance can be evaluated and Nequist plots can be drawn to get resistance-capacitance behavior [21]. Leakage current, breakdown voltage is very critical in design of electronic devices or barrier layers using functional ceramics. Piezoelectric properties are important to evaluate the behavior of functional ceramics under the action of external mechanical stress. Generally, Berlincourt method is employed to determine the piezoelectric coefficient and anti-resonant frequency. Quality factor and Mechano-electric coupling coefficients are analyzed using network analyzers [22].

Magnetic characterization is done using vibrating sample magnetometry (VSM) and superconducting quantum interference device magnetometry (SQUID). VSM gives variation of magnetic field with respect of temperature. SQUID is more sensitive compared to VSM and used for lower temperature and magnetic field ranges which are especially useful in glass like ceramics in which magnetic field strength is low and weak magnetic field gives little change in the temperature or vice versa [23]. Four probe instruments are used to measure variation of electrical resistance or resistivity as a function of temperature. Magnetic susceptibility is measured using AC susceptometer which are used for superconductor-based ceramics which operate at high temperature.

Microstructural analysis will give phase; porosity present in ceramics and also the presence and distribution of macro and micro cracks. X ray computed tomography (XCT) is another type of non-destructive testing which can be used to know the presence and location of defects in sintered ceramics like ultrasonic testing method [24].

Main Mechanical properties are hardness, fracture toughness, modulus of elasticity which can be evaluated using nanoindentation techniques,

Vickers hardness tests and fracture toughness testing matching using popular method known as indentation induced cracking (IIC) method. If ceramics has multilayers, then nanoindentation will be very useful in evaluating surface and subsurface characteristics [25]. Three-dimensional surface profiles, waviness, roughness etc can be evaluated using 3D profilometer or optical interferometer.

11.5 Fabrication Methods of Functional Ceramics

The fabrication of functional ceramics involves following steps:

a. Select the precursor material depending on the required properties and its application in actual device or component.
b. Preparation of powder is done using hydrothermal synthesis, co-precipitation, solid state reaction or sol-gel synthesis route. Proper size control of particles, its distribution, stoichiometry and composition of phase is important to get desired properties. If volatile material is to be eliminated, then drying and calcination is used as secondary processes.
c. Preparation of slurry or granular or paste like using powder and binders, plasticizers, surfactants, binders, initiators, viscosity has to be controlled.
d. Shaping and forming of ceramics. Proper suitable method to be chosen to form and shape the ceramics. Many fabrication processes are available dry pressing, isostatic pressing, slip casting, tape casting, extrusion and additive manufacturing processes. Advanced 3d printing techniques like stereolithography, direct ink writing, binder jetting, fused filament fabrication and selective laser sintering.
e. Drying and debinding. Green molds obtained during shaping and forming is dried using well controlled air drying. If the functional ceramics is done through additive manufacturing route, then polymer used in additive manufacturing must be removed by controlled heating. In both cases care should be taken to prevent warping.
f. Densification or Sintering. After it is dried or debinding, functional ceramics are densified carefully by controlling temperature, ramp rate, hold time and environment. Proper grain size, distribution and microstructure is to be ensured

during this process. Generally, furnace sintering, isostatic pressing, plasma sintering and microwave sintering are employed.

g. Secondary operations are to be carried out like machining, polishing and trimming. Sometimes additional machining is needed for electrode fixings. Thermal and stress relieving heat treatment are also needed depending upon the applications.
h. Testing after complete fabrication, functional composite ceramics are to be tested for their function and properties.

Additive manufacturing processes now is widely used method to manufacture the functional ceramics. Complex geometries can be printed using 3d printing with good accuracy and repeatability. There are various methods used. Some of them are stereolithography (SLA), Direct ink writing (DIW), fused filament method (FFM), binder jetting and selective laser sintering (SLS). Each of these methods offer unique advantages and some disadvantages. In stereolithography, first slurry of powder, binder, surfactant and additives is prepared and then mixed with resin, which is photosensitive. The slurry is cured layer by layer using digital light processing unit [26]. Once the process is completed, it is sent to debinding to remove polymer material from 3d printed parts. Secondary operations followed to give required properties.

Stereolithography (SLA) is one of the additive manufacturing techniques used for fabrication of functional ceramics. Flow properties and light absorption ability of slurry determines the success of the SLA. Number of ceramic particles, resin content, viscosity, rheology and density will affect the flowability of the slurry. If the solid particle content is more in slurry, then high density ceramic is obtained after sintering. Rheology and flow properties affect the ease of printability and modifiers like dispersants and rheology modifiers are added to adjust and optimize the flow properties and printability. Curing depth follows Beer–Lambert law, a good indicator of dimensional accuracy and adhesion between two layers. If debinding is not done properly, then there is a risk of cracking and warping of functional ceramics. Last stage sintering convers the green mold part into the densified functional ceramic with required microstructure, density and other properties. However, it is important to maintain correct sintering temperature profile and it must be done in controlled atmosphere to get desired properties.

Direct ink writing is a novel technique used for fabrication of functional ceramics. First ceramic ink which is shear thinning in nature is used.

As shear force is applied, material gets thinner. As the ceramics ink is sheared in fine nozzle, it becomes thin and then it is deposited or written like pen according to the programmed tool path. Binder and solvent are added to get the desired rheological and fluid flow properties. Using this process, multi material printing is possible with good flexibility and less special Equipment or attachments [27]. Once the part is printed, it undergoes either drying or debinding depending upon the material processing; and the removal of polymers is done effectively. New innovations include robot assisted printing and freeze form extrusion.

Powder based methods are quite common in fabrication of ceramic based functional materials. Ceramic powders are prepared carefully along with correct stoichiometric composition of alloying and other additives. Then binding element is deposited wherever the particles are to be deposited, and then particles are deposited. This process is repeated until the required layer thickness is obtained. After completing the deposition of all layers, the part is densified using sintering process. This process is rapid and can be used for mass production of parts. However, it is important to note that additive manufacturing can produce only one part at a time. After sintering is done some infiltration of secondary materials are used in matrix material to act as reinforcing element to enhance directional properties of functional ceramics [28]. Selective laser sintering (SLS) and selective laser melting (SLM)are the two laser methods. SLS and SLM are difficult in preparation of functional ceramics as normally, ceramics have high melting temperature and very low thermal conductivity.

During early part of the research in additive manufacturing, fused filament fabrication (FFF) was developed for polymer-based thermoplastics. Then, with the advent of high temperature nozzle development and coating, this process is extended for ceramics manufacturing as well. Nozzles are heated to the required temperature to control the rheological and flow characteristics and deposited layer by layer [29]. Organic binder is removed after printing and then part is sintered according to the sintering temperature profile to give required densification and other properties.

New trend in additive manufacturing is the development of hybrid processes. Physics, mechanics and mechanism of sintering technology is well known, and it is combined with Spark Plasma Sintering (SPS) and microwave sintering. SPS uses a pressure system in which pressure is applied in unidirectional to compress green ceramic mold. Lead time in SPS is less and can be integrated with additive manufacturing XY table movement to produce optimized densified functional ceramics with good grain boundary control and less defect density [30].

Additive manufacturing can print highly complex parts or devices and can be customized for a particular application. In medical field, customized design is important for a specific disease or medical problem and can be tailor made using 3d printing. Densified parts, accurate, defect free and controlled property functional ceramics are essential with micro, macro cracks, warping and unwanted residual stress in the device or parts. In 3D printing, particles and many alloying elements can be dispersed in base polymer to get the desired layer by layer device with good product integration and material efficiency.

11.6 Conclusions

Functional ceramics new class of advanced materials whose properties can be tailor made using preparation of powder or slurry with the required alloy composition can be fabricated using advanced manufacturing conventional and non-conventional processes. They are unconventional material which can respond to metaphysical field such as electric, magnetic, optical, thermal and mechanical stress filed. The material can be tailor made to the specific applications. Holistic approach is followed in this chapter to fully explore the potential of the functional ceramics for diverse applications. Classification of functional ceramic is covered in this chapter with a special emphasis on mechanism on which functional ceramics work such as ferro electric, piezo electric pyro electric, electro optics etc. Complex properties of functional ceramics such as dielectric constants and mechanism-based constants are discussed in this chapter with mathematical modeling and formulations. Furthermore, detailed characterization of these properties is also given in detail. Optimization of properties and alloy contents and procedures formally finalizes the fabrication route. Several fabrication details including additive manufacturing process are discussed in this chapter. In summary, this chapter provides a comprehensive view on functional ceramics, classification, properties, characterization, and fabrication technologies.

11.7 Future Research Opportunities

Functional ceramics is truly multiphysics and multidisciplinary domain and represents many opportunities for innovation and application. Multi ferroic ceramics which gives combined electrical, magnetic hybrid behavior is gaining importance especially in memory devices and sensors including bioinspired devices and products. Tunable bandgap materials are next generation materials for semiconductor fabrication devices and highly recommended for flexible and wearable devices, phonics, biomedical, military and other applications. Smart ceramic ink which has high density, phase purity and defect free can be printed using additive manufacturing techniques. Multi material printing allows functionally graded ceramics for specific applications with real time monitoring of microstructure and defects. Another emerging area of research is in advanced microscopy which can detect grain boundary compositional fluctuations as in nanoelectronics, nano photonics and advanced green energy storage devices. Bioactive, biodegradable ceramics with lab on chip could bring breakthrough innovation in field of biomedical sciences and transformative technological fields.

AI disclosure statement: The author acknowledges the use of AI-based tools for minor language editing or grammatical correction. However, all intellectual contributions, analyses, and conclusions presented in this chapter are entirely the author's own.

References

1. Kamutzki, A., Weingarth, D.W., Clemens, F., A comparison of synthesis approaches towards functional polycrystalline silicate ceramics, *arXiv preprint, arXiv:2205.11905*, 2022.
2. Jamaludin, S.S., Rosli, N.R.N., Shamsudin, M.I., Ismail, M.H., A review on the fabrication techniques of functionally graded materials. *Sci. Res. Essays*, 8, 556, 2013.
3. Semler, C.E. and Lanagan, M.T., Classification and applications of advanced ceramics. *J. Mater. Sci.*, 52, 6032, 2017.
4. Hirai, T., Functional ceramics. *J. Am. Ceram. Soc.*, 74, 1259, 1991.
5. Hirayama, Y., Concept and recent advances of advanced ceramics: Classification by function and property. *J. Adv. Ceram.*, 11, 437, 2022.
6. Najafabadi, M.M., Thermal conductivity and mechanical properties of functional ceramics. *Ceram. Int.*, 47, 8092, 2021.
7. Ramesh, R., Multiferroics: Progress and prospects in thin films. *Nat. Mater.*, 6, 21, 2007.
8. Wang, J., Neaton, J.B., Zheng, H., *et al.*, Epitaxial $BiFeO_3$ multiferroic thin film heterostructures. *Science*, 299, 1719, 2003.
9. Chen, Z. and Payne, D.A., Low-field dielectric behavior of sol–gel derived $BaTiO_3$ thin films. *J. Am. Ceram. Soc.*, 78, 295, 1995.
10. Scott, J.F., Applications of modern ferroelectrics. *Science*, 315, 954, 2007.

11. Waser, R. and Aono, M., Nanoionics-based resistive switching memories. *Nat. Mater.*, 6, 833, 2007.
12. Subodh, G. and Reddy, R.R., Electrical conductivity of rare-earth doped ceria ceramics. *J. Mater. Sci. - Mater. Electron.*, 25, 2956, 2014.
13. Ohta, H., Kim, S.W., Mune, Y., *et al.*, Giant thermoelectric Seebeck coefficient of a two-dimensional electron gas in $SrTiO_3$. *Nat. Mater.*, 6, 129, 2007.
14. Suryanarayana, C. and Norton, M.G., Thermal behavior of polycrystalline ceramics, in: *X-Ray Diffr. Pract. Approach*, vol. 12, p. 102, Springer USA, Boston, 1998.
15. Safari, A. and Akdogan, E.K., Piezoelectric and acoustic behavior of ceramic composites. *J. Electroceram.*, 20, 253, 2008.
16. Tsuda, K. and Tanaka, M., STEM-EELS and HAADF characterization of functional ceramics at atomic resolution. *Microscopy*, 63, 55, 2014.
17. Frost, H.J. and Ashby, M.F., Deformation mechanism maps: Plasticity and creep of metals and ceramics. *Acta Metall.*, 30, 1041, 1982.
18. Rodríguez-Carvajal, J., Recent advances in magnetic structure determination by neutron powder diffraction. *Physica B*, 192, 55, 1993.
19. Edirisinghe, M.J. and Dorey, R.A., Electrophoretic deposition of nanoceramic layers. *J. Eur. Ceram. Soc.*, 32, 2455, 2012.
20. Pradeep, T., Spectroscopic techniques in nanoscale ceramics. *Curr. Sci.*, 92, 1470, 2007.
21. Shackelford, J.F. and Alexander, W., Overview of ceramic characterization. *J. Mater. Eng. Perform.*, 18, 1121, 2009.
22. Yan, T.T. and Ricci, J.L., Mechanical and structural characterization of ceramics using nanoindentation and SEM. *J. Biomed. Mater. Res. A*, 81, 745, 2007.
23. Zhang, H., Li, J., Chen, Y., *et al.*, Advanced characterization of ferroelectric ceramics using synchrotron X-ray diffraction. *J. Appl. Cryst.*, 44, 117, 2011.
24. Glazer, A.M., Application of electron microscopy in ceramics. *J. Mater. Sci.*, 40, 1973, 2005.
25. McIntyre, G.J., Neutron and X-ray diffraction techniques for ceramics. *Acta Cryst. A*, 69, 45, 2013.
26. Chen, F., Huang, L., Zhang, D., Fabrication of 3D-printed functional ceramics using digital light processing. *J. Eur. Ceram. Soc.*, 40, 4505, 2020.
27. Bai, Y., Wang, Z., Zhang, Y., *et al.*, Additive manufacturing of ceramics: Issues and opportunities. *Ceram. Int.*, 47, 3004, 2021.
28. Nisbet, D.R., Forsythe, J.S., Shen, W., *et al.*, Biocompatibility and fabrication of 3D-printed ceramic scaffolds. *Acta Biomater.*, 9, 7005, 2013.
29. Chen, C.C., Lin, Y.H., Tang, C.M., *et al.*, 3D printing of piezoelectric ceramic scaffolds for bone tissue engineering. *Adv. Healthcare Mater.*, 7, 1701024, 2018.
30. Vaidhyanathan, R., Sebastian, M.T., Jose, R., *et al.*, Microwave sintering of functional ceramics: Synthesis and properties. *Mater. Sci. Eng. B*, 144, 43, 2007.

12

Circular Economy-Driven Approaches in Sustainable Manufacturing of Advanced Materials

Idhayaraja S.[1], Ajithkumar Sitharaj[2*], Ganesh N.[3], Arul Pandiyan[3] and Sandip Kunar[4]

[1]Department of Manufacturing Engineering, PSG College of Technology, Coimbatore, Tamil Nadu, India
[2]Department of Mechanical Engineering, KPR Institute of Engineering and Technology, Coimbatore, Tamil Nadu, India
[3]Department of Mechanical Engineering, Sri Shakthi Institute of Engineering and Technology, Coimbatore, Tamil Nadu, India
[4]Department of Mechanical Engineering, Aditya University, Surampalem, India

Abstract

This chapter analyzes circular economy-driven strategies in the sustainable production of advanced materials, with the objective of minimizing resource depletion and environmental impact. It presents an integrated framework that incorporates efficient production systems, sustainable material selection, and sophisticated waste management to facilitate regenerative industry practices. Materials are evaluated according to their renewability, recyclability, embodied energy, toxicity, and performance. Design for disassembly and reuse is prioritized to prolong product lifespan and recuperate end-of-life value. Lifecycle assessment tools are examined for measuring environmental impact throughout the value chain. Mechanical and chemical recycling techniques are essential for preserving material functionality and economic value. Additive manufacturing is recognized as a crucial facilitator of resource efficiency and closed-loop recycling. The chapter emphasizes the significance of digital technologies, including IoT, digital twins, and data-driven supply chains, in facilitating adaptive and transparent systems. Collectively, these strategies facilitate the development of

**Corresponding author*: ersajithkumarbe@gmail.com

Gurudas Mandal, Swarup Kumar Ghosh, Snehanshu Pal and Sandip Kunar (eds.) Sustainable Manufacturing of Advanced Materials, (245–266) © 2026 Scrivener Publishing LLC

robust, sustainable manufacturing ecosystems that adhere to circular economy principles.

***Keywords*:** Circular economy, sustainable manufacturing, advanced materials, resource efficiency, additive manufacturing, closed-loop supply chain

12.1 Introduction

The sustainable manufacturing of advanced materials is increasingly informed by the principles of circular economy, which seek to sever the link between economic growth and the consumption of resources and environmental impacts. A circular economy framework for the integrated sustainable manufacturing establishes the basis for an increasingly circular economy by departing from the linear "take-make-dispose" format of production systems and incorporating regenerative approaches across production systems. As such, a circular economy framework for integrated sustainable manufacturing incorporates a conscious approach to using renewable and non-renewable resources, so that there is a balance between using bio-based polymers, natural fibers, and renewable energy inputs alongside responsible stewardship and recapture of non-renewable, yet critical, materials like high-performance alloys and engineering thermoplastics [1]. The circular economy framework also allows for a design to process flows through production, distribution, and use that supports efficiency, traceability, and the retention of value in circular solutions, while waste management in the biological and technical loops would allow for safely reintroduced or transformed new resources [2].

Sustainable material selection methods are at the heart of this vision, with well-defined criteria for sustainable material selection having a strong focus on evaluating renewability, recyclability, embodied energy, toxicity, and performance. Design for disassembly and reuse to facilitate the recovery of components at end-of-life and lifecycle assessment of material selections provide science-based avenues for evaluating and improving environmental performance throughout every step. The recycling valorization of advanced materials is important for sustaining the functional and economic values of high-performance materials through advanced recovery models. Mechanical and chemical recycling are complementary strategies to recover metals, composites, and polymers from complex products and production waste streams [3].

Additive manufacturing for a circular economy can further enhance resource efficiency and result in closed-loop production. These layer-by-layer fabrication technologies bring new dimensions in lightweight, optimized component design with minimal waste and significant material efficiency [4]. Resource efficiency in additive processes supports near-net-shape production, while recycling additive manufacturing feedstocks promotes the recycling of powders, filaments and resins thereby reducing reliance on virgin feedstocks. The digitalization enabling closed-loop production allows IoT-enabled real-time monitoring and predictive data analytics to enable continuous feedback about materials and processes, and digital twin systems for holistic lifecycle optimization enable manufacturers to simulate and track productivity, and subsequently improve performance over time [5]. The development and implementation of closed-loop supply chain modeling and adoption closes the loop, where products, parts and materials can be systematically recovered, remanufactured and reintroduced into a production lifecycle [6]. These methods realize an interconnected ecology promoting sustainability, increasing resilience from resource constraints, and enabling industries to meet rising expectations for environmental stewardship and circular innovation.

12.2 Circular Economy Framework for Integrated Sustainable Production

This framework presents a holistic approach to advancing industrial systems away from a linear model of extraction, production, and disposal toward regenerative, circular models that minimize the environmental and waste impacts of retaining materials and products for as long as possible. It includes a systematic approach to the use of renewable resources such as bio-based polymers, sustainably sourced natural fibers, and renewable energy inputs, as well as the use of renewable and direct procurement of critical non-renewable resources such as high-performance alloys and engineered composites.

It is anticipated that by employing resource-efficient manufacturing strategies that leverage recycling technologies such as additive manufacturing (near-net-shape production) or precision machining (low-scrap production), and recover cost-effective materials through closed-loop production techniques, industries can use less energy and generate less waste and environmental damage [7]. Figure 12.1 depicts how electrochemical

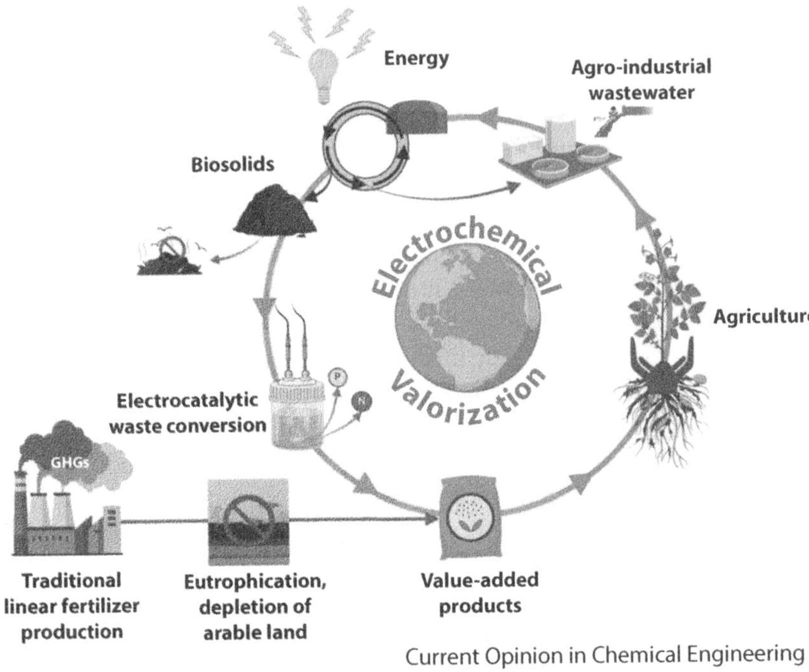

Figure 12.1 Electrochemical valorization of agro-industrial wastewater for sustainable fertilizer production [8]. It is available under an open-access license.

valorization using renewable or low carbon energy can close the loop on converting agro-industrial wastewater and biosolids to fertilizers and renewable energy. While traditional fertilizer production and use are linear, leveraged on fossil fuels with GHG emissions and the depletion of arable land, electrocatalytic waste conversion can sustainably recover agro-industrial nutrients like nitrogen and phosphorus. Not only can this process meet some of the global demand for nutrients, but it has the potential to meet a substantial portion of those demands; for example, US municipal biosolids can meet 9% of the nitrogen and 32% of the phosphorus needed to meet global nutrient demand [8].

Digital monitoring systems, including IoT-enabled sensors and cloud-based analytic platforms, provide instantaneous data on material flows, energy consumption and related emissions in real time to allow adaptive control of production processes. If closed-loop supply chains, enhanced through advanced logistics platforms and reverse logistics networks, allow products and parts to be collected, remanufactured, and reintroduced into the production cycle multiple times, the process becomes circular. This framework, promotor of semi-circular flows, encourages product and process design for minimal impact on the environment considering recyclability, reparability, and disassemblability in the initial product conceptualization stage [9].

This extended framework enhances the progress towards sustainable development goals through reduced greenhouse gas emissions, the preservation of natural resources, and innovation in cleaner production technologies. Ultimately, it culminates as a resilient manufacturing platform during a time of limited resources, pressing environmental issues, and changing market expectations and demands in the area of sustainable products and services.

12.2.1 Renewable and Non-Renewable Resources

At the heart of circular production is the proactive choice of consumption and management of both renewable and non-renewable resources aimed at retaining material and embodied energy value while mitigating environmental impact. Renewable resources afford the use of bio-based polymers like polylactic acid (PLA) and poly hydroxyl alkanoates (PHA) and naturally sourced fibers like hemp and flax. These resources were selected because they have regenerative capacity and are biodegradable. Renewable energy sources such as solar and biomass help ensure the footprints of production; related activities are lessened as well. Non-renewable resources or materials include high strength alloys, rare earth materials, and advanced engineering type plastics. Similarly, non-renewable resources afford circularity *via* computational value chain thinking through normative engineering strategies allowing for alloy recasting and solvent based polymer, and chemical depolymerizing recovery [10].

Material selection frameworks should incorporate criticality assessments, environmental impact (e.g., chemical footprints of materials) modeling, and indicators of circularity in discussions associated with sourcing activities to balance the performance and sustainability of materials. Tools such as lifecycle analysis and supply risk assessments are helpful in making decisions about prioritizing durable, recyclable material with less reliance on virgin extraction. Each of these behaviors create resilience to environmental and resource scarcity and align with regulatory direction encouraging sustainable production practices [11].

12.2.2 Process Flow across Production, Distribution, and Use

An integrated circular production flow necessitates advanced manufacturing processes and closed-loop designs. Additive manufacturing techniques, like selective laser melting and fused filament fabrication, facilitate near-net-shape production along with minimal waste and maximum efficiency of resources. Concurrently, advanced subtractive processes employing real-time monitoring will promote less energy use and scrap. Advanced logistics platforms and IoT tracking can further promote lower transport emissions and facilitate the re-entry of products and packaging [12].

During the use phase, products are designed with modular and reparability features. When parts can be removed and are replaceable, sensors and other embedded technology promote predictive maintenance that allows the repair of products, supporting long life cycles. Digital twins (performance models) that monitor products through the reclamation to remanufacturing to servicing to disposal process, can provide information to aid decisions on refurbishment or upgrades of products. New business models, like leasing and a product-as-a-service model, encourage the manufacturer to develop the product for long life and reuse instead of disposal in the waste stream. Because all phases of the production and distribution supply chain and the use are connected, this will lead to a regenerative system for production supply chains that maximizes value and minimizes waste [13].

12.2.3 Waste Management through Biological and Technical Loops

A resilient circular economy is predicated on effective waste management systems that enable resource recovery through biological and technical cycles. The biological loop manages the organic fraction of waste, i.e., food scraps, agricultural residues and biodegradable plastics, that involve recovering materials through anaerobic digestion (for energy) and composting (to make usable organic compost). The technical loop handles the durable fraction of waste, e.g., metals, composites, engineering polymers, through different recovery methods such as mechanical recycling, chemical depolymerization, and remanufacturing [14].

AI sorting systems with machine vision can autonomously recognize and distinguish different materials from products, allowing for relatively clean recovery of materials. Deconstruction techniques such as robotic disassembly, the full or modular deconstruction of products are a means to recover valuable reusable components and or materials. Policies to extend producer responsibility and develop deposit-return schemes are helpful ways to have producers/manufacturers take-back materials, the goal here is not to take back all materials to close the loop, which lessens environmental impacts, conserves resources and enables resilient supply chains with sustainable industrial systems that prioritize and reward sustainable value and responsibility.

12.3 Sustainable Material Selection Strategies

Sustainable material selection strategies are about systematically identifying, assessing, and selecting materials that allow products to perform as designed while minimizing environmental impacts over their entire lifecycle [15]. These strategies focus on renewable or high recyclability materials, as well as on materials with no hazardous or toxic byproducts. Figure 12.2 presents an integrated approach to managing carbon-based materials by integrating synthetic and natural material flows into a single recycling system. The proposed co-recycling approach allows complex input streams to be processed in the same recycling system by advancing thermochemical recycling technology to process adding value to mixed or contaminated carbon-based materials and including materials of different qualities. High-quality and mid-grade products can be reused/preserved for production; low-grade residues and losses, including downgraded natural and synthetic wastes, can be upcycled into valuable products. Thus, this integrated approach minimizes losses of carbon into landfills or off-gassing, continues to reduce dependence on fossil resources, and facilitates lakeside management of resource-based carbon, a key component for understanding circular economy [16]. To follow, employing thermochemical recycling and technologies (pyrolysis, gasification, etc.) can lower overall energy used as well as potentially enable near-zero on carbon emissions for energy in a decarbonized energy system, showcasing the crucial role of sustainable resource management.

Durability and ease of disassembly is also a big factor when selecting materials, particularly in conjunction with extending the lifespan of a product along with recovery at end-of-life. High-performance recycled thermoplastics, natural fiber composites, and low-toxicity flame retardants are increasingly being considered in automotive, and electronics applications and these materials aim to balance mechanical strength along with safety and environmental performance [17].

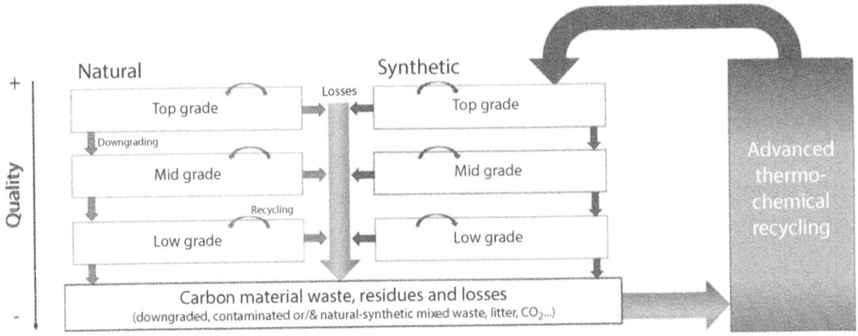

Figure 12.2 Co-recycling system integrating natural and synthetic carbon-based materials through advanced thermochemical recycling [16].

Advanced, data-driven platforms, including material informatics, digital twins and lifecycle databases, are a great way to optimize design and engineering trades between environmental footprint and mechanical properties, processing and economic costs. Paired with simulation and optimization, is the engine software that can estimate product performance with various macro- and micro-material states, along with decisions based on evidence and consistent with sustainability benefits. Recently, blockchain-enabled traceability systems have been made attractive to enhance transparency of sourcing and to ensure compliance with various legislation like REACH and RoHS [18].

By strategic alignment of material solutions for circular economy priorities, regulatory requirements, and customer expectations, organizations achieve lifecycle impact reduction and build resilience to supply chain disruptions, raw material shortages, and changing consumer demand. This integrated approach enables environmentally sustainable manufacturing innovation, and organizations are positioned to lead when resources become scarcer.

12.3.1 Criteria for Sustainable Material Selection

Selecting material for a sustainable production involves reviewing multiple criteria that weigh environmental, technical and economic objective outcomes. Renewability assesses if materials can be replenished over time; in terms of sustainable and bio-based materials, we favor plant-based polymers such as polylactic acid (PLA) and natural fibers such as hemp and flax so that we can avoid depleting non-renewable sources. Recyclability measures how much we can recover and reprocess materials that still provide some level of performance and promote closed-loop systems that keep valuable materials in circulation [19].

Embodied energy and carbon footprint measures, the cumulative energy and emissions attributable to the sum of extraction, processing, manufacturing and transportation (overall lifecycle) of the materials to promote options that have less impact such as recycled metals and bio-based plastics. Toxicity and safety measures the potential effects to human health and ecosystems of the product to monitor compliance with regulations like REACH and RoHS while avoiding dangerous substances. Durability and mechanical performance provide sufficient measures for designers to make sure material does not fail early or unexpectedly under load, so they meet strength and reliability objectives while essentially increasing the useful lifetime of the product to reduce waste [20]. Economic feasibility on a material resource level includes upfront costs but considers potential recovery, reuse and processing efficiencies that could bring savings. Finally, multi criteria decision frameworks allow to bring all these considerations together, using lifecycle assessment tools, scoring models and materials databases as means to undertake evidence-based decisions that meet sustainability objectives, regulatory demands, and an ever-increasing market demand.

12.3.2 Design for Disassembly and Reuse

Design for disassembly and reuse is a core principle of circular manufacturing that empowers products to be disassembled in an efficient manner to recover valuable components and materials at the end of life. At a minimum, successful DfD design strategies begin with modular design. Modules are often designed as functional units, where a component could be replaced, repaired, or upgraded without affecting other parts of the assembly. Methods of fastening, such as snap-fit joints, screw fastening, and interlocking connectors, facilitate easy mechanical separation, ensuring that parts can be retrieved without damaging parts or contaminating materials. Also, selecting compatible materials and avoiding adhesives or composite layers that do not permit recycling, will improve disassembly success [21].

Design for reuse takes the next step by allowing for recovered components to be reusable, or instead they are still of sufficient integrity to be remanufactured and included into a new product, thus reducing the extraction of virgin materials and curbing waste. For example, electronic devices that are inclusive of standardized modules and accessible enclosures can be refurbished and/or upgraded to prolong their usable life.

Incorporating DfD and reuse design principles earlier in the development of an electronic product will decrease lifecycle costs and compliance with take back legislation, as well as open doors to new business models such as lease or product as a service [22].

12.3.3 Lifecycle Assessment of Material Choices

Lifecycle Assessment (LCA) offers a systematic, science-based approach to rigorous thinking about environmental impacts related to materials in consideration, throughout the entire value chain, throughout the entire value chain. An LCA quantifies resource consumption, emissions, and environmental impacts from raw material extraction, processing, manufacturing, transportation and distribution, use and disposal. Researching the LCA may show that recycled aluminum uses up to 95 percent less energy and generates less greenhouse gas emissions compared to primary aluminum. While biopolymers created from renewable feedstocks may significantly reduce dependence on fossil fuels, this may come with trade-offs, such as land use impacts, or water use impacts in growing crops [23].

Advanced LCA software and digital tools can support scenario modeling, which allows designers to explore a number of material alternatives to further defined targets on sustainability. The integration of LCA into the material selection process and product development can allow organizations to identify opportunitics for improvements, substantiate claims for the environment, compliance, and transparency to customers and stakeholders. Ultimately, lifecycle assessment is an effective decision-support tool to ensure materials meet intended circular production systems and longer-term sustainability goals [24].

12.4 Additive Manufacturing for Circular Economy

Additive manufacturing (AM), generally known as 3D printing, is an exceptionally flexible production method that align with circular economy strategies by directly enabling the production of complex parts with minimal material and energy consumption. Where traditional subtractive techniques involve removing (cutting or machining) substantial quantities of raw material, AM adds material only where it is needed, layer by layer, to produce nearly net-shape components with superior resource efficiency [25].

This allows manufacturers to develop products for long-term use and easy reuse, since parts can be formed on-demand to substitute worn components instead of throwing away the entire systems. Similarly, the flexibility of AM processes facilitates using recycled materials and renewable feedstocks, such as recycled metal powders, reprocessed polymers, and even bio-based materials. Figure 12.3 presents the key factors for recycling and upcycling plastic waste that are currently prohibitive. These include high-performance thermoset plastics that, due to crosslinking, cannot be reprocessed; high costs and issues associated with classifying and separating these materials; food waste residues that proliferate and hinder cleaning and reuse; and additives and coatings the complicate, and can even hinder, recycling [26]. These challenges will require innovations such as self-cleaning plastic films, sustainable coating technologies, and sorting systems with greater efficiencies. Combining these could enable higher-value recovery of polymers, diminish the environmental footprint, and facilitate circular economies in plastic manufacturing.

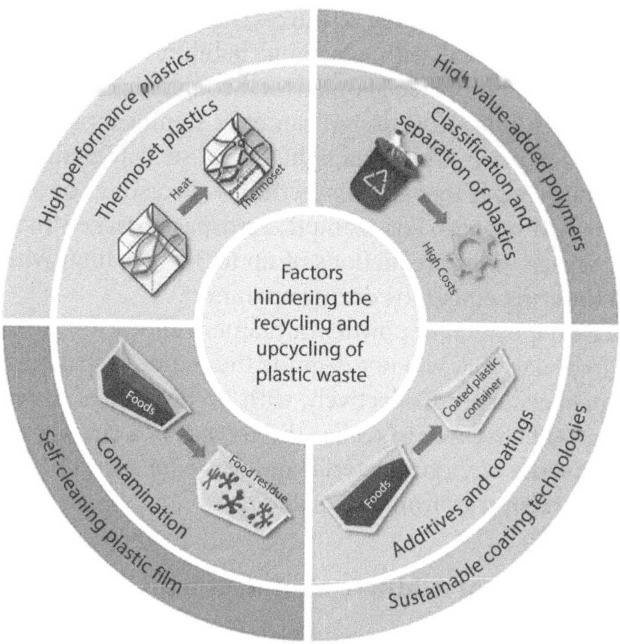

Figure 12.3 Factors hindering the recycling and upcycling of plastic waste [26].

AM also supports distributed manufacturing which allows production to occur closer to where consumption takes place, resulting in decreasing logistics complexity and transportation emissions [27]. Thus, adopting additive manufacturing systems advances sustainability goals and establishes shorter lead times and decreased storage costs, while also providing more opportunities for product customization. AM as a hybrid of digital design tools, advanced materials, and localized production, enables wasted material from the manufacturing of tradition models while prompting regenerative models that prolong the recirculation of materials and the environmental footprint of industrial-level activity.

12.4.1 Resource Efficiency in Additive Processes

One of the central advantages of additive manufacturing technologies is resource efficiency [28]. SLM and EBM (for metals) and FFF (for polymers), for example, are processes where the material is applied only as required to produce each layer of a part. Contrarily, there is often as much as 70-80% of a part reduced to scrap in traditional machining; with additive processes, material utilization rates over 90% are achievable [29].

This efficiency not only saves resources, but it also reduces production costs, and waste disposal with costs and liabilities reduced as well. The capability to design optimized geometries through digital means, topology optimization, and generative design can offer significant savings in materials with parts that are strong and lightweight, such as using lattice structures or organic shapes that are impractical with castings or forgings [30], as part of the AM process. For example, aerospace brackets manufactured using AM can see weight reductions of up to 30-50% improving fuel efficiency and lowering emissions during operation.

The precise deposition of powder or filament allows for the use of recycled input materials that will not degrade the required structural integrity or dimensional accuracy. Collectively, with AS because of its material efficiency, it is an effective way to realize the essence of a circular economy by minimizing the use of virgin material and waste on a landfill and allowing for high-value recoveries from the material.

12.4.2 Recycling of Additive Manufacturing Feedstocks

Recycling and reusing additive manufacturing feedstocks is a crucial way to further improve the sustainability of AM processes [31]. In metal AM technologies that use powder (e.g., SLM and EBM), unused powders after a build can be collected, sieved, and blended with virgin powder to ensure consistent quality and particle size distribution.

These powders can be recycled across multiple manufacturing cycles, creating savings in raw material costs and reducing material waste. Likewise, thermoplastic filaments and pellets used in extrusion-based processes (e.g., FFF) can be collected from support structures, failed prints, or scrap material, shredded, and reprocessed into new spools of filament. Because there are performance standards that must be met, recycled feedstocks are tested for their properties (e.g. particle size, flowability, chemical composition, and moisture content) [32]. Aside from mechanical recycling approaches, new chemical recycling approaches are being developed to degrade thermosettable composites and engineering polymers back to desired monomers or valuable oligomers (partially depolymerized chain). The aim of such methods is to recover both the resin and reinforcement fibers (i.e. carbon or glass). For instance, examples of solvolysis methods exist that only dissolve the matrix resins from composite parts while leaving the fiber untouched allowing reuse to produce new parts with the same properties. Establishing recycling practices based on advanced recycling can provide organizations with a closed-loop supply chain approach to materials circularity, reduce reliance on virgin resources and improve economic resilience to market uncertainty [33].

12.5 Digitalization Enabling Closed-Loop Production

Digitalization is reshaping manufacturing by allowing for closed-loop production systems, in which data inspires seamless connection between design, production, use, and recovery in a unified ecosystem. Instead of thinking in isolated phases, processes are interconnected through a combination of digital infrastructure, intelligent software, and connected devices, giving companies a way to observe, trace, and optimize materials, products, and processes while they are produced, used, and recovered. This continuous real-time flow of information puts the manufacturer in a position to improve operational efficiency, product quality, and reduce resource consumption by facilitating accurate decisions based on current information. Machine Learning and Artificial Intelligence provide insights into production and supply chain data to identify hidden patterns and forecasting. For example, predictive analytics provide the opportunity to optimize material flows to circulate recyclable materials further into the production cycle with limited loss of value [34].

Digitalization is also crucial to keeping up with developing pledges to sustainability regulations and voluntary standards, such as ISO 14001 or extended producer responsibility frameworks, and to enabling firms to demonstrate, they are conducting environmental stewardship. Because of the digital transparency provided by digitalization, companies can build trust with customers, investors, and regulators by having traceability of materials, carbon footprint reporting, and substantiation of recycled content claims [35]. Figure 12.4 depicts the current gaps and future opportunities to integrate Circular Economy and Sustainability in an Industry 4.0 context, including that there is still no overarching circular economy framework that integrates buildings, photovoltaics (PV), batteries and electric vehicles (EV), and properly describes how they interrelate in renewable energy driven life cycle decarbonization [36].

The roles of renewable-powered buildings and battery-powered vehicle systems in sustainability is still unclear, as is how advanced digital technologies are being applied like Internet of Things (IoT) and data-driven analytics to create operations in circular systems [37]. Therefore, the research calls for consideration of circular economy practices in different sectors while evaluating energy and carbon flows, examining the recovery of waste and conversion of waste to energy, and the use of digital tools to enable advanced circular systems. Overall, the framework includes the potential to enhance cascade reuse, reverse logistics, low-carbon refurbishment, and recycling by extending an integrated approach to carbon neutrality and sustainable development.

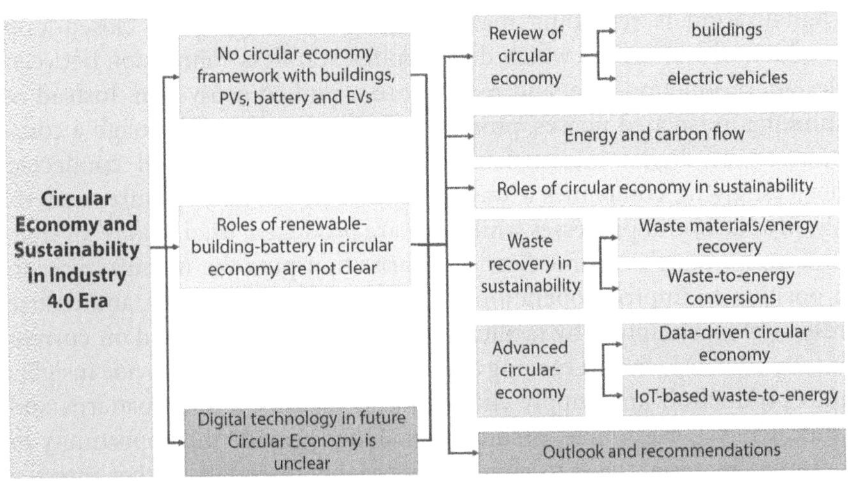

Figure 12.4 Circular economy and sustainability integration in the industry 4.0 era: Challenges and framework for low-carbon transition [36].

The digital platforms open up entirely new business models, such as product-as-a-service, pay-per-use, and leasing frameworks that not only encourage longer product life, but also incentivize manufacturers to design for durability and re-usability. Equipment manufacturers can now remotely monitor the condition of leased equipment and service or upgrade components to better extend their life and reduce waste. Digitalization provides the backbone to achieve closed-loop, circular value chains on scale, integrating physical processes with intelligent systems that achieve manufacturing processes that are, more efficient, but profoundly more sustainable, and resilient [38].

12.5.1 IoT-Enabled Real-Time Monitoring and Predictive Data Analytics

Real-time monitoring solutions enabled by IoT use interconnected sensors, RFID tags, and smart devices to aggregate detailed operational data at each stage of the production and distribution process. While sensors within machines are used to monitor temperature, vibration, and energy usage; product tags can track specific data about location, condition, and usage. All this valuable data feeds into centralized platforms, where predictive analytics algorithms can analyze activity trends and possibly identify issues before maintenance is needed, avoiding downtime or defects [39]. One of the significant benefits is that predictive maintenance allows manufacturers to intervene precisely when needed rather than simply scheduling future servicing, thereby making every action serve as a process to lengthen the life of their equipment and avoid any failures or delays.

IoT has also increased awareness of material flows and product returns in closed-loop systems. The monitoring of product returns by IoT can also help to shed light on recycling and remanufacturing processes to help companies ensure that recovered materials adhere to quality specifications and assist in determining the routing of returns for reuse. The combination of IoT monitoring with analytics helps companies reduce waste and increase efficiency, while enabling them to adapt quickly to demand changes or supply-chain shocks [40].

12.5.2 Digital Twin Systems for Comprehensive Lifecycle Optimization

A digital twin is a digital replica of a physical product, asset, or whole production line, that is updated in real-time from real-world data. The twins are created from design specifications and are populated with data from IoT sensors, process data, and historical performance data [41]. Digital twins let manufacturers simulate products and their performance under various conditions, tune and optimize performance, and predict maintenance needs. One example would be a digital twin of a wind turbine, which can estimate wear based on real-time loads, enabling maintenance to take place based on condition rather than time, improving operating life.

In manufacturing, digital twins can be used to simulate workflows and material flows within your process to help spot bottleneck scenarios and provide opportunities for optimized resource use. In closed-loop systems, digital twins monitor the entire lifecycle of a product from material acquisition to end-of-life recovery providing information about recyclability, environmental footprint, and economic contribution. This full lifecycle view allows data-driven decision-making to reduce emissions, increase durability, and enable a design to be refurbished or recycled easily. Closing the feedback loop between the physical and virtual worlds through digital twins will allow companies to drive continuing improvements in sustainability and competitive advantage over time.

12.5.3 Closed-Loop Reverse Logistics Framework

The diagram provides a holistic reverse logistics model that proactively represents the flow of waste, used products, and returned products back into productive use versus disposal. The Model starts with collection in which materials are received from several sources such as OEMs (original equipment manufacturers), retailers, and third parties. Materials received in this step will be categorized as waste or classified as used products and returned products. The reverse logistics system then flows onto separation and inspection *via* inbound logistics and quality control (QC) people, who inspect and evaluate each commodity in terms of condition to determine whether the commodity is suitable for reuse, repair, remanufacturing or recycling [42]. Suitable reusable, recoverable items move from inspection to storage, which by means of warehousing and inventory management, keeps track of inventory for reusable systems and obsolete systems waiting for further evaluation.

Once the products are ready for processing, they proceed into disassembly. Trained disassemblers and sorters disassemble the product to remove reusable components while sorting out proprietary or contaminated materials. Reusable components that meet fit-for-use specifications go towards further recovery, and contaminated components go to controlled treatment. Products that need further disassembly go to shredding and grinding.

Shredder operators use various specialized equipment to convert components into shredded materials and parts that can be reclaimed and reused [43]. This stage of reclamation ensures that even complex products or damaged products can provide recoverable resources.

All the outputs after disassembly and shred, flow to outbound logistics, the heart of transportation recovery material to the right destination. Resaleable, recyclable or remanufacturable materials are sent to recycling facilities, remanufacturing, or secondary markets. Waste streams are transferred by waste management organizations for disposal or non-recycled methods. Transporters work together to organize the transport of all the materials to properly handle and swiftly move forward with processing. Feedback loops exist across the system - to re-stash reusable systems and components into initial storage, or reintegration into production, complete their material loop [44].

The following impact model is indicative of circular economy practices, as it ensures that valuable materials are retained in circulation as long as possible, reduces the environmental burden of production, conserves resources, and fulfills obligations to regulation such as expansive producer responsibility [45]. The model provides a unified approach to inspection, disassembly, shredding, and outbound distribution, and it is an emerging workflow that will assist organizations with their economic and sustainability targets by promoting clear, trackable and transferable materials in the reverse supply chain.

12.6 Conclusion

Circular economy principles are a major departure from the traditional linear "take-make-dispose" approach and promote regenerative systems thinking where materials and products circulate for as long as possible. By combining materials derived from renewable resources, such as bio-based polymers and natural fibers, and responsibly managing non-renewable inputs such high-performance alloys and engineering plastics, manufacturers reduce, reuse and recycle materials into regenerative systems that dual purpose material and energy resource consumption. Sustainable material selection strategies are important to this transition as design strategies consider a breadth of sustainability factors including recyclability, embodied energy, end-of-life durability and destiny, and toxicity to make responsible design decisions. Design for disassembly (and reuse) ensures products can be taken apart as effectively, and easily as possible and then recycling developing componentry again, and service life assessments utilize a science-based process to quantify and model the environmental performance for the entire life of the product.

Recycling valorization further amplifies this process by utilizing both mechanical and chemical recycling to retrieve valuable metals, composites, and polymers from both end-of-life goods and production waste streams. Additive manufacturing fosters near-net-shape production, limits material waste, and relies on reclaimed feedstocks and on-demand production close to the point of consumption. Digitalization enhances closed-loop systems by extending the capabilities with IoT-enabled tracking, predictive analytics, and digital twins to monitor and improve materials, processes, and product performance. In combination with closed-loop supply chain systems that recover and recirculate materials, these systems are able to equip the industrial sector with more resilient, leaner, and sustainable production systems that meet the growing demands for circular innovation and environmental stewardship.

AI disclosure statement: The author acknowledges the use of AI-based tools for minor language editing or grammatical correction. However, all intellectual contributions, analyses, and conclusions presented in this chapter are entirely the author's own.

References

1. Pomponi, F. and Moncaster, A., Circular economy for the built environment: A research framework. *J. Cleaner Prod.*, 143, 1, 710–718, 2017.
2. Marinina, O., Kirsanova, N., Nevskaya, M., Circular Economy Models in Industry: Developing a Conceptual Framework. *Energies (Basel)*, 15, 24, 9376, 2022.
3. Mesa, J., González-Quiroga, A., Maury, H., Developing an indicator for material selection based on durability and environmental footprint: A Circular Economy perspective. *Resour. Conserv. Recycl.*, 160, 1, 104887, 2020.
4. Ajithkumar, S. and Arulmurugan, B., Investigation of Single and Double Pulse Techniques on Microstructure and Mechanical Anisotropy of Inconel 686 Component Fabricated by Wire Arc Directed Energy Deposition. *Met. Mater. Int.*, 31, 1648–1673, 2025.
5. Colorado, H.A., Velásquez, E.I.G., Monteiro, S.N., Sustainability of additive manufacturing: the circular economy of materials and environmental perspectives. *J. Mater. Res. Technol.*, 9, 4, 8221–8234, 2020.
6. Hapuwatte, B.M. and Jawahir, I.S., Closed-loop sustainable product design for circular economy. *J. Ind. Ecol.*, 25, 6, 1430–1446, 2021.

7. Lewandowski, M., Designing the business models for circular economy-towards the conceptual framework. *Sustainability (Switzerland)*, 8, 1, 43, 2016.
8. Ferreira-Garcia, D., Haque, S., Burke, B., Furst, A.L., Botte, G.G., Electrochemical organic waste conversion: a route toward food security and a circular economy. *Curr. Opin. Chem. Eng.*, 49, 1, 101156, 2025.
9. Voulgaridis, K., Lagkas, T., Angelopoulos, C.M., Nikoletseas, S.E., IoT and digital circular economy: Principles, applications, and challenges. *Comput. Networks*, 219, 1, 109456, 2022.
10. Ioannidis, F., Kosmidou, K., Papanastasiou, D., Public awareness of renewable energy sources and Circular Economy in Greece. *Renewable Energy, 206*, 1, 1086–1096, 2023.
11. Klemeš, J.J., Varbanov, P.S., Walmsley, T.G., Foley, A., Process Integration and Circular Economy for Renewable and Sustainable Energy Systems. *Renewable Sustainable Energy Rev., 206*, 1086–1096, 2019.
12. Schaubroeck, T., Gibon, T., Igos, E., Benetto, E., Sustainability assessment of circular economy over time: Modelling of finite and variable loops & impact distribution among related products. *Resour. Conserv. Recycl., 168*, 1, 105319, 2021.
13. Virtanen, M., Manskinen, K., Uusitalo, V., Syvänne, J., Cura, K., Regional material flow tools to promote circular economy. *J. Cleaner Prod., 235*, 1020–1025, 2019.
14. Romero-Hernández, O. and Romero, S., Maximizing the value of waste: From waste management to the circular economy. *Thunderbird Int. Bus. Rev.*, 60, 5, 757–64, 2018.
15. Rajeshkumar, L., Ajithkumar, S., Arulmurugan, B., Bhuvaneswari, V., Life cycle assessment of mineral fibers and their composites, in: *Synthetic and Mineral Fibers, Their Composites and Applications*, pp. 631–55, Elsevier, 2024.
16. Cañete Vela, I., Berdugo Vilches, T., Berndes, G., Johnsson, F., Thunman, H., Co-recycling of natural and synthetic carbon materials for a sustainable circular economy. *J. Cleaner Prod.*, 365, 1, 132674, 2022.
17. Ajithkumar, S., Arulmurugan, B., Rajeshkumar, L., Prospects of natural fiber-reinforced polymer composites in engineering and commercial applications, in: *Applications of Composite Materials in Engineering*, pp. 335–64, Elsevier, 2025.
18. Awan, U., Shamim, S., Khan, Z., Zia, N.U., Shariq, S.M., Khan, M.N., Big data analytics capability and decision-making: The role of data-driven insight on circular economy performance. *Technol. Forecasting Social Change, 168*, 1, 120766, 2021.
19. Ribul, M., Lanot, A., Tommencioni Pisapia, C., Purnell, P., McQueen-Mason, S.J., Baurley, S., Mechanical, chemical, biological: Moving towards closed-loop bio-based recycling in a circular economy of sustainable textiles. *J. Cleaner Prod.*, 326, 1, 129325, 2021.

20. Hagelüken, C. and Goldmann, D., Recycling and circular economy towards a closed loop for metals in emerging clean technologies. *Miner. Econ.*, *35*, 3, 539–562, 2022.
21. O'Grady, T., Minunno, R., Chong, H.Y., Morrison, G.M., Design for disassembly, deconstruction and resilience: A circular economy index for the built environment. *Resour. Conserv. Recycl.*, *175*, 1, 105847, 2021.
22. Vanegas, P., Peeters, J.R., Cattrysse, D., Tecchio, P., Ardente, F., Mathieux, F., *et al.*, Ease of disassembly of products to support circular economy strategies. *Resour. Conserv. Recycl.*, *135*, 1, 323–334, 2018.
23. Lei, H., Li, L., Yang, W., Bian, Y., Li, C.Q., An analytical review on application of life cycle assessment in circular economy for built environment. *J. Build. Eng.*, *44*, 1, 103374, 2021.
24. van Stijn, A., Malabi Eberhardt, L.C., Wouterszoon Jansen, B., Meijer, A., A Circular Economy Life Cycle Assessment (CE-LCA) model for building components. *Resour. Conserv. Recycl.*, *174*, 1, 105683, 2021.
25. Ajithkumar, S., Arulmurugan, B., Balaji, D., Prasath, V.M.M., Idhayaraja, S., Hybrid Welding Processes, in: *Advanced Welding Technologies*, pp. 177–96, Wiley, 2025.
26. Al Rashid, A. and Koç, M., Additive manufacturing for sustainability and circular economy: needs, challenges, and opportunities for 3D printing of recycled polymeric waste. *Mater. Today Sustainability*, *24*, 1, 100529, 2023.
27. Sitharaj, A., Arulmurugan, B., Balaji, D., Idhayaraja, S., Additive Manufacturing of Titanium-based Alloys by Wire Arc Directed Energy Deposition (WA-DED), in: *Advances in Solid-State Welding and Processing of Metallic Materials*, pp. 400–17, CRC Press, Boca Raton, 2025.
28. Ajithkumar, S. and Arulmurugan, B., Microstructure and mechanical properties of Inconel 686 fabricated by gas metal arc welding-based wire arc directed energy deposition: impact of cryogenic treatments. *J. Mater. Sci.*, *59*, 40, 19273–19302, 2024.
29. De Pasquale, G., Luceri, F., Riccio, M., Experimental Characterization of SLM and EBM Cubic Lattice Structures for Lightweight Applications. *Exp. Mech.*, *59*, 4, 469–482, 2019.
30. Sitharaj, A., Arulmurugan, B., Dharani Kumar, S., Idhayaraja, S., Manishraj, D., Premkumar, N., Wire-Arc Directed Energy Deposition of Aluminum Alloys: A Review on Properties, Challenges, and Applications, pp. 99–115, 2025.
31. Nataraj, G. and Ramesh Babu, S., Enhancing mechanical properties of PLA-based bio composite filament reinforced with horse gram filler for 3D printing applications. *J. Vinyl Add. Tech.*, *31*, 2, 453–468, 2025.
32. Dhiman, S., Joshi, R.S., Singh, S., Gill, S.S., Singh, H., Kumar, R., *et al.*, Recycling of Ti6Al4V machining swarf into additive manufacturing feedstock powder to realise sustainable recycling goals. *J. Cleaner Prod.*, *348*, 1, 131342, 2022.

33. Shanmugam, V., Das, O., Neisiany, R.E., Babu, K., Singh, S., Hedenqvist, M.S., *et al.*, Polymer Recycling in Additive Manufacturing: an Opportunity for the Circular Economy. *Mater. Circ. Econ.*, *2*, 1, 11, 2020.
34. Ajithkumar, S., Arulmurugan, B., Meiyazhagan, R., Welding of nickel alloys interface by finite element modeling and artificial intelligence and envisage by patent landscape. *Int. J. Interact. Des. Manuf. (IJIDeM)*, 1–32, 2025.
35. Hernández-Chover, V., Castellet-Viciano, L., Bellver-Domingo, A., Hernández-Sancho, F., The Potential of Digitalization to Promote a Circular Economy in the Water Sector. *Water (Switzerland)*, *14*, 22, 3722, 2022.
36. Zhou, Y., AI-driven digital circular economy with material and energy sustainability for industry 4.0. *Energy AI*, 20, 100508, 2025.
37. Vellingiri, A., Mohanasundaram, K., Tamilselvan, K.S., Maheswar, R., Ganesh, N., Multiple Sensor based Human Detection Robots: A Review. *Int. J. Smart Sens. Intell. Syst.*, *16*, 1, 2023.
38. Burmaoglu, S., Ozdemir Gungor, D., Kirbac, A., Saritas, O., Future research avenues at the nexus of circular economy and digitalization. *Int. J. Prod. Perform. Manage.*, *72*, 8, 2247–2269, 2023.
39. Ajithkumar, S., Arulmurugan, B., Idhayaraja, S., Robotic Welding and Automation: Cutting-Edge Technology for Industry, in: *Advanced Welding Technologies*, pp. 497–512, Wiley, 2025.
40. Reuter, M.A., Digitalizing the Circular Economy: Circular Economy Engineering Defined by the Metallurgical Internet of Things. *Metall. Mater. Trans. B*, *47*, 6, 3194–3220, 2016.
41. Ajithkumar, S., Arulmurugan, B., Prasath, V.M.M., Raja, P.U.M., Aravindh, M., Advancements in Robotic Welding Sensing Technology, in: *Advanced Welding Technologies*, pp. 477–95, Wiley, 2025.
42. Gharibi, K. and Abdollahzadeh, S., A mixed-integer linear programming approach for circular economy-led closed-loop supply chains in green reverse logistics network design under uncertainty. *J. Enterp. Inf. Manage.*, *38*, 1, 1–31, 2021.
43. Mallick, P.K., Salling, K.B., Pigosso, D.C.A., McAloone, T.C., Towards a circular economy: Development of a support tool for designing reverse logistics systems. *J. Environ. Manage.*, *351*, 1, 119819, 2024.
44. Wilson, M., Paschen, J., Pitt, L., The circular economy meets artificial intelligence (AI): understanding the opportunities of AI for reverse logistics. *Manage. Environ. Qual. Int. J.*, *33*, 1, 9–25, 2022.
45. Shahidzadeh, M.H. and Shokouhyar, S., Toward the closed-loop sustainability development model: a reverse logistics multi-criteria decision-making analysis. *Environ. Dev. Sustain.*, *25*, 5, 4597–4689, 2023.

13

Sustainable Additive Manufacturing: Innovations in Advanced Materials and Processing Techniques

Manoj Mohan Prasath V.[1]**, Ajithkumar Sitharaj**[2*]**, Karuppusamy R.**[3] **and Sandip Kunar**[4]

[1]*Department of Mechanical Engineering, Karpagam Academy of Higher Education, Coimbatore, India*
[2]*Department of Mechanical Engineering, KPR Institute of Engineering and Technology, Coimbatore, India*
[3]*Department of Electronics and Communication Engineering, Hindustan Institute of Technology, Coimbatore, India*
[4]*Department of Mechanical Engineering, Aditya University, Surampalem, India*

Abstract

Additive manufacturing (AM) enables the fabrication of complex and high-precision components while significantly reducing material waste compared to conventional manufacturing methods. Using advanced materials, AM supports environmentally responsible practices by lowering raw material consumption and emissions. This review examines the environmental benefits of AM technologies and identifies significant application areas that contribute to sustainability across various mechanical and biomedical sectors. As industries and medical increasingly prioritize environmental responsibility, AM facilitates improvements in process efficiency, material utilization, and the production of customized, on demand components. It also influences product development by enabling design flexibility, user specific features, and lightweight structures. Despite existing challenges at different stages of the manufacturing life cycle, AM continues to impact innovation, reshape industrial business models, and alter value chain dynamics. The paper further discusses developments in hybrid systems and bio-fabrication, emphasizing the potential of AM in promoting sustainable and energy-efficient production practices.

**Corresponding author*: ersajithkumarbe@gmail.com

Gurudas Mandal, Swarup Kumar Ghosh, Snehanshu Pal and Sandip Kunar (eds.) Sustainable Manufacturing of Advanced Materials, (267–292) © 2026 Scrivener Publishing LLC

Keywords: Additive manufacturing, environmentally responsible, industrial sectors, flexibility, life cycle

13.1 Introduction

The manufacturing sector is experiencing a paradigm shift largely due to newer technologies allowing for cost-effective, resource efficient and small-scale manufacturing. Traditional methods of manufacturing are energy-intensive and lead to a damaging environmental impact due to their waste and emissions. New paradigms like servitization, personalization and pre-sumption in conjunction with the emergence of Additive Manufacturing (AM) technologies have encouraged industry to rethink their production methods [1]. AM technologies include 3D printing, 3D scanning and certain specialized software that can support decentralized and individual production. Compared to traditional manufacturing methods, the AM process offers some significant benefits in terms of sustainability, potentially making it a much more environmentally friendly process. This shift not only aids in reducing emissions, but is also supremely significant in mitigating climate change [2].

AM and other advanced manufacturing processes have the potential to disrupt the industry by facilitating more compact, localized, and collaborative value chains that can yield significant sustainability potential. AM is a layer-by-layer, additive process as opposed to traditional, subtractive approaches, demonstrating greater synergies with natural growth patterns while dramatically reducing material waste. This provides the opportunity to decouple the creation of economic and social value from the environmental consequences of traditional manufacturing [3]. As AM and other advanced manufacturing processes spread across industries, adequate and continued research and development will be needed for successful and long-lasting implementation. AM systems have significant potential to support global sustainability ambitions through reduction of waste, reused materials and recycled materials; however, there are practical, measurable steps needed to support these initiatives and collaboration, and coinvestigation must occur across the entire value chain including employees, customers, suppliers, communities, governments, NGOs, certification bodies and universities [4].

While the developments in hardware and models have advanced AM systems, polymer chemists and materials scientists have begun to focus on designing AM-specific materials (e.g., thermoplastics, thermosets,

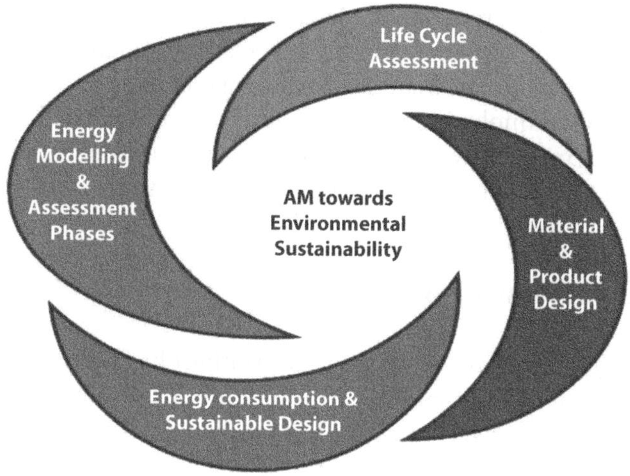

Figure 13.1 Environmental sustainability [8]. [Reproduced under open license].

elastomers, and composites). The call for quantifying whether AM is contributing to sustainability issues or helping to decrease these sustainability issues has been echoed, as we as a society try and come to grips with the growing concerns surrounding environmental issues, and plastic waste gathering on a global scale. Moving forward, it is clear that materials and processes for AM need to adhere to the principles of sustainable practices when addressing sustainability concerns [5, 6].

Generative architecture allows us to correlate many components into a singular product, eliminating the need for redundant material and labor for production processes and addressing the principal sustainability issues in industry. Furthermore, generative design allows users, designers, and engineers to take risks and be creative with material choices while still critically examining design and engineering processes and eliminating the need for repeated physical prototyping while saving time, decreasing waste, and conserving material usage [7]. AM is now being utilized across industry, since it can help solve issues we face related to biodiversity and environmental change demonstrated in Figure 13.1. 3D printing technologies assist in the development of properties that are relatable to personalized products, while supporting better quality standards derived from potential environmental performance. Both sustainable development and environmental requirements are a component of any growth plan related to the ongoing growth model. Reductions in energy usage and raw material use advancing sustainable development [8]. This technology builds objects layer by layer from digital files and 3D scanning data with materials to build from like powders, wire, etc. It is readily available for even home usage and can manipulate materials including metals, food, etc. This paper examined AM technologies, roles in industry, and associated sustainability. This paper reviewed what we can do now and the implications of these methods on the sustainability of manufacturing systems.

13.2 Overview of AM

AM technology (also referred to as 3D printing) has the ability to transform various aspects of manufacturing and production. It involves building objects layer by layer by using distinct digital 3D models, which can be made using CAD software or 3D scanning. The system is made up of

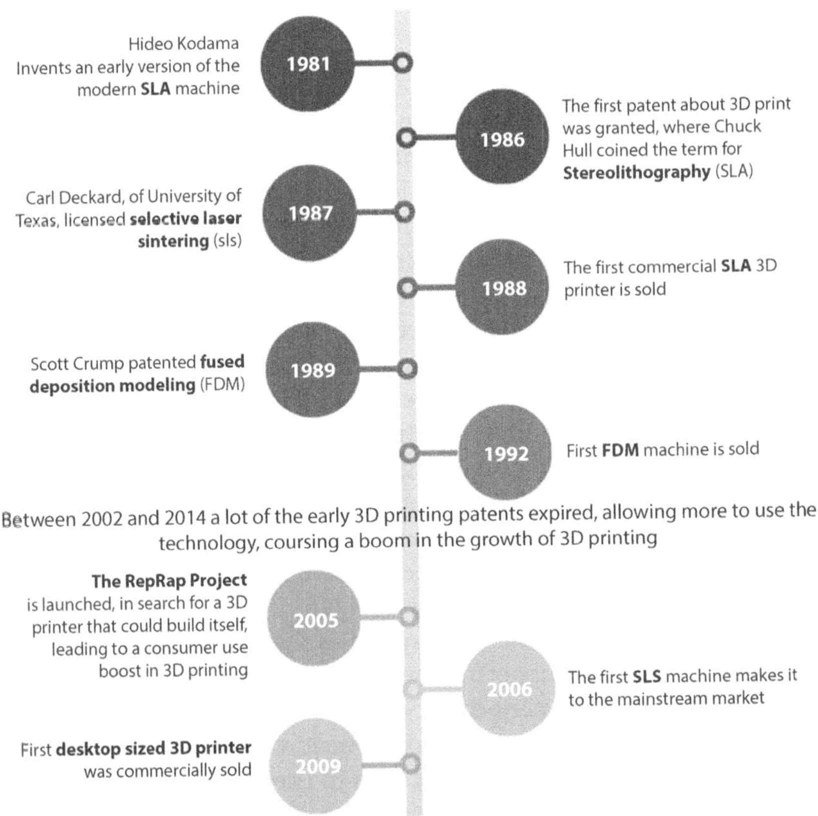

Figure 13.2 History of AM [9]. It is available under an open-access license.

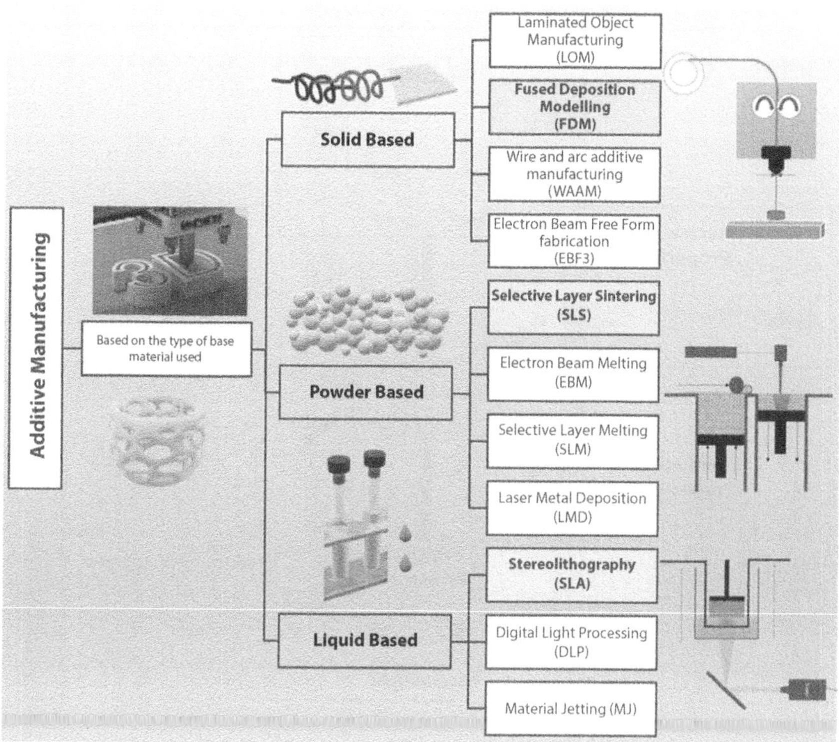

Figure 13.3 Classification of AM [10]. [Reproduced under open license].

a machine that has followed a specific approach to recreate models of the digital world in the physical world. The ISO/ASTM 52900-15 standard that categorizes AM processes into seven modalities: binder jetting, directed energy deposition, material extrusion, material jetting, powder bed fusion, sheet lamination and vat photopolymerization. As with many technologies, many of the principles of AM were developed independently by multiple inventors (circa 1994-2017) (Figure 13.2). More so, AM can also be categorized into three groups based on base material: liquid, powder, and solid (Figure 13.3) [9].

13.2.1 Evaluation of AM Materials

To assess whether 3D printing materials are usable for a desired application, a substantive analysis of their importance in mechanical, thermal, chemical, and environmental properties given in Figure 13.4 [10]. Mechanical strength, durability, and wear resistance are critical to load-bearing and

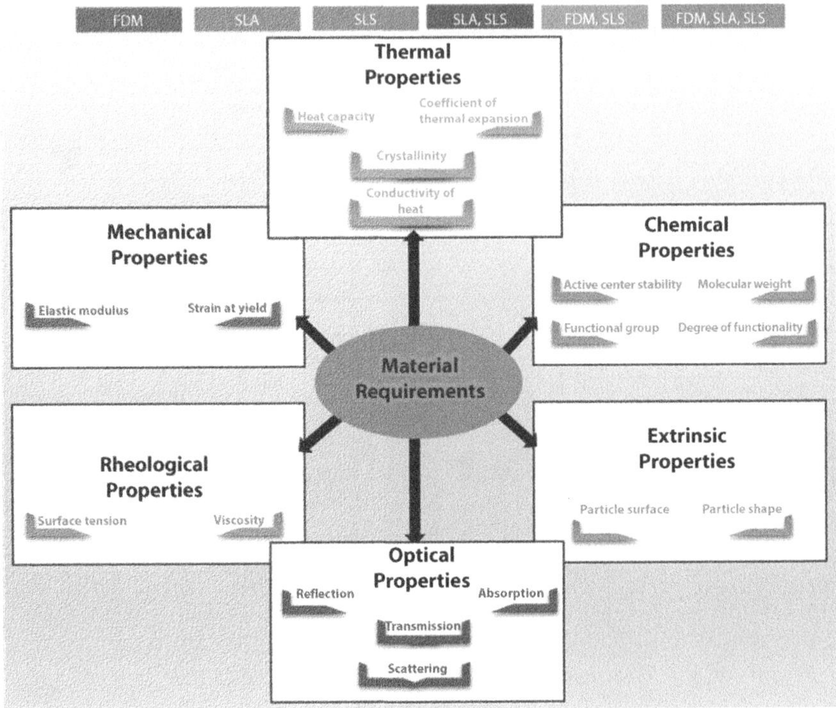

Figure 13.4 Properties of AM materials [10]. [Reproduced under open license].

functional parts. Various types of materials include PLA, ABS, PETG, and nylon with varied performance shown in Figure 13.5 [11]. As example thermal properties, heat resistance and melting point dictate a materials performance abilities in temperature; for instance, PEEK and polycarbonate perform well in heat.

Chemical resistance is just as significant, particularly in hostile environments where we rely on stable materials like nylon and PETG. Concerns related to printability, e.g., ease of extrusion, bed adhesion, warping, etc., contribute to production efficiency. Obviously, PLA is one of the most forgiving materials available. Surface resolution and finish are also important; applications requiring precision or aesthetic attributes are commonly resolved through resin-based materials. In applications where safety is paramount, e.g., medical or food products, we need to be able to guarantee biocompatibility and consider the type of materials we can use [12]. Where environmental aspects are also planned, biodegradability and recyclability are slowly becoming more significant, and PLA is an adapted approach.

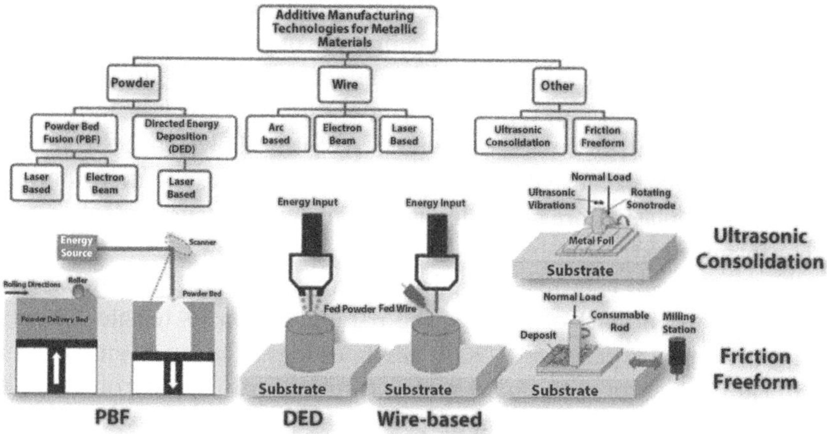

Figure 13.5 Materials used in AM technology [15]. [Reproduced under open license].

The cost of raw materials, and their availability, also influence a material decision, particularly in large-scale or cost-sensitive projects.

13.2.2 Different Types of AM Materials

3D printing and prototyping processes are revolutionizing manufacturing and many industries. One element that contributes to 3D printing's versatility is the number of materials that are available. The physical properties of the object are determined by the materials, and those physical properties will determine the strength, flexibility, durability, and overall look of the finished object, thus it is crucial to select the correct materials in order to achieve the best outcome [13]. Generally speaking, 3D printing materials can be classified into three major categories: ceramics, metals, and plastics. Each major material category includes numerous subcategories that can be applied for specific printing technologies or applications [14]. This study will analyze these classifications of materials and subcategories, discussing the familiar and unique properties, advantages, and disadvantages of 3D printing material properties.

13.2.2.1 Polymer Materials

Polymers are the most common materials used for 3D printing, and they are used because they are cheap, flexible, and versatile and can be used for a variety of AM methods. We see innovations in polymer-based 3D printing material science and manufacturing that range from rapid prototyping to complex medical applications. The major types of polymer materials found in 3D printing include composite polymer, thermoplastic, and thermosetting resin. There are many types of thermoplastic, but two of the most common are Acrylonitrile Butadiene Styrene (ABS) and Polylactic Acid (PLA). PLA is preferred because it is readily recycled and is easy to thermomold. These thermoplastics are generally used in material extrusion (MEX) and other forms of extrusion-based AM processes [16].

Fontana *et al.* [17], investigated the performance of PLA through an analysis of the impact of 3D printing parameters on its tensile strength using design of experiment (DOE) methodology. The study employed a 3x3 factorial design with one response (replication) printing of the PLA tough specimens. Fontana *et al.*, reported the layer thickness was shown to have a more significant influence on tensile than infill percentage of PLA. Fontana *et al.* developed a regression model to show ultimate tensile strength related to both, layer thickness and infill percentage. Tensile, flexural and fatigue test specimens were printed on a MakerBot Replicator 2X. Comparing the samples from these tests, a raster rotation demonstrating the maximum tensile strength (64 MPa) was observed at a 45° rotation, while the maximum flexural strength (102 MPa) was seen at a 0° raster rotation. A 90° raster orientation provided the minimum fatigue life. Fontana *et al.* suggested the mechanical properties of the raw PLA filament were comparable to the printed parts; therefore, they were potential candidates for recycling failed prints. Microscope testing was completed on empty or hollow parts, which showed internal voids attributed to the formational printing process with green PLA filaments [18].

In addition, work has been conducted that indicated that the printed parts of an open-source RepRap 3D printer are as, or better than printed parts manufactured with proprietary systems. A study of a Lulzbot TAZ 3D printer 3D printing PLA with various colors suggested that statistically significant correlations exist between extruder temperature and crystallinity, which are necessary for assessing the behavior of 3D printed material properties [19]. The mechanical behavior of 3D printed materials has been well characterized in the aerospace and biomedical fields but there is still a gap in understanding how the material responds to applied loads. A theoretical model based on transverse isotropy, classical lamination theory, and Hill-Tsai criterion was formulated to predict the tensile strength of FDM PLA. This model was validated by 3D printing specimens at seven angles and seven thicknesses along each angle and were used to predict trends in tensile strength. It was found that the ultimate tensile strength decreased with larger angles and thicker layers. The model could be used to extend toward FDM or SLA materials as well [20].

Ulkir *et al.* [21], discusses that both the material make-up and print parameters can affect the mechanical and physical properties of 3D printed thermoplastics for aerospace applications. One study examined the nozzle temperature on ABS materials for use in drones and other model aircraft. This resulted in an overall decline in tensile strength (by 41.52%). Most importantly, the study found that higher nozzle temperatures had an effect on the viscosity of the materials, and subsequently mass, hardness and density were also lower. Another study examined ABS nanocomposite filaments where the ABS was reinforced with graphene nanoplatelets (GnP) and/or carbon nanotubes (CNTs). The study found that GnP negatively affected tensile and flexural properties, while CNTs improved tensile and flexural properties, but increased brittleness. It was found that higher contents of GnP increased the dielectric constant [22]. A third study used twin-screw extrusion to develop thermoplastic starch (TPS)/ABS biomass alloys to produce black or white filaments, which had superior mechanical and thermal attributes when compared to commercial ABS, with potential for use in 3D printing [23].

In research involving PETG-based composites (i.e., PETG+CF and PETG+KF), the Taguchi method and ANOVA were used to improve bending properties of the composites by adjusting the printing parameters [24]. In addition to this research, studies on 3D printed PETG composites examined the effect of fiber reinforcement on the properties of printed parts for automotive and aeronautical applications. Although compressive strength was reduced due to fiber reinforcement, much more so with Kevlar, the compressive modulus was improved by 12.4% when utilizing carbon fiber. Additionally, fiber filled PETG composites had more stress relaxation and displacements over time [25]. In another study, the tensile and compressive properties of PLA, PETG, PC, and ASA were tested under simulated marine conditions. PC had the greatest resistance to loading post-degradation and PLA had similar properties compared to PETG prior to degradation. Additionally, PC and PETG had the greatest toughness and durability under marine conditions compared to other materials, indicating their suitability for marine environments [26].

13.2.2.2 Ceramic Materials

The scaffolds are characterized by biocompatibility testing, strength, durability, and biocompatible interactions with the host and implant, before implantation. The 3D printed ceramic implants (scaffold) can possibly deliver drugs combined with regeneration of bones (osteogenic properties) and can also deliver drugs and allow for regeneration of cartilage (chondrogenic properties). The ceramic scaffolds are 3D printed as a green body, which is then sintered at high temperature to create a solid structure. For the scaffolds, key elements in the pre-processing are: simulation testing, material selection, design, and scanning. In the scaffold post-processing, key elements include strengthening the implant, and residual removal. Both natural bioceramics and synthetic bioceramics are possible, Table 13.1 demonstrates their performance characteristics and uses [27]. Additionally, the studies explore the mechanical behavior and anisotropy of 3D printed ceramics using various AM processes, including primarily stereolithography (SLA) and digital light projection (DLP).

Li *et al.* [28], analyzed 3D-printed alumina ceramics fabricated by stereolithography, looking at how the presence of additives affected their mechanical properties. The authors reported that increasing the content of TiO_2 improved the flexural strength of sintered alumina from 139.2 MPa to 216.7 MPa. However, there was a significant increase in anisotropic conditions with the addition of 1.5 wt% TiO. With increasing amounts of $CaCO_3$, both shrinkage and flexural strength decreased, likely due to large amounts of residual porosity. While the addition of MgO to alumina reduces anisotropic differences, this work helpfully delineates how flexural strength, shrinkage, and anisotropic attributes are subject to design alterations when printing compacts of alumina ceramic by 3D printer. Similarly, another study created ZrO_2 (zirconia) ceramic bars with stereolithography 3D printing, sintering, and creating assorted dimensions (widths) of zirconia ceramics. When measuring anisotropic attributes showing surface quality, roughness along the fabricated direction was characterized relatively low with 0.41 μm and 1.07 μm respectively making the subject properties further efforts to improve upon. Dimensional accuracy of the ceramics was determined by measuring warpage along with flatness metrics. The ceramic bars achieved flexural strength, fracture toughness, hardness, and density values up to 99.3%. Ultimately the authors determined the anisotropic nature of the behavior to exist in the variances of the scanning paths and difference in the dimensions when printing [29].

Further research focused on the possible use of digital light projection (DLP) 3D printing to fabricate polymer-derived ceramics (PDCs) with complex geometries. The authors sought to study the mechanical anisotropy of these ceramics printed at 2 orientations (0° and 90°). The authors tested both the green-body precursors and final sintered ceramics, and differences in mechanical properties were found dependent on build direction; in this case, the 90°-oriented ceramics were stronger but had a less surviving rate during pyrolysis than the 0°-oriented. The authors reported that build orientation is an important design variable to consider regarding the structural integrity. More work needs to be done to better characterize defect formation in 3D-printed PDCs [30], and their study also emphasized how 3D printing build orientation, and metallic additive and composition properties modulated mechanical performance and the anisotropy of the printed ceramic components.

Table 13.1 Bioceramics and their application [27].

Materials	Description	Application
b-Tricalcium phosphate	Good porosity, low cracking or deformation, low shrinkage, biodegradability, and bio absorption	Hard tissue repair
Silica carbide	Outstanding strength, favorable compressive characteristics, fracture behavior, and microscopic characteristics	Light weight design of structural ceramics
Ferrite	Magnetic behavior, high biocompatibility, and affordability	Magnetic hyperthermia and bone tissue regeneration
Alumina	Chemical inertness, biocompatibility, affordability, and accessibility	Joint replacements, bone spacers, dental implants
Hydroxyapatite	Excellent biocompatibility, compression performance, and cell growth	Clinical repair of bone defects
Zirconium oxide	Good mechanical properties	Bone repair, bone tissue engineering
Calcium silicate	Outstanding osteoconductivity, bioactivity, and biodegradability	Hard tissue regeneration and as coating for metal implants
Barium titanate	Good biocompatibility	Large bone defects
Calcium carbonate	High osteoconductivity, bioactivity, and good biocompatibility	Bone tissue engineering, osteoconductive bone substitutes

13.2.2.3 Metal Materials

The process of Metal 3D printing entails the pigment layer by layer of feedstock to form complex geometries from these materials. Over the past 10-20 years, methods have been refined and commercialized and 3D printing in metals has been widely adopted and used in a variety of sectors: industry, academia and R&D [31]. This article provides a comprehensive overview of metal 3D printing processes by addressing microstructure and mechanical properties, as well as the technical, and qualification, challenges and trends [32]. Yelamanchi *et al.* [33], investigated the composition and mechanical properties of fiber metal laminates (FMLs) with 3D-printed composites for possible aerospace applications where FMLs are utilized for their unique mechanical performance. The authors fabricated continuous carbon fiber reinforced composites from a Markforged system, sandwiched between layers of aluminum alloy. The 3D-printed FMLs were subjected to tensile tests, low-velocity impact test, and high-velocity impact tests. The results suggest that the 3D-printed FMLs exhibit higher tensile strength and impact resistance than standalone composite materials, indicating the potential of 3D printing for laminates that are hybrid.

In a similar vein, Živčák *et al.* [34], chose to consider the mechanical performance of 3D-printed maraging steel samples made by direct metal laser sintering (DMLS). They measured reduced elongation and ultimate tensile strength due to post-process hardening, yet the levels of nickel and titanium were high. The printed samples revealed more brittle neck regions on cross-section analysis; however, this outcome did not produce a significant compromise in production quality. Strength of Ti-6Al-4V products earlier enhanced *via* furnace hardening. Other research findings that have focused on aluminum powder-reinforced polymer matrix composites (PMCs) examined mechanical performance and performance in many areas found that toughness improved with the addition of aluminum suggesting that the addition of aluminum powder in incremental levels will mark a pathway toward the development of polymer-based composites with extreme strength improvements for high-end applications [35].

13.3 A Sustainable AM Technology

Creating a sustainable AM environment takes into account social impacts, economic sustainability, environmental sustainability and process safety. The benefits of 3D printing include lower costs, the least amount of material waste and the ability to recycle while also creating lighter, more energy efficient, and environmentally friendly products. Generally speaking, AM has the potential to alter the industrial landscape by improving resource efficiency and providing opportunities for safe, sustainable production and consumption processes. AM influences business models substantially by providing decentralized and flexible manufacturing processes, which also encourages sustainable development and sustainability goals. While AM can substitute for traditional methods of manufacturing, AM represents its greatest economic value when it substitutes for traditional "customized" production methods. The better the organizations can understand the value of AM the more organizations will implement AM whether directly or through service providers. While the successful realization of the potential benefits from AM is tied to specialized management skills and competencies, and although government policies and investments, as well prior experiences of success associated with rapid prototyping, could assist in facilitating "ease of adoption", organizations must also consider the advances to their organizational design potential in AM. This includes, but is not limited to, AM's digital design fill storage, on demand spare part production, modular design, and a well thought out approach to repair, remanufacturing and refurbishing [36].

13.3.1 Sustainable AM Manufacturing Methods

The objective of sustainable AM approaches is minimizing environmental impacts, at the same time as eliminating waste and improving resource efficiency, throughout the product lifecycle. While the literature indicates a good amount of research has been done surrounding AM technologies, it demonstrates a limited amount of consideration of the wider societal effects. AM has greater design flexibility at a greater level of resource, while traditional welded products use higher input heat, waste more material, and have limited capability of applying complex design geometries [37]. A variety of AM processes and techniques are covered, in addition to electron beam lithography. They include inkjet printing (IJP), stereolithography (SLA), laminated object manufacturing (LOM), selective laser sintering (SLS), and fused deposition modeling (FDM). The review considers some emerging AM techniques such as three-dimensional (3D) printing (3DP), laser engineered net shaping (LENS), and electron beam melting (EBM), in comparison to traditional manufacturing methods. The review outlines positive aspects of AM as presented in Table 13.2, identifying AM as promising opportunity for improved material and resource utilization, enhanced flexibility in production, and increased design adaptation in product development, all of which help in the promotion of more sustainable manufacturing approaches.

Many studies have examined the socio-environmental implications of AM. Huang *et al.* [38], analyzed AM technology from a technical standpoint by comparing it to traditional manufacturing, highlighting advantages, such as flexibility of production, efficiency of materials and resources, and customization of parts. The authors also mentioned disadvantages,

Table 13.2 Advantages of sustainable AM manufacturing methods.

Sustainable AM methods	Advantage	Application
Fused Deposition Modeling (FDM)/ Fused Filament Fabrication (FFF)	Low material waste Uses biodegradable/recycled materials (e.g., PLA, rPET) Low energy consumption	Prototyping Custom jigs/fixtures Educational models
Selective Laser Sintering (SLS)	No support structures needed Recyclable powder Strong, functional parts	Aerospace and automotive parts Functional prototypes Medical devices
Binder Jetting	Low-temperature process (less energy) Works with recycled metal powders Minimal post processing	Sand casting molds Complex metal parts Ceramic components
Direct Energy Deposition (DED)	Efficient metal usage Ideal for repair/ remanufacturing	Turbine blade repair Oil & gas equipment Heavy machinery parts
Stereolithography (SLA)	High precision with minimal waste Biodegradable and low toxicity resins available	Dental/surgical models Jewelry Artistic and wearable components
Material Extrusion with Recycled Filament	Encourages circular economy Reduces plastic waste Local, on-demand production	Custom consumer products spare parts Educational and research use

such as size limits, quality defects, and high costs. Woodson *et al.* [39], discussed the tumbling potential of 3D printing for sustainable industrial transformation using AM technology. However, Woodson *et al.* noted challenges to sustainable commercial viability of AM technology, including inflexible supply chains, health and safety concerns, legal and social challenges, and developmental or usability barriers. Priarone and Ingarao *et al.* [40], compared traditional manufacturing to a hybrid system of AM and finish machining and found that using integrated systems was more energy efficient and produced lower CO_2 emissions than traditional manufacturing.

13.3.2 Sustainable Manufacturing in Environment

Environmental concerns have become increasingly relevant to the manufacturing industry because of the reliance on finite raw materials and energy usage. Applicable to AM techniques such as Direct Metal Deposition (DMD), AM is more sustainable than current practices because it is expected to create less waste, improve resource optimization, and utilize lightweight raw materials and unique AM techniques. As well, researchers have offered design strategies such as A-DfAM and C-DfAM that show consideration for early design with respect to environmental data, cost impact, and improved product quality. As part of this initiative, researchers are exploring the use of lightweight unprocessed raw materials, such as aluminum, or unprocessed iron oxide to promote sustainability and reduce the environmental impact.

A platform that is a fully developed Design for AM (DfAM) framework combines design needs and manufacturing limitations while allowing insights into the development of eco-efficient products. It also compared AM with typical manufacturing processes, and projected future developments in AM, especially in healthcare applications, [41]. It examined considerations around operator health and safety, emissions when materials are processed using extrusion-type technologies, and considerations around the reusability of recycled powders. It evaluated combinations of materials like stainless steel micropowder mixed with cement paste, and Wire Arc AM (WAAM), as assessed through a multi-criteria decision-making process to improve general and commercial sustainability and efficient production [42]. Ongoing research directions should assess the application of artificial intelligence and numerical modeling processes to achieve optimal process parameters, predict materials behaviors and enhance the sustainability framework in relation to AM technologies [43].

13.3.3 Life Cycle Assessment

Life Cycle Assessment (LCA) is an essential tool for assessing the environmental impact of production processes by examining the input and output flows over a product's life cycle [44]. Many studies indicate that powder production and fabrication account for nearly 90% of the environmental impact of traditional machining [45]. Some more in-depth research has been done to determine the main drivers of environmental degradation, such as energy use, materials waste, and tooling [46]. The application of 3D printing in construction is being assessed through case studies of each project phase, and the case studies have provided insights. LCA has been conducted to determine the environmental performance of AM (AM) processes based on mask-image-projection-based stereolithography, focusing on ecosystem and human health impacts.

For crucial purposes like aeronautical turbines, specific LCA frameworks with decision criteria were created around manufacturing decisions. There was also a methodology presented to improve the LCA of AM materials focused on reducing health and environmental risks [47]. LCA studies of the production of a mold core using both traditional and AM processes compared raw material usage, emissions of greenhouse gases, and information about the operational aspects of traditional and AM processes. AM has shown some promise for producing custom pieces and small batch production with great advantages during the usage phase [48]. Some LCA methods and software has been developed to evaluate performance in terms of finance and environment but also including studies or trials to evaluate efficiency of processes and effectiveness of coatings. All of the researchers collected some advanced data on the AM processes, energy usage, and sustainability aspects regarding AM materials, including those from WAAM and wide range of other AM [49].

Energy efficiency analyses have been carried out for processes including LENS, powder bed fusion, and paste extrusion including investigations into powder reuse and waste disposal methods. Environmental evaluations of AM-fabricated medical implants, exemplified by Ti-6Al-4V knee components illustrate the need for sustainable manufacturing practices. Life cycle assessments (LCAs) comparing AM product versions demonstrate that assessments have a meaningful role in assisting industries and policymakers with making informed choices, creating pathways to develop green methods of manufacturing and environmentally responsible innovations [50]. Moreover, the positive attributes of natural fiber-renforced polymer composites in engineering and commercial application have been highlighted maintaining lower environmental impacts, recyclability and significant potential for sustainable product [51].

13.4 Innovation in AM Technology

AM has evolved significantly beginning in the late 1970s, through several innovative inventions. In 1977, Wyn Kelly Swainson demonstrated an early concept of fusion with plastic materials using lasers. In 1981, Dr. Hideo Kodama proposed a technique to build 3D plastic models using photopolymerization. These developments formed the foundation of AM. In 1986, Chuck Hull developed Stereolithography (SLA). This system represented the first practical AM system, consisting of a UV light source to solidify resin, and Chuck Hull started 3D Systems Corporation with it. In 1989, Scott Crump invented Fused Deposition Modeling (FDM) by extruding melted thermoplastics in layers, after which Stratasys was formed. In 1992, Carl Deckard and Joe Beaman proposed Selective Laser Sintering (SLS) to build solid objects from powdered materials using a laser. In the late 1990s, 3D printers were commercially available and opened the doors to accessibility. In 1999, a group of researchers in the medical field at the Wake Forest Institute for Regenerative Medicine created a laboratory-grown bladder using AM methods, which was a new field for biomedical engineers.

In 2002, Object Geometries introduced the PolyJet printing method, which jets and cures liquid photopolymers to create models with a high resolution. The RepRap project (2004), started by Adrian Bowyer, sought to create open-source self-replicating 3D printers, reducing the barrier to entry for the technology. In 2005, designer Patrick Jouin produced the first fully functional piece of furniture (using AM). In 2006, Arcam commercialized the Electron Beam Melting (EBM) process that allowed metal parts to be printed in a vacuum using electron beams.

In 2009, the launch of the BfB RapMan printer represented the first approachable material extrusion kit designed for home use. There was slow uptake of Binder Jetting, which was developed by MIT in 1993, until the 2010's when it began to find traction as an alternative with a quicker, more scalable production option for complex parts. Carbon started scaling AM into resin printing with their 2015 launch of a technology (CLIP) that offered much quicker and smoother resin-based printing. Most recently, developments into multi-material printing and bioprinting in 2023 have led to the creation of tissue constructs and 3D-printed organs being made for transplants. Collectively, AM innovations have increased technical capabilities as well as unique use cases, across healthcare, aerospace, manufacturing, and consumer applications [52].

13.4.1 Recent Innovations

New advances in AM are now evident with transformational technologies such as, High-Speed Sintering technology (HSS), Continuous Liquid Interface Production (CLIP), Laser Powder Bed Fusion (L-PBF), Directed Energy Deposition (DED) or Multi-Material and Functionally Graded AM [53]. HSS is a type of powder bed fusion technology that uses infrared energy to selectively sinter polymer powder which allows for faster build rates but also makes it a suitable candidate for manufacturing functional plastic parts at an industrial scale. CLIP provides high speed printing with high quality surface finishes by employing the use of a combination of light and oxygen to cure liquid resin to solid parts in a 3D printing process. L-PBF is the most precise way of manufacturing, by using a laser to fuse fine metallic powder's part layers to part layers, and creates complex metallic structures with the utmost accuracy. DED has adapted both powder and

Table 13.3 Mechanical field of AM [55].

Filed	Application	Description
Mechanical	Lightweight structural components	Used in aerospace and automotive for weight reduction and fuel efficiency
	Customized tooling, jigs, and fixtures	Improves assembly line precision and reduces lead time
	Heat exchangers and turbine blades	Complex internal geometries for enhanced thermal performance
	Functional prototypes	Rapid design iteration and testing
	Topology-optimized parts	Achieves material efficiency and strength through design optimization
	Functionally graded components	Enables gradual change in material properties within a single part
	Repair and remanufacturing	DED used for repairing high-value metal components

wire-based feedstock systems, adding material capabilities and part performance and surface finish. Many options have emerged for multi-material and functionally graded AM and therefore parts that consist of many compositions/materials and properties can be designed and manufactured within one part for advanced, application-specific designs across fields of study [54]. Tables 13.3 and 13.4. show the characteristics of a sustainable AM process in the mechanical field and biomedical field of applications.

Table 13.4 Biomedical field of AM [55].

Application area	Description	Materials used	Benefits
Prosthetics and Orthotics	Custom-fit prosthetic limbs and orthotic devices designed for individual patients.	PLA, ABS, TPU, Nylon	Lightweight, cost-effective, highly customizable
Cranial and Maxillofacial Implants	Patient-specific implants for skull, jaw, and facial reconstruction.	Titanium, PEEK, PMMA	Anatomical accuracy, biocompatibility, reduced surgery time
Dental Implants and Aligners	Crowns, bridges, surgical guides, dentures, and orthodontic aligners.	Resin, Zirconia, Metal alloys	Precision fit, faster production, personalized solutions
Bone Scaffolds	Porous structures promoting bone tissue growth and regeneration.	Bioceramics, Titanium, PLA, Hydroxyapatite	Enhanced healing, biodegradable, supports bone integration
Surgical Instruments	Customized tools designed for specific surgical procedures.	Stainless steel, Titanium, Nylon	Sterilizable, ergonomic, tailored for surgeons

(*Continued*)

Table 13.4 Biomedical field of AM [55]. (*Continued*)

Application area	Description	Materials used	Benefits
Organ and Tissue Models	Anatomical replicas used for surgical planning, education, and training.	Silicone, Resin, Hydrogels	Accurate visualization, risk-free practice, patient education
Bioprinting of Tissues	Layer-by-layer deposition of living cells to create tissue-like structures.	Bioinks (cells + hydrogels)	Potential for organ regeneration and drug testing
Hearing Aids	Personalized shells and internal components for optimal fit and comfort.	Acrylic resins	Fast turnaround, perfect anatomical fit, high comfort
Drug Delivery Systems	3D-printed pills and capsules with controlled release mechanisms.	Biodegradable polymers	Personalized dosing, controlled release, improved compliance

13.5 Conclusion

AM has gained traction in industries like aerospace, automotive, healthcare, and food production— thanks to the advantages of customization and efficiency in production. The development of materials, processes, and design methodologies has been even more helpful to the advancement of AM within the manufacturing systems. AM promotes sustainability by reducing input production input through lightweight structures, providing longevity of products, modular designs, local production, and recycling in closed-loop value chains. Sustainability when implementing AM can be split into 4 broad categories: redesigning products and processes, material input processing, make-to-order manufacturing, and closed-loop systems.

AM demonstrates further potential in operational processes in mechanical and biomedical applications, promoting reduced harm to the environment. Although AM shows promise, there remain issues with operational costs, available materials for inputs and process efficiency that need to be addressed before any further vigorous possibilities are explored to promote social, environmental, and industrial sustainability.

AI disclosure statement: The author acknowledges the use of AI-based tools for minor language editing or grammatical correction. However, all intellectual contributions, analyses, and conclusions presented in this chapter are entirely the author's own.

References

1. Ravi, G., Murugan, N., Arulmani, R., Microstructure and mechanical properties of Inconel-625 slab component fabricated by wire arc additive manufacturing. *Mater. Sci. Technol. (United Kingdom)*, 36, 16, 1785–1795, 2020.
2. Ajithkumar, S. and Arulmurugan, B., Investigation of Single and Double Pulse Techniques on Microstructure and Mechanical Anisotropy of Inconel 686 Component Fabricated by Wire Arc Directed Energy Deposition. *Met. Mater. Int.*, 31, 1648–1673, 2024.
3. Qi, X., Li, Y., Li, F., Du, J., Niu, J., Fan, W., *et al.*, Regulating strength and ductility of additively manufactured Inconel 718 alloy *via* adding nano-TiC and deep cryogenic treatment. *Mater. Sci. Eng., A*, 910, 1, 46852, 2024.
4. Prajapati, D.K. and Kumar, R., Additive Manufacturing Sustainability in Industries. *Adv. Sci. Eng. Med.*, 12, 894–9, 2020.
5. Alghamdi, S.S., John, S., Choudhury, N.R., Dutta, N.K., Additive manufacturing of polymer materials: Progress, promise and challenges. *Polymers (Basel)*, 13, 5, 753, 2021.
6. Sitharaj, A., Arulmurugan, B., Veeraraghavan, V., Idhayaraja, S., Vivekanandh, S.R., Kavin, M., *et al.*, Hybridization in Wire Arc Directed Energy Deposition of Titanium Alloys: A Review, pp. 441–65, 2005.
7. Wang, H., Du, W., Zhao, Y., Wang, Y., Hao, R., Yang, M., Joints for treelike column structures based on generative design and additive manufacturing. *J. Constr. Steel Res.*, 184, 2021.
8. Gopal, M., Lemu, H.G., Gutema, E.M., Sustainable Additive Manufacturing and Environmental Implications: Literature Review. *Sustainability (Switzerland)*, 15, 1, 504, 2023.
9. Wong, K.V. and Hernandez, A., A Review of Additive Manufacturing. *ISRN Mech. Eng.*, 2012, 1, 1–10, 2012.

10. Kafle, A., Luis, E., Silwal, R., Pan, H.M., Shrestha, P.L., Bastola, A.K., 3d/4d printing of polymers: Fused deposition modelling (fdm), selective laser sintering (sls), and stereolithography (sla). *Polymers (Basel)*, 13, 18, 3101, 2021.
11. Kopar, M. and Yildiz, A.R., Experimental investigation of mechanical properties of PLA, ABS, and PETG 3-d printing materials using fused deposition modeling technique. *Materialpruefung/Mater. Test.*, 65, 12, 1795–1804, 2023.
12. Vinyas, M., Athul, S.J., Harursampath, D., Nguyen Thoi, T., Experimental evaluation of the mechanical and thermal properties of 3D printed PLA and its composites. *Mater. Res. Express*, 6, 11, 115301, 2019.
13. Kumari, S., Vishwakarma, P., Abhishek, K., Advancements in 3D Printing Materials for Diverse Industries: A Review and Future Prospects. *E3S Web Conf.*, 552, 01038, 2024.
14. Sitharaj, A., Arulmurugan, B., Dharani Kumar, S., Idhayaraja, S., Manishraj, D., Premkumar, N., Wire-Arc Directed Energy Deposition of Aluminum Alloys: A Review on Properties, Challenges, and Applications, pp. 99–115, 2025.
15. Srivastava, M., Rathee, S., Patel, V., Kumar, A., Koppad, P.G., A review of various materials for additive manufacturing: Recent trends and processing issues. *J. Mater. Res. Technol.*, 21, 1, 2612–2641, 2022.
16. Ligon, S.C., Liska, R., Stampfl, J., Gurr, M., Mülhaupt, R., Polymers for 3D Printing and Customized Additive Manufacturing. *Chem. Rev.*, 117, 15, 10212–10290, 2017.
17. Fontana, L., Minetola, P., Iuliano, L., Rifuggiato, S., Khandpur, M.S., Stiuso, V., An investigation of the influence of 3d printing parameters on the tensile strength of PLA material. *Mater. Today Proc.*, 57, 2022.
18. Letcher, T. and Waytashek, M., Material property testing of 3D-printed specimen in pla on an entry-level 3D printer. *ASME International Mechanical Engineering Congress and Exposition, Proceedings (IMECE)*, vol. 2A, 2014.
19. Wittbrodt, B. and Pearce, J.M., The effects of PLA color on material properties of 3-D printed components. *Addit. Manuf.*, 8, 2015.
20. Yao, T., Deng, Z., Zhang, K., Li, S., A method to predict the ultimate tensile strength of 3D printing polylactic acid (PLA) materials with different printing orientations. *Compos. B Eng.*, 163, 2019.
21. Ulkir, O., Ertugrul, I., Ersoy, S., Yağımlı, B., The Effects of Printing Temperature on the Mechanical Properties of 3D-Printed Acrylonitrile Butadiene Styrene. *Appl. Sci.*, 14, 3376, 2024.
22. Vidakis, N., Maniadi, A., Petousis, M., Vamvakaki, M., Kenanakis, G., Koudoumas, E., Mechanical and Electrical Properties Investigation of 3D-Printed Acrylonitrile–Butadiene–Styrene Graphene and Carbon Nanocomposites. *J. Mater. Eng. Perform.*, 29, 2020.
23. Kuo, C.C., Liu, L.C., Teng, W.F., Chang, H.Y., Chien, F.M., Liao, S.J., *et al.*, Preparation of starch/acrylonitrile-butadiene-styrene copolymers (ABS) biomass alloys and their feasible evaluation for 3D printing applications. *Compos. B Eng.*, 86, 36–39, 2016.

24. Valvez, S., Silva, A.P., Reis, P.N.B., Optimization of Printing Parameters to Maximize the Mechanical Properties of 3D-Printed PETG-Based Parts. *Polymers (Basel)*, 14, 13, 2564, 2022.
25. Valvez, S., Silva, A.P., Reis, P.N.B., Compressive Behaviour of 3D-Printed PETG Composites. *Aerospace*, 9, 2022.
26. Ramírez-Revilla, S., Camacho-Valencia, D., Gonzales-Condori, E.G., Márquez, G., Evaluation and comparison of the degradability and compressive and tensile properties of 3D printing polymeric materials: PLA, PETG, PC, and ASA. *MRS Commun.*, 13, 2022.
27. Ly, M., Spinelli, S., Hays, S., Zhu, D., 3D Printing of Ceramic Biomaterials. *Eng. Regener.*, 3, 2022.
28. Li, H., Elsayed, H., Colombo, P., Enhanced mechanical properties of 3D printed alumina ceramics by using sintering aids. *Ceram. Int.*, 49, 2023.
29. Xing, H., Zou, B., Li, S., Fu, X., Study on surface quality, precision and mechanical properties of 3D printed ZrO2 ceramic components by laser scanning stereolithography. *Ceram. Int.*, 43, 2017.
30. Brinckmann, S.A., Young, J.C., Fertig, R.S., Frick, C.P., Effect of print direction on mechanical properties of 3D printed polymer-derived ceramics and their precursors. *Mater. Lett.: X*, 17, 2023.
31. Ajithkumar, S., Arulmurugan, B., Idhayaraja, S., Robotic Welding and Automation: Cutting-Edge Technology for Industry, in: *Advanced Welding Technologies*, pp. 497–512, Wiley, 2025.
32. Das, S., Bourell, D.L., Babu, S.S., Metallic materials for 3D printing. *MRS Bull.*, 41, 2016.
33. Yelamanchi, B., Macdonald, E., Gonzalez-Canche, N.G., Carrillo, J.G., Cortes, P., The mechanical properties of fiber metal laminates based on 3d printed composites. *Materials*, 13, 2020.
34. Živčák, J., Nováková-Marcinčínová, E., Nováková-Marcinčínová, L., Balint, T., Puškár, M., Increasing mechanical properties of 3D printed samples by direct metal laser sintering using heat treatment process. *J. Mar. Sci. Eng.*, 9, 2021.
35. Ford, S. and Despeisse, M., Additive manufacturing and sustainability: an exploratory study of the advantages and challenges. *J. Cleaner Prod.*, 137, 2016.
36. Sitharaj, A., Arulmurugan, B., Balaji, D., Idhayaraja, S., Additive Manufacturing of Titanium-based Alloys by Wire Arc Directed Energy Deposition (WA-DED), in: *Advances in Solid-State Welding and Processing of Metallic Materials*, pp. 400–17, CRC Press, Boca Raton, 2025.
37. Ajithkumar, S., Arulmurugan, B., Balaji, D., Prasath, V.M.M., Idhayaraja, S., Hybrid Welding Processes, in: *Advanced Welding Technologies*, pp. 177–96, Wiley, 2025.
38. Huang, S.H., Liu, P., Mokasdar, A., Hou, L., Additive manufacturing and its societal impact: A literature review. *Int. J. Adv. Manuf. Technol.*, 67, 2013.

39. Woodson, T.S., 3D printing for sustainable industrial transformation. *Development (Basingstoke)*, 58, 2015.
40. Priarone, P.C. and Ingarao, G., Towards criteria for sustainable process selection: On the modelling of pure subtractive versus additive/subtractive integrated manufacturing approaches. *J. Cleaner Prod.*, 144, 2017.
41. Priyadarshini, J., Singh, R.K., Mishra, R., Dora, M., Application of additive manufacturing for a sustainable healthcare sector: Mapping current research and establishing future research agenda. *Technol. Forecasting Social Change*, 194, 122686, 2023.
42. Bekker, A.C.M., Verlinden, J.C., Galimberti, G., Challenges in assessing the sustainability of wire + arc additive manufacturing for large structures. *Solid Freeform Fabrication 2016: Proceedings of the 27th Annual International Solid Freeform Fabrication Symposium - An Additive Manufacturing Conference, SFF 2016*, 2016.
43. Ajithkumar, S., Arulmurugan, B., Meiyazhagan, R., Welding of nickel alloys interface by finite element modeling and artificial intelligence and envisage by patent landscape. *Int. J. Interact. Des. Manuf. (IJIDeM)*, 2025.
44. Rajeshkumar, L., Ajithkumar, S., Arulmurugan, B., Bhuvaneswari, V., Life cycle assessment of mineral fibers and their composites, in: *Synthetic and Mineral Fibers, Their Composites and Applications*, pp. 631–55, Elsevier, 2024.
45. Faludi, J., Bayley, C., Bhogal, S., Iribarne, M., Comparing environmental impacts of additive manufacturing vs traditional machining *via* life-cycle assessment. *Rapid Prototyping J.*, 21, 2015.
46. Plantamura, O.I., Is 3D Printed House Sustainable? *Cisbat*, vol. 9, 2015.
47. Kokare, S., Oliveira, J.P., Godina, R., Life cycle assessment of additive manufacturing processes: A review. *J. Manuf. Syst.*, 68, 2023.
48. Tang, Y., Mak, K., Zhao, Y.F., A framework to reduce product environmental impact through design optimization for additive manufacturing. *J. Cleaner Prod.*, 137, 2016.
49. Angelini, E., De Benedetti, B., Grassini, S., Marino, M., Life cycle assessment of coating treatments for automotive magnesium parts. *Mater. Sci. Forum*, 426, 2003.
50. Dostatni, E., Dudkowiak, A., Rojek, I., Mikołajewski, D., Environmental analysis of a product manufactured with the use of an additive technology – AI-based vs. traditional approaches. *Bull. Pol. Acad. Sci. Tech. Sci.*, 71, 1, e144478, 2023.
51. Ajithkumar, S., Arulmurugan, B., Rajeshkumar, L., Prospects of natural fiber-reinforced polymer composites in engineering and commercial applications, in: *Applications of Composite Materials in Engineering*, pp. 335–64, Elsevier, 2025.
52. Pei, E., Kabir, I.R., Leutenecker-Twelsiek, B., History of AM, in: *Springer Handbooks of Additive Manufacturing*, pp. 3–29, Cham: Springer International Publishing, 2023.

53. Ajithkumar, S. and Arulmurugan, B., Microstructure and mechanical properties of Inconel 686 fabricated by gas metal arc welding-based wire arc directed energy deposition: impact of cryogenic treatments. *J. Mater. Sci., 59*, 40, 19273–19302, 2024.
54. Fidan, I., Huseynov, O., Ali, M.A., Alkunte, S., Rajeshirke, M., Gupta, A., *et al.*, Recent Inventions in Additive Manufacturing: Holistic Review. *Inventions*, 8, 4, 103, 2023.
55. Khanlar, L.N., Rios, A.S., Tahmaseb, A., Zandinejad, A., Additive manufacturing of zirconia ceramic and its application in clinical dentistry: A review. *Dent. J. (Basel)*, 9, 9, 104, 2021.

14

Semi-Solid Metal Casting: A Sustainable Approach for High-Performance Component Fabrication

Arul Pandian Palanisamy[1]*, Ganesh Nataraj[1] and Ajithkumar Sitharaj[2]

[1]Department of Mechanical Engineering, Sri Shakthi Institute of Engineering and Technology, Coimbatore, Tamil Nadu, India
[2]Department of Mechanical Engineering, KPR Institute of Engineering and Technology, Coimbatore, Tamil Nadu, India

Abstract
Semi-solid metal casting (SSMC) is a novel manufacturing process that functions in the temperature regime between the solidus and liquidus condition of metals where both solid and liquid phases coexist. This special situation generates thixotropic behavior, which allows improved control of flow and dimensional accuracy. By taking advantage of both traditional casting and forging, SSMC has great advantages, including improved porosity reduction, increased mechanical properties, and enhanced microstructural homogeneity. One of the strongest points of SSMC is the fact that it aligns with the goals of sustainable manufacturing. It uses lower temperatures to process; and hence, it has lower energy and emission requirements. Furthermore, the near-net-shape nature of the process lessens material wastage and machining needs, further increasing its environmental effectiveness. This chapter explores the basic concepts, process types (thixocasting and rheocasting), and materials applied in SSMC, with an emphasis on the use of aluminum and magnesium alloys. The chapter also brings into focus the sustainability indicators like lifecycle analysis and die life optimization. Applications of SSMC cut across different high-performance industries like automotive, aerospace, and biomedical engineering. Regardless of its benefits, limitations include high equipment cost and process complexity. The chapter ends with future directions in material development, hybrid manufacturing integration, and wider application of SSMC in environmentally friendly industrial operations.

**Corresponding author*: arulpsg02@gmail.com

Gurudas Mandal, Swarup Kumar Ghosh, Snehanshu Pal and Sandip Kunar (eds.) Sustainable Manufacturing of Advanced Materials, (293–310) © 2026 Scrivener Publishing LLC

Keywords: Semi-solid casting, sustainability, high-performance components, rheocasting, thixocasting

14.1 Introduction

Over the past few years, the global move toward sustainable production practices has gathered momentum because of increased environmental worries, depletion of resources, and strict regulation frameworks. The manufacturing sector, which usually gobbles up a lot of energy and raw materials, is under greater pressure to embrace processes that bring down carbon emissions, material use, and environmental footprint. As part of this shift, the investigation of advanced and sustainable metal processing methods has emerged as a research priority in academic and industrial fields.

Semi-solid metal casting (SSMC) has been identified as a viable option within this framework. Situated between traditional casting and forging operations, SSMC takes advantage of the semi-solid condition of metal alloys—where solid and liquid phases coexist—to obtain desirable mechanical and microstructural properties in the finished product. Such an unusual state, generally realized by heating the alloy in the solidus to liquidus range, facilitates thixotropic flow behavior, improves mold filling, less turbulence, and fewer casting defects than in the fully liquid casting methods.

The SSMC process is generally categorized into two basic routes (shown in Figure 14.1): thixocasting with pre-cast and heat-treated billets with

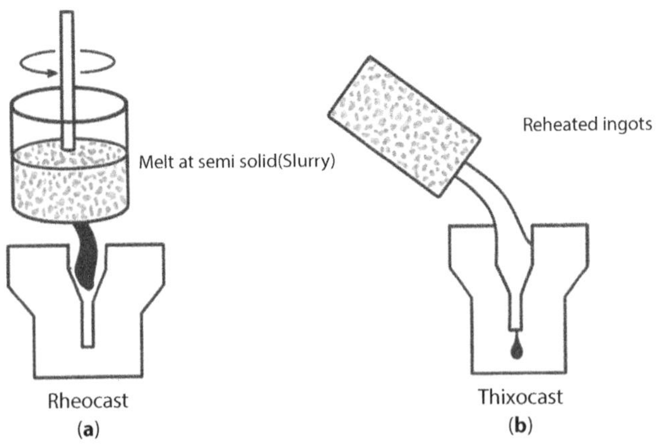

Figure 14.1 Types of semi-solid casting methods: (a) Rheocast (b) Thixocast.

microstructure control; and rheocasting with direct production of the semi-solid slurry from molten metal. Both methods are intended to create near-net-shape parts with enhanced dimensional integrity, less porosity, and finer grain microstructures. Such properties are most beneficial in applications demanding high mechanical strength, close tolerance, and improved fatigue resistance, including automotive suspension components, aerospace brackets, and orthopedic implants [1, 2].

From a sustainability standpoint, SSMC has some inherent benefits over traditional manufacturing methods. The reduced operating temperatures involved in SSMC cut energy consumption substantially, leading to lower costs of production and carbon footprint. Also, the lessened necessity of secondary processes like machining means less material loss and longer life dies. These features render SSMC as a prime contender for assimilation into sustainable production systems and circular economy models [3, 4].

A further major advantage of SSMC is its application with lightweight metals like aluminum and magnesium, which find greater usage in automotive and aerospace industries for reducing weight and enhancing fuel efficiency. The fact that the process can form parts with better mechanical performance from these types of materials further increases the environmental and economic benefits of the process [5].

Despite its merits, the application of SSMC on a wider industrial level encounters some challenges. They are high initial equipment cost, restricted alloy choice, and stringent temperature and process control requirements. In addition, standardization of process parameters and the dissemination of knowledge to a greater extent are necessary to achieve the full potential of SSMC in commercial manufacturing operations [6].

This chapter gives a systematic summary of semi-solid metal casting, starting with its basic principles and categories, and going on to materials, process routes, and indicators of sustainability. Its applications in high-performance industries are presented, followed by present constraints and research directions for overcoming current limitations. Through this discourse, this chapter seeks to bring out the revolutionary potential of SSMC as a sustainable and high-performance manufacturing method of the future.

14.2 Fundamentals of Semi-Solid Metal Casting

Semi-solid metal casting (SSMC) is a near-net-shape manufacturing process that functions in the temperature regime between the solidus and liquidus of a metal alloy, producing a combination of solid and liquid phases.

Under this novel state lies a thixotropic slurry—a fluid that deforms under shear stress but maintains its shape when at rest—enabling better mold filling, lower porosity, and better dimensional stability [7, 8]. The decrease in turbulence in flow reduces gas entrapment and formation of oxides, thereby leading to the improvement in the mechanical strength of the end component [9].

(i) Classification

SSMC processing is generally categorized into thixocasting and rheocasting based on the method of preparation of the semi-solid slurry. Thixocasting requires specially prepared non-dendritic billets, which are heated back to semi-solid condition and pressure-injected into dies. The process has very good repeatability and structure consistency but is more expensive as a result of billet preparation [10]. Conversely, rheocasting creates the semi-solid slurry directly from molten metal through controlled agitation and cooling. This process is economical and versatile but needs tight control over solid fraction and grain morphology for quality assurance [11, 12].

(ii) Advantages

- Improved dimensional accuracy
- Reduced porosity and shrinkage
- Enhanced mechanical properties
- Lower forming forces and die wear

14.3 Process Routes and Materials

The process of semi-solid metal casting normally involves three primary steps: material preparation, reheating, and die casting. In the material preparation phase, various techniques are applied to obtain the required semi-solid microstructure, such as mechanical stirring, electromagnetic stirring, cooling slope technique, spray forming, and magneto-hydrodynamic (MHD) stirring. These methods are employed to produce a fine, non-dendritic microstructure with controlled solid fraction. Reheating is subsequently done to achieve the semi-solid state of the material and then high-pressure injection into a die to produce the final product [13, 14].

(i) Suitable Alloys

Aluminum and magnesium alloys have the widest application in SSMC because of their superior castability, light weight, and strength-to-weight ratio. Aluminum alloys like A356, A357, and AA7075 are extensively employed in thixocasting and rheocasting owing to their suitability in the semi-solid condition as well as in their opportunities for use in the automotive and aeronautical sectors. Magnesium alloys such as AZ91D are also preferred because of its lower weight and sufficient strength, which makes them suitable for weight-constrained components [15, 16].

Copper and steel alloys have been examined for semi-solid processing but possess challenges like limited semi-solid temperature range, elevated melting point, and oxidation. Improvements in alloy design and sophisticated processing techniques are being continuously developed to expand the material scope for SSMC [17].

(ii) Solid Fraction and Grain Morphology

The control of solid fraction and grain morphology is critical to the quality and performance of SSMC products. Typically, the solid fraction is maintained between 30% and 60%, depending on the desired properties and process parameters. A lower solid fraction increases fluidity but may compromise shape retention, while a higher fraction enhances mechanical integrity but may reduce mold filling capability [18].

Achieving a fine, spherical (non-dendritic) grain structure is essential for consistent flow behavior and defect-free casting. Mechanical and electromagnetic stirring techniques are commonly used to fragment and spheroidize dendritic grains during the slurry preparation stage. Cooling slope casting and ultrasonic treatment have also been employed to refine grain morphology and enhance uniformity [19, 20].

Ultrasonic vibration during solidification improves slurry homogeneity by breaking up dendrites and promoting the formation of fine, globular grains. The combination of shear and thermal conditions during slurry preparation influences the final microstructure and mechanical properties of the component. The resulting structure enhances fluidity, reduces porosity, and contributes to improved mechanical performance [21].

Additionally, advancements in process modeling and real-time monitoring systems are being integrated to optimize grain size distribution and solid fraction during casting. These developments allow for better control over slurry rheology and mold filling behavior, ultimately leading to improved product quality and process repeatability [22, 23]. Recent studies have also investigated the synergistic effects of multi-scale grain refinement techniques, combining electromagnetic stirring with ultrasonic vibration to achieve more uniform and finer grain structures. These hybrid techniques show promising results in improving the mechanical properties and surface finish of the final product [24]. Furthermore, researchers are exploring novel slurry preparation methods, such as chemical inoculation and controlled cooling paths, to expand the range of alloys processable by SSMC [25].

The integration of artificial intelligence (AI) and machine learning (ML) algorithms into the process control systems has opened new possibilities for predictive modeling and adaptive process optimization. AI-driven frameworks can analyze real-time sensor data to regulate temperature, stirring rate, and solid fraction, thereby minimizing defects and ensuring product consistency [26]. Overall, the combination of alloy selection, precise control over solid fraction, and advanced grain refinement strategies plays a pivotal role in the successful implementation of SSMC across various industrial sectors. Continued advancements in materials science, process engineering, and digital manufacturing will further enhance the reliability and scalability of SSMC in the future [27].

14.4 Sustainability Aspects

The implementation of sustainability in contemporary manufacturing has been driving innovation toward cleaner and more efficient processes. Semi-solid metal casting (SSMC) is an example of this trend with several environmental and economic advantages. SSMC is better than conventional casting and forging technologies as it encourages the conservation of energy, use of resources, and lesser environmental footprint, qualifying it as a green manufacturing technology.

One of the highest-profile sustainability advantages of SSMC is that it is energy efficient. Since the process has only a limited temperature range between solidus and liquidus, the amount of energy used to keep the semi-solid state is much lower than in traditional casting or melting [28]. This means that less energy is used and there are fewer greenhouse gases emitted, which are crucial industrial sustainability metrics.

SSMC also provides better material utilization. The near-net-shape component production capability significantly minimizes secondary machining needs, reducing raw material loss and scrap generation. In conventional casting or machining, much of the material is usually removed and wasted in post-processing. SSMC shapes parts in closer proximity to their final dimensions, thus saving material and lowering operational expenditure [29].

Cleaner production is another critical aspect of SSMC. Less oxides and impurities are created with the lower process temperature and less turbulence in slurry injection. The limited application of lubricants and coolants also reduces environmental contamination. The process is less industrial waste-producing, reduces dangerous emissions, and reduces the use of hazardous chemicals to acceptable levels, meeting ISO 14000 environmental management standards [30].

Longer die life helps in sustainability as it decreases tooling expenses and maintenance requirements. Because SSMC entails lower temperature and pressure than hot forging or high-pressure die casting, thermal and mechanical fatigue of the dies is significantly alleviated. This results in longer tool service life, less replacement, and reduced downtime, which are major contributors to overall manufacturing sustainability [31].

Life Cycle Assessment (LCA) research comparing SSMC with conventional manufacturing processes always reveal a reduced environmental impact. The studies compare the energy inputs, emissions, and wastes of a product throughout its entire life cycle from raw material to end-of-life disposal. Findings from recent LCA studies show that SSMC can decrease the overall carbon footprint of a component by 20–40% because of savings in energy, less scrap, and negligible post-processing [32].

Additionally, SSMC promotes lightweighting efforts, especially with the utilization of aluminum and magnesium alloys due to their low weight and recyclability. Lightweight components promote fuel efficiency in automotive, and aerospace uses, which leads to indirect environmental benefits during the product life cycle [33]. Finally, SSMC offers a holistic response to sustainable production. Its built-in benefits, e.g., energy conservation, decreased material loss, less emissions, and longer tool life—make it an excellent option for companies that wish to conform to international environmental objectives. With sustainability growing more integral to industrial planning, the implementation of SSMC can be a crucial step toward both economic effectiveness and environmental accountability [34].

14.5 Mechanical and Microstructural Characteristics

The microstructural and mechanical properties of SSMC parts are remarkably better than those produced by conventional casting processes. These are due to controlled solidification, minimized turbulence, and special slurry behavior in semi-solid processing. Parts produced using SSMC exhibit superior tensile and yield strength, better fatigue life, and highly refined microstructure—properties that make them favorable for safety-critical and high-load industrial applications [35].

The most significant of the mechanical benefits of SSMC is the attainment of high yield and tensile strength in the components. This results from the uniform, fine grain structure and lower casting defects including cold shuts and shrinkage porosity. Low turbulence flow of the semi-solid slurry reduces gas and inclusion entrapment, which are normally stress concentrators and decrease the mechanical integrity. Experimental data on aluminum alloy A356 indicate that thixocast components have a tensile strength up to 20–30% greater than for gravity-cast equivalents [36].

Fatigue resistance is also an important mechanical property that is improved by SSMC. The finer microstructure with reduced inclusions and increased grain boundary bonding leads to enhanced cyclic loading resistance and crack growth resistance. Because fatigue failure tends to initiate from surface defects and internal porosity, near-defect-free structure of SSMC components significantly enhances their fatigue life. This renders SSMC extremely suitable for dynamic components like suspension arms, engine mounts, and aerospace structural brackets [37].

Microstructurally, SSMC produces fine, globular (non-dendritic) grain microstructures, as opposed to traditional castings that typically exhibit coarse, dendritic microstructure. The change is brought about using shear stirring, electromagnetic stirring, or cooling slope methods during the slurry preparation step. The spherical grain microstructure produces isotropic mechanical response and maintains properties invariant with changing orientations of the component [38].

Moreover, segregation and porosity are greatly minimized in SSMC. Segregation, normally the result of non-uniform cooling rates in conventional casting, results in concentration gradients of alloying elements. The uniform solidification rate in SSMC reduces the occurrence of this problem, resulting in more uniform compositions. Porosity, the result of gas entrapment or shrinkage during solidification, is also minimized because of the lower fluid flow velocity and better feeding in the semi-solid state [39].

Mechanical properties of SSMC parts can also be controlled by optimizing processing parameters including solid fraction, cooling rate, and injection pressure. For example, a solid fraction between 40–60% guarantees proper flowability as well as mechanical performance. Additionally, accurate control of temperature during reheating and die filling avoids over-shearing and overheating of the slurry, which degrades the final microstructure [40].

Heat treatment also increases the mechanical properties of SSMC components. T6 heat treatment of thixocast aluminum components, for instance, can greatly improve hardness and strength through precipitation hardening. Research indicates that peak-aged A357 alloy components have better wear resistance and load-carrying capacity following heat treatment and are comparable to wrought components in most applications [41].

Fractography of SSMC fracture test pieces shows high toughness and energy-absorbing ability through ductile fracture surfaces with deep dimples. The findings are consistent with tensile and impact test results, validating the high reliability of semi-solid processed materials [42].

SSMC's microstructural control also leads to improved machinability and surface finish over conventional castings. The fine-grain, homogeneous microstructure results in less tool wear and tighter tolerances for secondary operations. This is especially beneficial for parts with high dimensional accuracy and surface integrity, including transmission cases and pump housing [43].

Additionally, the conjunction of improved mechanical properties and better microstructural characteristics in SSMC render it very appropriate for lightweight structural applications. Replacement of heavier components with semi-solid processed magnesium or aluminum parts in automotive and aerospace sectors leads to weight savings, fuel savings, and emissions decrease without reduction in strength or safety levels [44].

Evolving research in semi-solid processing is also considering the inclusion of reinforced metal matrix composites (MMCs) to enhance mechanical behavior further. Adding ceramic particles like SiC or Al2O3 into the SSMC matrix can achieve hybrid composites with superior wear resistance, rigidity, and thermal stability. These advancements open the potential application scope of SSMC to more stringent mechanical environments [45].

Overall, the synergism of high mechanical strength, fatigue resistance, microstructural uniformity, and process versatility makes SSMC a viable substitute for conventional manufacturing processes. With ongoing enhancements in alloy design, control of slurry, and post-processing technologies, SSMC will likely dominate the manufacture of next-generation high-performance components in industries ranging from consumer electronics to defense and aerospace [46].

14.6 Applications in High-Performance Components

Semi-solid metal casting (SSMC) process has achieved extensive usage in those industries where mechanical performance, weight savings, and process efficiency are of maximum concern. Because of its capability to cast near-net-shaped components with high dimensional accuracy and superior mechanical properties, SSMC is progressively being incorporated into the manufacturing plans of automotive, aerospace, biomedical, defense, and energy industries. These industries require materials and processes that perform to high standards while conforming to worldwide sustainability targets [47].

(i) Automotive Applications

The vehicle industry was one of the first to embrace SSMC technology. Thixocasting and rheocasting processes have successfully been used in manufacturing parts such as brake calipers, engine brackets, steering knuckles, suspension arms, and transmission cases. The parts are advantageously provided through semi-solid processing with a high strength-to-weight ratio and porosity reduction, which are immensely important for vehicular safety and performance [48].

For example, brake calipers fabricated with SSMC exhibit enhanced fatigue strength and improved dimensional stability relative to gravity-cast ones. The finer microstructure obtained in SSMC improves brake performance and cyclic loading durability. Additionally, the potential for the employment of lightweight aluminum alloys leads to vehicle weight reduction overall, which improves fuel economy and minimizes emissions [49].

Automakers have also documented substantial material waste and machining time savings as a result of SSMC's near-net-shape forming characteristics. Case histories involving major auto makers like BMW and General Motors show that parts produced by semi-solid methods can meet or exceed the mechanical performance of forged or machined parts while reducing production cost and energy use [50].

(ii) Aerospace Applications

In the aerospace industry, the requirement for lightweight and high-strength parts is even more essential because of the fuel efficiency and load-carrying requirements. SSMC has been found to be effective for manufacturing aerospace brackets, actuator housings, seat structures, and mounting interfaces. The parts need to endure harsh mechanical loads and temperature fluctuations without loss of integrity [51].

With SSMC, manufacturers can produce complex geometries with less porosity and uniform grain structures, leading to improved fatigue life and structural reliability. Aerospace brackets made of semi-solid processed aluminum alloys, for instance, have consistent mechanical performance and better dimensional stability under service stresses. Moreover, the fewer number of steps in manufacturing reduces production time and minimum quality control problems [52]. Continuing research and development activities are investigating the use of magnesium alloys and advanced composites in aerospace SSMC to decrease component weight further. These developments have the potential to advance the contribution of semi-solid casting to future aircraft and space vehicles [53].

(iii) Biomedical Applications

Biomedical engineering is one more area wherein the special strengths of SSMC are increasingly being investigated. Technology facilitates the production of high-strength, biocompatible parts including orthopedic implants, dental structures, surgical devices, and artificial joints [54].

The delicate, non-dendritic microstructure and porosity of SSMC components lead to improved load-carrying capacity and increased fatigue resistance—critical requirements for biomedical implants that undergo repeated mechanical loading in the human body. Additionally, the accuracy in dimensions of SSMC minimizes the requirement of heavy post-processing, which is crucial for individualized implants and complex geometries [55].

Recent research has indicated that cobalt-chromium and titanium semi-solid processed alloys can be employed as hip and knee implants with mechanical performance comparable to, or better than, conventionally produced equivalents. Further, SSMC research with biodegradable magnesium-based alloys promises temporary implants that degrade naturally within the body, alleviating the necessity for reoperation [56].

(iv) Defense and Energy Applications

SSMC also finds increasing applications in defense and energy industries where structural integrity, light weight, and high performance under severe conditions are essential. Some of the military and defense applications of structural elements are drone parts, weapon housing, structural components, and protective gear frames. These components usually demand a synergy of impact resistance, fatigue strength, and corrosion protection [57].

In the energy industry, SSMC has been used to produce wind turbine components, heat exchangers, fuel cell enclosures, and electrical enclosures. These processes require materials with the ability to manage high thermal and mechanical stress while holding dimensional accuracy. The energy and environmental sustainability of the SSMC process are beneficial to the clean energy revolution, providing a technological advantage over conventional processes [58].

(v) Comparative Case Studies

Several case studies have established that SSMC components frequently surpass materials made using traditional processes. In comparative testing of thixocast engine brackets with die-casting and forging equivalents, the SSMC parts exhibited greater dimensional repeatability, lower weight, and comparable or better tensile strength [59].

Correspondingly, rheocast suspension arms were fatigue-tested under conditions simulating road loading and showed improved fatigue life and reduced failure rates than the equivalent gravity-cast parts. These results emphasize the process capability to manufacture trustworthy, long-lived, and economical mission-critical parts [60].

In summary, the uses of semi-solid metal casting are ever increasing in sectors. Its potential to integrate material efficiency, mechanical excellence, and environmental sustainability makes it a 21st-century strategic technology for manufacturing high-performance components. With the development of material science and processing technologies, the future horizon of SSMC will most likely encompass more sophisticated alloys, hybrid materials, and tailored designs according to varied application requirements [61].

14.7 Challenges and Limitations

As much as semi-solid metal casting (SSMC) provides many technical and environmental advantages, numerous constraints limit its global manufacturing adoption. It is critical to understand these limitations for the creation of feasible solutions and increasing the industrial use of this innovative technology.

One of the most important hindrances is the high setup and equipment price. There is specialized equipment needed to create and manipulate the semi-solid state of the alloy, such as induction heating systems, temperature control systems, and exact injection mechanisms. Initial investments are much more expensive than for standard casting setups, and SSMC becomes less appealing to small- and medium-scale businesses [62].

The other challenge is the limited choice of alloys. While SSMC has been successful for aluminum and magnesium alloys, the application to others such as copper, titanium, and high-alloy steel is still in the research stage. The narrow range of semi-solid temperature and the intricate phase behavior of most alloys limit their applicability to semi-solid processing [63].

Accurate temperature control during slurry preparation and injection is essential to maintain the proper solid fraction and defect-free components.

Small excursions can lead to erroneous flow behavior, incomplete fill-out, or structure inconsistencies. This requires the incorporation of high-precision sensors and closed-loop feedback systems, making the process more complex and costly [64].

Scaling up the process to extremely large components is a limitation too. The viscosity and flow behavior of the semi-solid slurry are hard to control over distances or in enormous molds, resulting in complications such as premature solidification or insufficient mold filling. This limits the application of SSMC to small- and medium-sized components unless novel solutions are explored to resolve flow control and thermal management problems [65].

To overcome such limitations, current research is aimed at creating new alloy compositions with broader semi-solid processing windows and enhanced flow characteristics. Concurrently, sophisticated process modeling and simulation software are employed to balance temperature distribution and flow characteristics. Additionally, the use of real-time monitoring systems, artificial intelligence (AI), and digital twins introduces new opportunities for predictive control and improved quality control [66].

By overcoming such technical and economic hurdles, the future use of SSMC may be more scalable, cost-efficient, and more available for a wider variety of industrial processes.

As semi-solid metal casting (SSMC) progresses, its future is in advanced materials, smart processing systems, and sustainability-based innovations. Efforts at research are aimed at increasing process capabilities and making the technology more versatile for different industrial requirements.

One of the promising avenues is the development of high-performance alloys designed specifically for semi-solid processing. Those with more extensive semi-solid temperature ranges, improved flow behavior, and increased microstructural stability are being researched to overcome present material constraints and expand the processing window of SSMC [67].

The blending of hybrid manufacturing methods, including the combination of SSMC with additive manufacturing processes, presents promising potential for fabricating complex, multi-material, or functionally graded parts. The combination has the potential to provide mass customization while retaining the mechanical and structural advantages of semi-solid processing [68].

Smart manufacturing technologies, such as artificial intelligence (AI), machine learning (ML), and Internet of Things (IoT) systems, are being integrated to provide real-time monitoring and control. These systems provide predictive modeling, adaptive process modification, and less human intervention, resulting in increased consistency and quality assurance [69].

Consistent with worldwide environmental objectives, future work is also targeting the employment of sustainable raw materials, like recycled metals and bio-based feedstock, to reduce environmental footprints. These materials, assuming compatibility with SSMC, would significantly improve the greenness of the process and diminish dependence on virgin materials [70]. By widening the process window, unifying automation, and utilizing intelligent systems, SSMC has the potential to become a cornerstone technology in the next generation of high-performance and sustainable manufacturing.

14.8 Conclusions

Semi-solid metal casting (SSMC) provides a rationale solution for those industries looking for sustainable and high-performance manufacturing technologies. Through processing in the semi-solid temperature range, it overcomes the limitations of conventional casting and forging to produce parts with improved mechanical properties, less porosity, and refined microstructures. The feasibility of producing near-net-shape components substantially lowers material loss, machining needs, and energy consumption, making SSMC compatible with the concepts of green manufacturing.

The flexibility of SSMC is spread over a range of industries, such as automotive, aerospace, biomedical, defense, and energy. Its uses are further growing with enhancements in material compatibility and advancements in process technology. The combination of intelligent control systems, process simulation, and automation should enhance streamlining production and enhance quality consistency further.

Although there are limitations to its current application, e.g., expensive equipment, narrow alloy selection, and scalability—research into continuing development of alloys, digital control of processes, and hybrid manufacturing is making its wider adoption possible. SSMC is likewise getting momentum in materials research with sustainability, applying recycled metals and sustainable feedstocks. In summary, SSMC is a revolutionary manufacturing process with immense potential to achieve future performance, economic, and environmental requirements. Further interaction among researchers, engineers, and industrial parties will be important in order to surpass current limitations and realize the full advantages of the novel process.

AI disclosure statement: The author acknowledges the use of AI-based tools for minor language editing or grammatical correction. However, all intellectual contributions, analyses, and conclusions presented in this chapter are entirely the author's own.

References

1. Flemings, M.C., Behavior of metal alloys in the semi-solid state. *Metall. Trans. A*, 22, 5, 957, 1991.
2. Kapranos, P., Kirkwood, D.H., Atkinson, H.V., Liu, T., Thixoforming high performance alloys. *Mater. Sci. Eng., A*, 289, 1–2, 30, 2000.
3. Fan, Z., Semi-solid metal processing. *Int. Mater. Rev.*, 47, 2, 49, 2002.
4. Atkinson, H.V., Future of semi-solid processing. *Mater. Sci. Technol.*, 21, 2, 253, 2005.
5. Haga, T., Kapranos, P., Kirkwood, D.H., Production of semisolid slurries for thixoforming using the cooling slope method. *J. Mater. Process. Technol.*, 178, 1–3, 489, 2006.
6. Liu, T., *et al.*, Challenges and developments in semi-solid metal forming. *J. Manuf. Process.*, 28, 606, 2017.
7. Liu, Y. and Atkinson, H.V., Microstructure evolution in semi-solid alloy processing. *J. Mater. Sci.*, 36, 3, 505, 2001.
8. Brown, S.G.R. and Doherty, R.D., Behavior of thixotropic metal slurries. *Acta Metall.*, 38, 6, 1085, 1990.
9. Flemings, M.C., Solidification processing. *Metall. Trans. B*, 26, 4, 933, 1995.
10. Haga, T. and Suzuki, S., Semi-solid forming of aluminum alloys using the cooling slope method. *Mater. Sci. Forum*, 331–337, 911, 2001.
11. Yang, X., *et al.*, Rheocasting using electromagnetic stirring. *J. Alloys Compd.*, 504, 1, L1, 2010.
12. Kumar, S. and Singh, R., Semi-solid processing of metal alloys. *Mater. Today Proc.*, 1, 2, 165, 2013.
13. Fan, Z. and Wang, Y., MHD stirring for microstructure control. *J. Mater. Sci.*, 40, 2, 419, 2005.
14. Shah, R., *et al.*, Design and optimization of semi-solid processing routes. *Metals*, 8, 6, 452, 2018.
15. Samuel, A.M. and Samuel, F.H., Cast aluminum alloys in automotive applications. *J. Mater. Process. Technol.*, 135, 1, 1, 2003.
16. Zhang, B., *et al.*, Properties of AZ91D magnesium alloy in semi-solid processing. *Mater. Charact.*, 96, 138, 2014.
17. Rao, K.P., *et al.*, Feasibility of semi-solid forming for steel and copper. *J. Mater. Eng. Perform.*, 21, 4, 554, 2012.
18. Lin, X., *et al.*, Influence of solid fraction on flow and formability. *Met. Mater. Int.*, 22, 3, 405, 2016.

19. Atkinson, H.V., *et al.*, Grain refinement in semi-solid slurry. *Solid State Phenom.*, 141–143, 91, 2009.
20. Tang, Y. and Zhang, L., Effects of ultrasonic treatment on aluminum semi-solid slurry. *Mater. Des.*, 67, 72, 2015.
21. Wan, D., *et al.*, Microstructure control by ultrasonic vibration. *Mater. Lett.*, 199, 57, 2017.
22. Roy, S., *et al.*, Simulation in semi-solid processing. *J. Mater. Process. Technol.*, 214, 6, 1315, 2014.
23. Chen, W. and Xu, Z., Machine vision and AI in SSMC. *Procedia Manuf.*, 48, 112, 2020.
24. Lu, S., *et al.*, Synergistic effects of EMS and ultrasonic refinement. *Materials*, 14, 18, 5112, 2021.
25. Prasad, S. and Ramesh, C.S., Alloy design for semi-solid casting. *J. Alloys Compd.*, 491, 1–2, 47, 2010.
26. Wu, D., *et al.*, Digital twin technology for SSMC. *J. Manuf. Syst.*, 62, 111, 2022.
27. Sharma, R. and Mahapatra, M.M., Advances in materials and process modeling. *Mater. Today Proc.*, 72, 3, 1081, 2023.
28. Herring, D.H., Energy efficiency in semi-solid processing. *Heat Treat. Prog.*, 16, 5, 20, 2016.
29. Liu, C., *et al.*, Material efficiency in SSMC. *J. Clean. Prod.*, 41, 1, 2013.
30. Yang, F. and Chen, S., Environmental performance of semi-solid processes. *Resour. Conserv. Recycl.*, 132, 87, 2018.
31. Thirumalai, A. and Srinivasan, R., Tool life in SSMC. *Int. J. Adv. Manuf. Technol.*, 105, 2989, 2019.
32. Gao, X., *et al.*, LCA of thixoformed aluminum parts. *Sustain. Mater. Technol.*, 29, e00303, 2021.
33. Xu, Z. and Zhou, B., Lightweighting *via* semi-solid casting. *Light Met. Age*, 73, 4, 24, 2015.
34. Qureshi, F. and Berto, F., Green manufacturing through SSMC. *Environ. Eng. Manage. J.*, 21, 1, 45, 2022.
35. Hassan, S.F., *et al.*, Strengthening mechanisms in semi-solid components. *Mater. Sci. Eng., A*, 464, 1–2, 174, 2007.
36. Rohatgi, P.K., *et al.*, Mechanical properties of SSMC alloys. *Mater. Des.*, 39, 508, 2012.
37. Huang, W., *et al.*, Fatigue behavior of thixocast parts. *Mater. Sci. Eng., A*, 528, 3, 1133, 2011.
38. Chen, J. and Atkinson, H.V., Grain morphology and flow behavior. *Mater. Charact.*, 60, 7, 726, 2009.
39. Yang, J., *et al.*, Porosity in SSMC components. *J. Mater. Process. Technol.*, 214, 10, 2230, 2014.
40. Tian, Y., *et al.*, Optimization of solid fraction in rheocasting. *Metals*, 10, 9, 1223, 2020.

41. Zhao, Y. and Zhai, Y., Heat treatment of A357 thixocast parts. *J. Alloys Compd.*, 685, 667, 2016.
42. Lee, S.H., *et al.*, Fractography of SSMC alloys. *Eng. Fract. Mech.*, 77, 8, 1213, 2010.
43. Moradi, M. and Torkamani, H., Surface finish and machinability of SSMC components. *J. Manuf. Process.*, 35, 377, 2018.
44. Mirbagheri, S.M.H., *et al.*, Lightweight design using SSMC. *Mater. Des.*, 52, 845, 2013.
45. Prakash, S. and Patel, B.R., Hybrid MMCs *via* semi-solid casting. *Compos. B Eng.*, 241, 110042, 2022.
46. Sharma, A., *et al.*, SSMC for next-generation high-performance parts. *Mater. Today Proc.*, 46, 8, 11712, 2021.
47. Lee, Y.C., *et al.*, Application of SSMC in transportation. *J. Mater. Eng. Perform.*, 16, 4, 482, 2007.
48. Maity, K., *et al.*, SSMC components in automotive applications. *Mater. Today Proc.*, 18, 4243, 2019.
49. Mahadevan, V. and Palanivel, R., Brake caliper enhancement *via* SSMC. *Int. J. Automot. Technol.*, 22, 5, 1115, 2021.
50. Patterson, T., BMW's semi-solid casting advances. *Automot. Eng. Mag.*, 125, 3, 42, 2017.
51. Huber, D. and Gartner, M., Aerospace structures from SSMC. *J. Aerosp. Mater.*, 35, 2, 67, 2015.
52. Karthikeyan, I. and Balasubramanian, V., Reliability of SSMC aerospace brackets. *Mater. Des.*, 31, 5, 2284, 2010.
53. Zhou, J., *et al.*, Magnesium composites in aerospace. *J. Magnes. Alloys*, 10, 1, 75, 2022.
54. Rajak, D.K., *et al.*, Biomedical-grade semi-solid castings. *J. Manuf. Process.*, 55, 152, 2020.
55. Ko, Y.H., *et al.*, Dimensional precision in biomedical SSMC. *Procedia CIRP*, 65, 123, 2017.
56. Zhang, Z., *et al.*, Magnesium-based biodegradable implants. *Bioact. Mater.*, 6, 10, 4041, 2021.
57. Pandey, A.K., *et al.*, Defense applications of SSMC alloys. *Mater. Today Proc.*, 19, 2500, 2019.
58. Yoon, J. and Lee, S., SSMC components in wind energy. *Renewable Energy*, 85, 779, 2016.
59. Tan, J.C., *et al.*, Engine brackets *via* SSMC: A comparative study. *Met. Mater. Int.*, 24, 2, 221, 2018.
60. Patel, R., *et al.*, Rheocast suspension systems under dynamic loading. *Int. J. Veh. Struct. Syst.*, 12, 4, 389, 2020.
61. Balu, A. and Raj, A.K., SSMC for future manufacturing. *Mater. Today Proc.*, 47, 1, 567, 2021.
62. Jain, N. and Aggarwal, A., Economic analysis of SSMC implementation. *Eng. Econ. J.*, 43, 3, 215, 2018.

63. Rajesh, M., *et al.*, Alloy design constraints in SSMC. *Mater. Today Proc.*, 38, 4, 2317, 2021.
64. Kannan, S. and Thirumalai, K., Importance of thermal control in SSMC. *Trans. Indian Inst. Met.*, 72, 10, 2593, 2019.
65. Bhandari, R., *et al.*, Scaling challenges in semi-solid casting. *J. Manuf. Sci. Eng.*, 144, 5, 051008, 2022.
66. Menon, S., *et al.*, AI and sensor-based control in SSMC. *Smart Manuf. Syst.*, 2, 1, 13, 2023.
67. Wang, H., *et al.*, Advanced alloy systems for semi-solid casting. *Mater. Res. Express*, 10, 4, 045901, 2023.
68. Park, J., *et al.*, Additive–SSMC hybrid processes. *Addit. Manuf.*, 55, 102799, 2022.
69. Thomas, P., *et al.*, Smart monitoring in foundry processes. *J. Intell. Manuf.*, 32, 6, 1487, 2021.
70. Lee, D.H. and Singh, R., Use of recycled feedstock in SSMC. *Resour. Conserv. Recycl.*, 162, 105054, 2020.

15

Designing High-Performance Alloys for Sustainable Structural Applications Using Molecular Dynamics Simulations

K. Vijay Reddy[1*], Sunil Kumar Pathak[1], Shailesh Kumar Singh[1,2] and Snehanshu Pal[3]

[1]Climate Change and Data Science (CCDS), CSIR-Indian Institute of Petroleum, Dehradun, India
[2]Academy of Scientific and Innovative Research (AcSIR), Ghaziabad, India
[3]Department of Metallurgy and Materials Engineering, Indian Institute of Engineering Science and Technology, Shibpur, Howrah, West Bengal, India

Abstract

This chapter discusses the application of molecular dynamics (MD) simulations in the design of advanced next-generation metallic systems by microstructural engineering, with focus on high-entropy alloys (HEAs) and nanocrystalline (NC) metals. Through the coupling of phase modulation and grain boundary engineering, the mechanical behavior of these systems was broadly examined at various temperatures. In case of HEAs, the combination of ordered B2 (BCC) grains in an FCC matrix showed a temperature-dependent change in deformation mechanisms and dislocation-mediated plasticity at low temperatures to diffusion-controlled processes at high temperatures. Similarly, in case of NC Cu system, the integration of amorphous intergranular films (AIFs) showed that the crack growth is affected by the film thickness and crack orientation. The specimen with a 0.5 nm AIF showed a significant increase in strength and also retarded fracture. These findings demonstrate how atomistic simulations offer useful knowledge of deformation behavior, which guides us to the synthesis of high-performance, sustainable materials. The research highlights the utility of MD simulations as a predictive and economic method in modern alloy design.

**Corresponding author*: vijay00vegeta@gmail.com

Gurudas Mandal, Swarup Kumar Ghosh, Snehanshu Pal and Sandip Kunar (eds.) Sustainable Manufacturing of Advanced Materials, (311–324) © 2026 Scrivener Publishing LLC

Keywords: Molecular dynamics simulation, high entropy alloys, nanocrystalline metals, sustainable materials

15.1 Introduction

The design and development of high-performance metallic alloys and systems are the foundation of modern engineering. This is particularly true in structural applications, where thermo-mechanical stability, and environmental resistance are essential [1, 2]. With a growing focus on sustainability, the metallurgical community faces two main challenges: first, it must develop alloys/materials that not only offer superior mechanical and functional performance, but also meet eco-friendly and energy-efficient manufacturing and operational practices. In this perspective, attaining this balance is through control and manipulation of the microstructural characteristics that determines the mechanical behavior at an atomic level [3, 4]. One of the strategies includes controlling the intrinsic structure and distribution of grain boundaries (GBs) that act as barriers to dislocation motion and regulate deformation processes like creep, fracture, and corrosion–thereby improving durability and performance [5, 6]. Specifically, tuning and manipulating the crystallographic structure, and thereby the energy aids in improving the strength without sacrificing much of ductility.

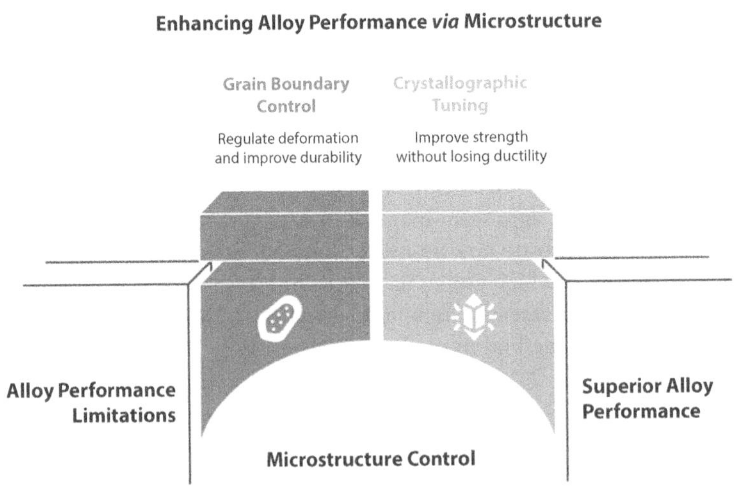

Figure 15.1 Schematic illustration showing that controlling and tuning the grain boundaries and crystallographic phases acts as a bridge in the enhancement of material performance (Author's own).

Grain boundaries (GBs) with specific symmetries (defined by coincidence site lattice (CSL) theory), including coherent twins, are decisive in facilitating plastic deformation without the nucleation of failure [7, 8]. Moreover, at atomic scales, ultrafine grains and high-density interfaces alter the conventional deformation mechanisms and causes enhancement of strength and toughness [8]. On the other hand, tailoring the GB structure (i.e., segregation, controlling excess free volume and energy) through atomic-scale engineering also allows a precise control of the interfacial properties, which aids in further enhancement of the mechanical properties [9]. Apart from GB engineering, modulating the crystallographic structures and their fractions is also an effective technique to enhance the mechanical performance without varying the overall composition of the system. For instance, addition of fine dispersed hard phase particles in the ductile matrix can impede the movement of dislocations and result in increase in strength through Orowan mechanism [10]. Similarly, tuning the size and distribution of ordered phases (i.e., B2 or L12 phases) can determine the yielding of the strain hardening of the material. Together, these inflections in the system can lead to developing high-performance alloys accompanied with sustainability and enduring reliability.

To have an in-depth understanding and insights to optimize these mechanisms, molecular dynamics (MD) simulations become an important computational tool, as it provides a discrete atomistic view of the underlying mechanisms that control the properties of the system [11–14]. Unlike other large-scale continuum simulation modeling, which are dependent on the empirical rules and other mechanical assumptions, MD is based on the first principles rules (used as interatomic potential files) coupled with the atomic trajectories of individual atoms (determined by solving Newton's equation of motion) that results in the physical mechanism. Recent studies have shown the potential of MD simulations in understanding the deformation behavior of super-alloys and high entropy alloys (HEAs) such as AlCoCrFeNi HEAs, Cu/Ni nanocrystalline systems, and other relevant metallic system [15]. For example, our recent study has revealed that tailoring the GB structure with special characteristics aids in thermal stability and mitigation of Hall-Petch breakdown of nanocrystalline (NC) Cu specimen [8]. Similarly, another research focused on the intrinsic spectral properties of GBs *via* excess free volume, energy, and segregation and revealed how the variation of these parameters affects the overall GB migration and grain growth [9]. Modifying the GB characteristics also influence the properties during mechanical processing of nanostructured metallic systems enabling us to manipulate the texture [16]. Apart from the technical atomistic aspect, MD simulation also aids in sustainability as it can detect and study and analyze novel alloys and material systems without undergoing extensive experimental trails. This capability of atomistic simulation minimizes the time requirement for material development as well as resource consumption, aligning our objective towards sustainable development of materials.

In summary, MD simulation can be a useful and essential theoretical tool that can help in actively guiding alloy design and optimizing the operating parameters. In this chapter, the authors have briefly elucidated the elementary simulation setup, the findings, and the underlying physical mechanism. This chapter will give a glance into the depth of how MD simulations can be implemented to design and engineer novel advanced alloys with enhanced physical and structural properties.

15.2 MD Simulation Setup

15.2.1 Designing the Simulation System for High Entropy Alloys

Here, MD simulations were implemented to study the influence of phase modulation and GB engineering on the deformation behavior of nanostructured metallic systems. The simulations were performed using the open-source Large-scale Atomic/Molecular Massively Parallel Simulator (LAMMPS) package [17], and the post-processing visualizations were carried out using OVITO software [18]. Specifically, for the investigation of the phase modulation (FCC and BCC), a FeNiCrCoAl high entropy alloy system has been considered. In this case, an embedded atom method (EAM) potential created by Farkas and Caro was employed to describe the interaction between the atoms [19]. Specific ordered BCC phase (B2) was built using first principle Density Functional Theory (DFT) tool and then the structure was merged in the MD simulation box domain [20, 21]. The DFT tool was used to define stoichiometries and position of elements to create a stable B2 phase. For the specimen dimension, a cell comprising of dual phases and a dimension of 20 nm × 20 nm × 20 nm is designed, where the FCC matrix was embedded with ordered B2 phase uniformly. The phase fraction of the B2 phase was varied by increasing the number of B2 grains (e.g. 2, 4, 7) to analyze its effect on the mechanical behavior of the system (refer to Figure 15.2). Apart from, designing the material system, there are important equilibration protocols (such as thermostat, ensembles, operating temperature and pressure, timesteps, output variables) that needs to be set before the initiation of the simulation. Here, the system was

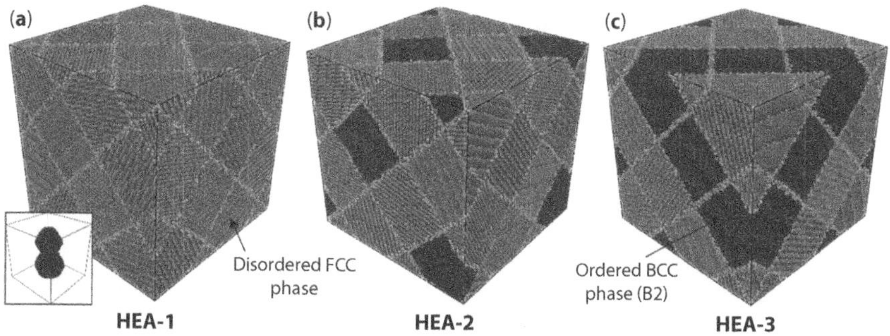

Figure 15.2 Polyhedral Template Matching (PTM) snapshots illustrate the three HEA specimens, each with different Al content. (a) HEA-1 with 2 BCC grains is embedded at the centre of the specimen (as shown in the inset), (b) HEA-2 contains 4 BCC grains, and (c) HEA-3 features 7 BCC grains [15]. (Reproduced under permission).

energy minimized using conjugate gradient method followed by equilibration under NPT (constant pressure and temperature) ensemble for three different temperatures [22, 23]. To study the mechanical deformation, the specimens were subjected to uniaxial tensile deformation along the Z-axis at a constant strain rate of $1 \times 10^8\ s^{-1}$ under canonical (NVT) ensemble conditions. The timesteps for these simulations were set to be 2 fs, and periodic boundary conditions were applied along all three directions.

15.2.2 Simulating Novel Interfaces in Metallic Systems

To explore the effects of interfacial structure on crack behavior and plastic strain localization, a nanocrystalline Cu system is modeled and the grain boundaries were modified with special interfaces, i.e., amorphous intergranular films (AIF). These systems were created by embedding the $Cu_{64}Zr_{36}$ amorphous layers (about 0.5 to 2 nm thickness) at the GB regions to mimic a crystalline bulk with metallic glass (MG) interfacial films. After equilibrating and energy minimization using LAMMPS [17], a crack was implanted inside the specimen to study its growth under uniaxial tensile deformation and cyclic deformation (refer to Figure 15.3). The visualization of the mechanisms is once again done through OVITO software [18]. Altogether, the two simulations provide a comprehensive architecture of designing novel materials by the manipulation of phases and interfaces to enhance fracture toughness, strength, and plastic strain of the nanoscale metallic systems. Interested researchers can explore more in the literature, where an in-detailed step by step procedure of designing these structures are provided [15, 24].

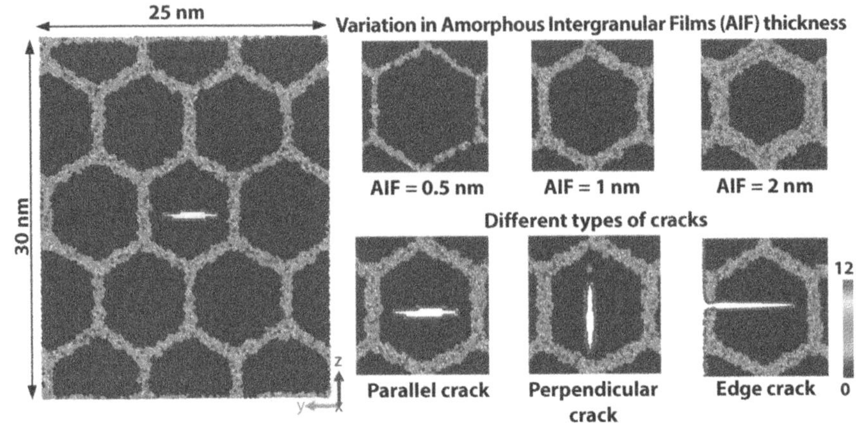

Figure 15.3 (a) Centro-symmetry parameter (CSP) snapshot shown for a representative specimen with an AIF thickness of 1 nm along with the dimensions; (b) snapshots depict specimens with different AIF thicknesses, highlighting the orientations and positions of the embedded cracks [24]. (Reproduced under permission).

15.3 Results and Discussion

15.3.1 Influence of Phase Modulation on Mechanical Behavior of the System

The tensile stress-strain plots were studied and presented here to understand the effect of B2 phase on the strength and toughness of the HEA specimens. Figure 15.4 (a–c) presents the stress–strain curves for the three HEA specimens (HEA-1, HEA-2, and HEA-3) tested at low (100 K), ambient (300 K), and high (1173 K) temperatures, respectively. At 100 K, all specimens exhibit significant plastic deformation prior to fracture, with HEA-1 showing the highest ultimate tensile strength (UTS) of approximately 10 GPa (refer Figure 15.4 (a)). However, HEA-1 also undergoes early and rapid plastic deformation at approximately 3.5 GPa, as highlighted in the inset of Figure 15.4 (a). In contrast, HEA-2 and HEA-3 demonstrate greater strength but reduced ductility at the same temperature. This variation in deformation behavior is likely influenced by differences in B2 phase content, a topic elaborated in later sections. At 300 K (refer to Figure 15.4 (b)), the UTS values for all three specimens converge in the range of 8 to 9 GPa. While HEA-2 and HEA-3 show only slight changes compared to the 100 K case, HEA-1 displays a noticeable reduction in UTS, which can be attributed to increased atomic thermal vibrations at higher temperatures.

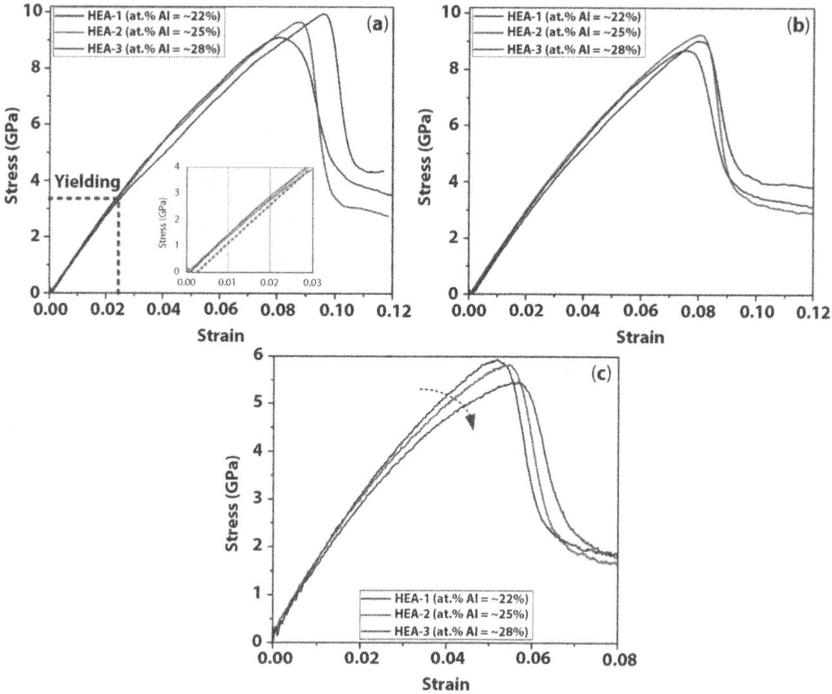

Figure 15.4 Stress-strain curves showing the tensile deformation behaviour of HEA specimens (HEA-1, HEA-2, and HEA-3) at different temperatures: (a) 100 K, (b) 300 K, and (c) 1173 K. The inset in (a) highlights the early onset of rapid plastic deformation in the HEA-1 specimen [15]. (Reproduced under permission).

These vibrations facilitate dislocation mobility and enhance their interactions under applied strain, accelerating plastic flow—a behavior consistent with findings in similar alloy systems. Despite the temperature increase, HEA-1 continues to show early onset of plastic deformation, mirroring its behavior at 100 K. At 1173 K, as shown in Figure 15.4 (c), the tensile strength of all specimens drops significantly. Notably, a higher B2 phase content now correlates with reduced strength and accelerated plastic deformation, especially as Al content increases. This reversal in trend suggests a transition from dislocation-driven to diffusion-dominated deformation mechanisms at elevated temperatures, particularly along phase interfaces.

The authors have further analyzed the fracture behavior and in this analysis, they have examined the shear strain distribution within the HEA specimens during the tensile deformation, with particular focus on the strain partitioning between the FCC and BCC phases across different temperatures. Literature studies have indicated that the atomic strain distribution in FeNiCrCoAl high-entropy alloys is strongly influenced by both the underlying phase (FCC or BCC) and the testing temperature [25]. To capture these effects, a comparative analysis of shear strain was conducted in all three HEA specimens at low (100 K) and high (1173 K) temperatures at a simulation time of 120 ps, as illustrated in Figure 15.5. At 100 K, deformation is characterized by the formation of sharp and intense shear bands within the FCC regions, corresponding to stacking faults generated by Shockley partial dislocations. Additionally, the FCC–BCC interfaces exhibited moderately elevated strain, although their contribution to the overall deformation remains limited. Notably, the B2 (BCC) grains accommodate plastic strain through the development of multiple shear bands oriented along the (110) planes-a consistent observation across all specimens (Figure 15.5 (a1–a3)). At elevated temperatures (1173 K), the shear strain becomes more localized near the FCC–BCC interfaces and within the BCC regions, while the FCC matrix shows increased atomic mobility due to enhanced thermal activation (refer to Figure 15.5 (b1–b3)). These observations are consistent with literature reports indicating

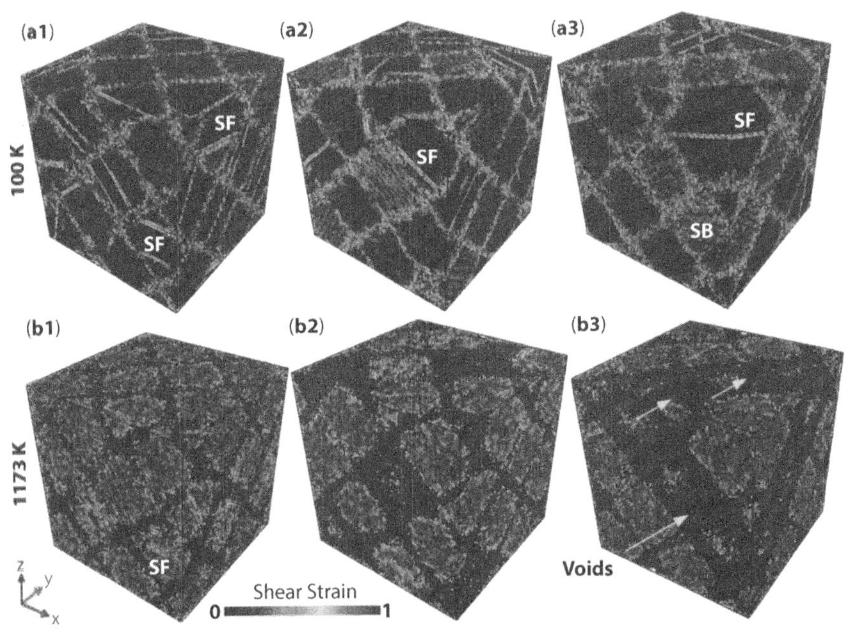

Figure 15.5 Atomic strain snapshots at t=120 ps showing the strain distribution in HEA specimens at two temperatures: 100 K and 1173 K. (a1, b1) corresponds to HEA-1, (a2, b2) to HEA-2, and (a3, b3) to HEA-3. The color bar indicates the range of atomic shear strain values from low to high [15]. (Reproduced under permission).

that increasing the Al content enhances hardness while reducing ductility and plastic strain accommodation [26–28]. Overall, the results underscore the dominant role of the BCC phase in governing the mechanical response of HEAs during the high-temperature deformation.

15.3.2 Effect of Interface Modulation on Crack Retardation Behavior

Figure 15.6 presents the stress–strain curves for nanocrystalline (NC) cu specimens and those with varying AIF thicknesses, each containing cracks of different orientations subjected to uniaxial tensile deformation. The results indicate that the ultimate tensile strength (UTS) of the specimens is significantly influenced by crack orientation. Specifically, specimens with

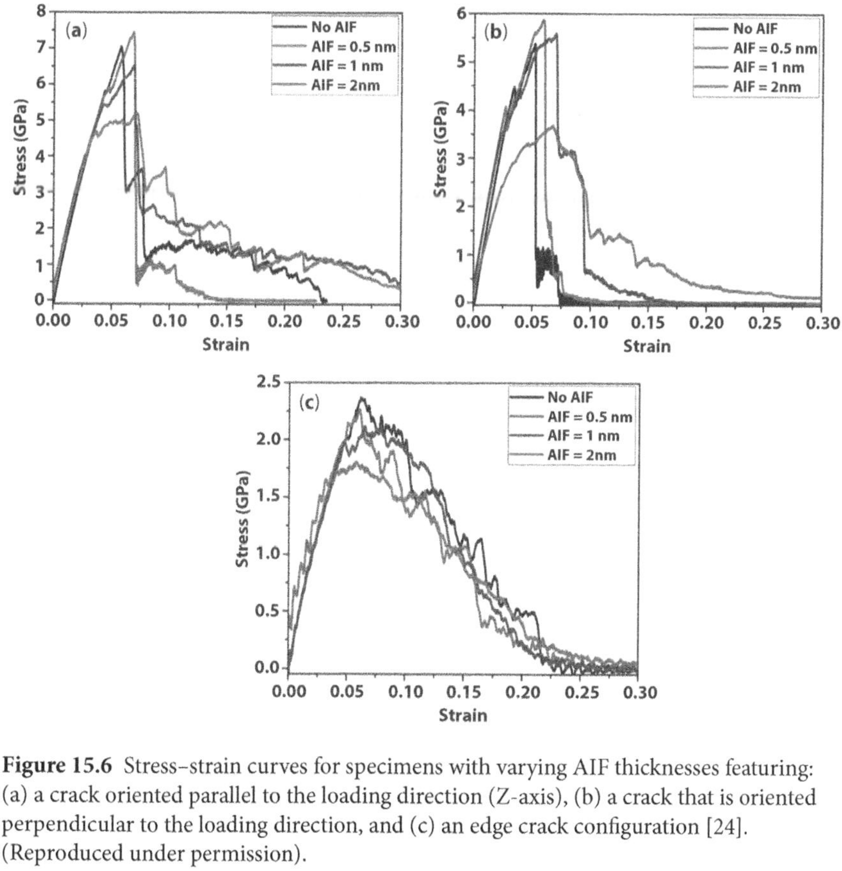

Figure 15.6 Stress–strain curves for specimens with varying AIF thicknesses featuring: (a) a crack oriented parallel to the loading direction (Z-axis), (b) a crack that is oriented perpendicular to the loading direction, and (c) an edge crack configuration [24]. (Reproduced under permission).

parallel cracks (refer to Figure 15.6 (a)) exhibited the highest UTS, while those with edge cracks (refer Figure 15.6 (c)) showed the lowest. This orientation-dependent behavior is consistent with observations reported in the literature for brittle and quasi-brittle solids [29, 30]. For the parallel crack configuration, the specimen with a 0.5 nm thick AIF demonstrated the highest tensile strength. This behavior can be attributed to its similarity with NC cu exhibiting heavy grain boundary solute segregation, which suppresses grain coalescence and thus enhances strength. However, despite the higher UTS, this specimen experienced rapid crack propagation and failed at a strain of approximately 15%.

As the AIF thickness increased to 1 nm and 2 nm, the tensile strength decreased below that of the pure NC cu specimen. Nonetheless, these thicker AIF layers significantly delayed fracture, with the specimens sustaining up to ~25% tensile strain before failure. This suggests that while thicker AIFs reduce tensile strength, they effectively retard crack propagation in the parallel orientation. A similar trend was observed for specimens with perpendicular cracks (refer Figure 15.6 (b)), where the 0.5 nm AIF enhanced strength relative to NC cu, but further increases in AIF thickness led to a reduction in UTS. Notably, the post-UTS stress drop occurred at a higher strain (~10%) in the 0.5 nm AIF specimen compared to ~7.5% in the NC cu sample, indicating delayed fracture due to hindered crack growth. Overall, the results demonstrate that an AIF thickness of 0.5 nm is more effective in slowing parallel crack growth than perpendicular cracks. For edge cracks (as observed in Figure 15.6 (c)), a slight variation was observed: the 0.5 nm AIF specimen exhibited lower tensile strength than NC cu and further increases in AIF thickness continued to reduce the UTS. However, like internal cracks, fracture was delayed up to ~25% tensile strain in these cases, confirming that AIFs still contribute to crack retardation even when strength is compromised.

15.4 Conclusions

This chapter has illustrated how molecular dynamics (MD) simulation is an effective atomistic tool for designing and comprehending the deformation mechanisms of high-entropy alloys (HEAs) and nanocrystalline (NC) metallic systems. By modulating the phases and grain boundary (GB) engineering, enhanced strength, ductility, and fracture, resistance can be attained without changing the overall chemical composition. In FeNiCrCoAl HEAs, the incorporation of ordered B2 (BCC) grains in an

FCC matrix exhibited a multi-faceted interaction between strength and ductility with mechanical performance being highly sensitive to phase fraction and temperature. Under low temperatures, the FCC phase supported plasticity *via* stacking faults, whereas the resistance to deformation was controlled through the B2 grains. At high temperature, deformation mechanism changed to diffusion-controlled behavior, and localization of strain in the BCC phase resulted in void nucleation and subsequent fracture. Similarly, interface modulation by amorphous intergranular films (AIFs) in NC cu systems demonstrated orientation-dependent crack resistance. It was observed that the specimen with 0.5 nm AIF thickness showed enhanced strength and postponed crack propagation, particularly under parallel crack orientations, whereas thicker films improved ductility by slowing down fracture at higher strains. Collectively, these results highlight the significance of microstructural tuning, phase distribution and interfacial properties, toward high-performance and thermos-mechanical stability in metallic materials. MD simulations not only provide insight into underlying atomistic mechanisms but also support sustainable materials science by mitigating the extent of experimental requirements. The methods outlined in this chapter provide a robust computational framework to allow future alloy design approaches.

AI disclosure statement: The author acknowledges the use of AI-based tools for minor language editing or grammatical correction. However, all intellectual contributions, analyses, and conclusions presented in this chapter are entirely the author's own.

References

1. Fragassa, C., Lesiuk, G., Epp, J., High-performance applications of metals and alloys: Material properties, behaviour modeling, optimal design and advanced processes. *Metals*, 13, 8, 1485, 2023.
2. Yang, L., Jiang, X., Sun, H., Shao, Z., Fang, Y., Shu, R., Effects of alloying, heat treatment and nanoreinforcement on mechanical properties and damping performances of Cu–Al-based alloys: A review. *Nanotechnol. Rev.*, 10, 1, 1560–1591, 2021.
3. Gleiter, H., Nanostructured materials: basic concepts and microstructure. *Acta Mater.*, 48, 1, 1–29, 2000.
4. Li, X., Lu, L., Li, J., Zhang, X., Gao, H., Mechanical properties and deformation mechanisms of gradient nanostructured metals and alloys. *Nat. Rev. Mater.*, 5, 9, 706–723, 2020.

5. Zhang, B.B., Tang, Y.G., Mei, Q.S., Li, X.Y., Lu, K., Inhibiting creep in nanograined alloys with stable grain boundary networks. *Science*, 378, 6620, 659–663, 2022.
6. Hao, D., Wang, Q., Ma, H., Wu, D., Kang, J., Dong, H., Su, R., The influence of grain boundary character distribution on the high-temperature creep behavior and damage mechanism of Inconel 718. *Mater. Des.*, 248, 113525, 2024.
7. Reddy, K.V. and Pal, S., Effect of grain boundary complexions on the deformation behavior of Ni bicrystal during bending creep. *J. Mol. Model.*, 24, 4, 87, 2018.
8. Pal, S., Reddy, K.V., Deng, C., Improving thermal stability and Hall-Petch breakdown relationship in nanocrystalline Cu: A molecular dynamics simulation study. *Mater. Lett.*, 324, 132821, 2022.
9. Pal, S., Reddy, K.V., Yu, T., Xiao, J., Deng, C., The spectrum of atomic excess free volume in grain boundaries. *J. Mater. Sci.*, 56, 19, 11511–11528, 2021.
10. Scattergood, R.O., Koch, C.C., Murty, K.L., Brenner, D., Strengthening mechanisms in nanocrystalline alloys. *Mater. Sci. Eng., A*, 493, 1-2, 3–11, 2008.
11. Reddy, K.V. and Pal, S., Influence of rolling temperature on the structural evolution and residual stress generation of nanocrystalline Nickel during nano-rolling process. *Comput. Mater. Sci.*, 184, 109935, 2020.
12. Reddy, K.V. and Pal, S., Dynamic probing of structural evolution of single crystal Fe during rolling process using atomistic simulation. *Steel Res. Int.*, 90, 7, 1800636, 2019.
13. Reddy, K.V., Meraj, M., Pal, S., Molecular dynamics simulation based investigation of strain induced crystallization of nickel metallic glass. *Mater. Chem. Phys.*, 237, 121831, 2019.
14. Pal, S. and Reddy, K.V., *Molecular dynamics for materials modeling: A practical approach using LAMMPS platform*, 168, CRC Press, Boca Raton, 27 March 2024. https://doi.org/10.1201/9781003323495.
15. Reddy, K.V., Kumar, P., Vashistha, S., Pal, S., Singh, S.K., Effect of structural modulation of B2 phase on the deformation mechanism in FeNiCrCoAl high entropy alloy: An atomistic insight. *Mater. Chem. Phys.*, 340, 130840, 2025.
16. Reddy, K.V. and Pal, S., Atomistic insight into the texture weakening and shear-shuffle twinning mechanism during cold-rolling of magnesium. *JOM*, 74, 4, 1387–1394, 2022.
17. Plimpton, S., Fast parallel algorithms for short-range molecular dynamics. *J. Comput. Phys.*, 117, 1, 1–19, 1995.
18. Stukowski, A., Visualization and analysis of atomistic simulation data with OVITO–the Open Visualization Tool. *Modell. Simul. Mater. Sci. Eng.*, 18, 1, 015012, 2009.
19. Farkas, D. and Caro, A., Model interatomic potentials for Fe–Ni–Cr–Co–Al high-entropy alloys. *J. Mater. Res.*, 35, 22, 3031–3040, 2020.
20. Van De Walle, A., Asta, M., Ceder, G., The alloy theoretic automated toolkit: A user guide. *Calphad*, 26, 4, 539–553, 2002.

21. Enkovaara, J., Rostgaard, C., Mortensen, J.J., Chen, J., Dułak, M., Ferrighi, L., Jacobsen, K.W., Electronic structure calculations with GPAW: A real-space implementation of the projectoraugmented-wave method. *J. Phys.: Condens. Matter*, 22, 25, 253202, 2010.
22. Evans, D.J. and Holian, B.L., The Nose–Hoover thermostat. *J. Chem. Phys.*, 83, 8, 4069–4074, 15 October 1985. https://doi.org/10.1063/1.449071.
23. Nakano, A., Parallel multilevel preconditioned conjugate-gradient approach to variable-charge molecular dynamics. *Comput. Phys. Commun.*, 104, 1-3, 59–69, 1997.
24. Pal, S., Reddy, K.V., Deng, C., On the role of Cu-Zr amorphous intergranular films on crack growth retardation in nanocrystalline Cu during monotonic and cyclic loading conditions. *Comput. Mater. Sci.*, 169, 109122, 2019.
25. Bönisch, M., Wu, Y., Sehitoglu, H., Twinning-induced strain hardening in dual-phase FeCoCrNiAl0. 5 at room and cryogenic temperature. *Sci. Rep.*, 8, 1, 10663, 2018.
26. Stepanov, N.D., Shaysultanov, D.G., Chernichenko, R.S., Tikhonovsky, M.A., Zherebtsov, S.V., Effect of Al on structure and mechanical properties of Fe-Mn-Cr-Ni-Al non-equiatomic high entropy alloys with high Fe content. *J. Alloys Compd.*, 770, 194–203, 2019.
27. Zhang, C., Zhang, F., Diao, H., Gao, M.C., Tang, Z., Poplawsky, J.D., Liaw, P.K., Understanding phase stability of Al-Co-Cr-Fe-Ni high entropy alloys. *Mater. Des.*, 109, 425–433, 2016.
28. Wang, W.R., Wang, W.L., Wang, S.C., Tsai, Y.C., Lai, C.H., Yeh, J.W., Effects of Al addition on the microstructure and mechanical property of AlxCoCrFeNi high-entropy alloys. *Intermetallics*, 26, 44–51, 2012.
29. Ma, L., Wang, X., Feng, X.Q., Yu, S.W., Numerical analysis of interaction and coalescence of numerous microcracks. *Eng. Fract. Mech.*, 72, 12, 1841–1865, 2005.
30. Nguyen, G.D., Korsunsky, A.M., Einav, I., A constitutive modelling framework featuring two scales of behaviour: Fundamentals and applications to quasi-brittle failure. *Eng. Fract. Mech.*, 115, 221–240, 2014.

16

Recyclable and Renewable Composite Materials in Additive Manufacturing

Ganesh Nataraj[1]*, Ramesh Babu S.[2], A. R. Ravikumar[3], Arul Pandian[1], Karthikeyan Rajendran[1] and Ajithkumar Sitharaj[2]

[1]Department of Mechanical Engineering, Sri Shakthi Institute of Engineering and Technology, Coimbatore, Tamil Nadu, India
[2]Department of Mechanical Engineering, KPR Institute of Engineering and Technology, Coimbatore, Tamil Nadu, India
[3]Department of Mechanical Engineering, School of Engineering, Mohan Babu University, Tirupati, India

Abstract
The research examines how recyclable and renewable composite materials serve as essential components for developing sustainable practices in additive manufacturing (AM). The research begins with a description of AM basics before emphasizing sustainable materials as essential for environmental protection through circular economy principles. The study provides detailed analysis of thermoplastic-based and thermoset-based recyclable composites through discussions about post-consumer recycled plastics and fiber reinforcements as well as material degradation complexities and vitrimers as emerging solutions. The research devotes substantial attention to renewable composite materials through detailed analysis of bio-based polymers (PLA, PHAs, Bio-PE, Bio-PP, cellulose-based) and natural fiber reinforcements (hemp, flax, jute, bamboo) which presents their characteristics and AM compatibility and environmental advantages and discusses their moisture absorption challenges. The research examines hybrid renewable composites together with the emerging field of materials derived from algae and fungi. The study provides a thorough examination of material preparation techniques alongside AM process optimization methods and necessary characterization techniques including Life Cycle Assessment (LCA) which validate sustainability claims. The practical applications of these sustainable AM composites are demonstrated across diverse sectors, including biomedical, automotive, packaging, construction, and aerospace, showcasing their

**Corresponding author*: ganmech2002@gmail.com

Gurudas Mandal, Swarup Kumar Ghosh, Snehanshu Pal and Sandip Kunar (eds.) Sustainable Manufacturing of Advanced Materials, (325–350) © 2026 Scrivener Publishing LLC

potential for creating customized, lightweight, and environmentally responsible products. The study identifies major technical barriers alongside economic challenges and regulatory hurdles which include material degradation and scalability problems and high costs and the industry-wide absence of standardization. The research ends by emphasizing the vast possibilities of sustainable AM while stressing the necessity of continuous joint work between researchers and industry leaders and policymakers to drive the development of a circular sustainable manufacturing system that supports economic growth and environmental protection.

***Keywords*:** Sustainable additive manufacturing, recyclable composites, renewable materials, 3D printing applications, material characterization

16.1 Introduction

Additive Manufacturing (AM) or 3D printing technology has developed substantially during the last three decades from its original application as a prototyping tool to become a standard manufacturing technology. The technology has expanded its applications into aerospace, automotive, biomedical, and consumer goods sectors because it enables the creation of complex geometries and customized products with minimal waste [1–3]. FDM, SLA, SLS, and MJF represent the primary AM processes which construct objects through digital model layer-by-layer, manufacturing using different materials and energy sources [4]. The manufacturing processes deliver multiple benefits to traditional production techniques through design flexibility and detailed structure creation and material conservation which makes AM a sustainable alternative [5, 6]. Aerospace and automotive sectors employ AM for both prototyping and manufacturing lightweight high-performance components which help achieve sustainability targets by minimizing material consumption and environmental emissions [7, 8]. The biomedical sector uses AM technology to produce personalized implants and prosthetics which results in improved patient results and customized solutions. The consumer goods industry uses AM to create customized products at high speed which satisfies various consumer requirements. The advantages of AM remain limited by high production expenses and restricted material compatibility and irregular build quality that researchers address through design algorithm improvements and AI-based process monitoring systems. The maturation of AM technology will integrate with different industries to create innovative solutions that enhance sustainability and boost manufacturing productivity [9]. Research indicates that natural fiber polymer composites provide sustainable lightweight solutions for various industries at reduced costs.

The present study investigates moisture resistance and durability to enhance operational capabilities and industrial adoption [10].

Additive manufacturing (AM) requires sustainable materials because traditional manufacturing methods together with virgin material usage produce major environmental damage and deplete resources and pollute the environment. The manufacturing industry depends on non-renewable resources while producing large amounts of waste which causes environmental deterioration and produces significant greenhouse gas emissions [11, 12]. The restricted recyclability and negative environmental effects of conventional AM materials made from petroleum-based plastics drive the need for sustainable alternatives [13, 14]. The circular economy concept provides a solution to these problems through material reduction and reuse and recycling practices which decrease waste generation and resource utilization [15, 16]. The manufacturing sector now implements this approach because regulatory requirements combine with customer preferences for green products and corporate sustainability initiatives [17, 18]. The combination of natural fillers with Green PLA-based composites enhances the mechanical properties of 3D printing materials. The environmentally friendly replacement material boosts engineering strength while promoting sustainability and waste utilization [19]. Bio-based polymers together with recycled composites have become popular choices because they reduce carbon emissions while providing better end-of-life recovery options which match circular economy principles [20, 21]. The adoption of sustainable materials in AM receives additional support from biomimicry and nanotechnology; technologies which enable the creation of eco-friendly materials with superior performance and durability. The implementation of a circular economy in AM encounters obstacles because it requires technological progress alongside investments in sustainable practices and industrial sector resistance to change. The implementation of sustainable materials in AM processes remains crucial for achieving sustainable environmental and economic benefits which will create a resilient manufacturing ecosystem for the long term.

Materials qualify as recyclable when they can be transformed into fresh products after their first use to reduce waste while preserving resources. The recycling industry uses mechanical processes together with chemical processes and thermal processes as its primary methods. The recycling process of mechanical reprocessing maintains material chemical structures while chemical recycling transforms materials into monomers for re-polymerization and thermal recycling produces energy through incineration [22, 23]. The recycling of AM materials faces difficulties because of polymer sorting requirements and contamination issues and the special challenges of reprocessing thermoset materials with their cross-linked structures [24, 25]. Renewable materials derived from natural biomass sources and agricultural waste provide sustainable alternatives for fossil-based materials. The materials decompose naturally through biodegradation and break down into non-toxic components when composted under specific conditions [26, 27]. Lignocellulosic biomass stands as a sustainable renewable resource for developing biodegradable and recyclable plastics which help decrease fossil fuel dependence [28]. The combination of environmentally friendly root fiber composites (RFCs) with fiber-matrix bonding surface treatments demonstrates substantial market potential. The development of processing techniques and fiber modification methods enables new industrial applications AJI. A sustainable AM ecosystem depends on the essential connection between recyclable and renewable materials which enables waste reduction through recycling and environmental benefits from renewable resources. The integrated system supports a circular economy by maintaining continuous material cycles which decreases environmental effects and increases economic stability [29]. The AM industry can establish a sustainable future by using recyclable and renewable materials which reduce its carbon emissions [30].

16.2 Recyclable Composite Materials in Additive Manufacturing

This section delves into the foundational concepts and practical applications of using recyclable materials within additive manufacturing. We will explore various material categories, the principles guiding their reusability, and the innovative techniques driving their integration into 3D printing workflows. An overview of these categories is provided in Figure 16.1.

16.2.1 Principles of Recyclability in AM

AM product designers must adopt proactive measures to establish recyclability at the initial stages of product conceptualization. Design for recyclability (DfR) refers to the process of choosing materials which both can be recycled and are part of established recycling systems. The selection of thermoplastics over thermosets represents a key approach in AM design because thermoplastics can undergo multiple cycles of re-melting and reprocessing. The application of DfR principles in AM includes two main aspects which are the reduction of non-separable material combinations

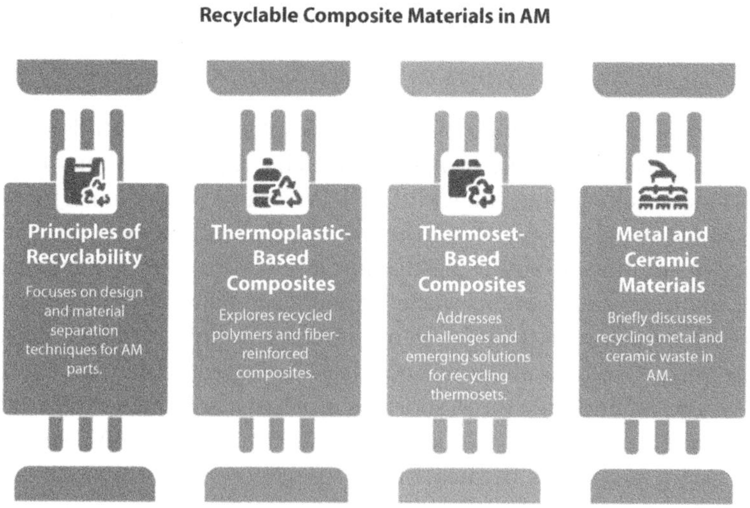

Figure 16.1 Recyclable composite materials in AM.

and the creation of geometries that enable simple product disassembly and material separation after use. The unique layered design of AM creates an opportunity to incorporate features that support recycling by implementing material breakage points or sorting system identifiers. The fundamental purpose of DfR in AM involves transforming end-of-life waste materials into new feedstock that can be used for AM production and other manufacturing processes.

The increasing use of multi-material printing systems in advanced AM systems creates a major challenge for achieving total recyclability. Multi-material AM generates complex parts with customized properties through a single printing process, but it complicates the recycling process. The complex intermixing of different materials with distinct melting points and chemical characteristics and recycling methods creates substantial difficulties for effective separation methods. The combination of a high-performance polymer with distinct reinforcing fibers or two different polymers printed together within a single component becomes impossible to recycle using traditional mechanical methods. A solution to this issue demands the creation of specialized material separation methods which focus on AM composite materials through selective dissolution and density separation following cryogenic grinding or advanced chemical depolymerization techniques. The lack of efficient and large-scale separation technologies for multi-material designs produces unrecyclable waste that contradicts the sustainability targets of additive manufacturing.

16.2.2 Thermoplastic-Based Recyclable Composites

16.2.2.1 Recycled Polymers as Matrix Materials

The foundation of sustainable additive manufacturing depends on thermoplastic-based recyclable composites which use post-consumer recycled (PCR) plastics including rPET, rPP, rHDPE and rABS as matrix materials for AM filaments and powders. The use of recycled polymers provides an attractive solution to virgin plastics because it decreases dependence on fossil resources and prevents waste from entering landfills. The process of adding these materials to AM systems presents multiple obstacles. Material degradation stands as the main obstacle during recycling because thermal processing multiple times shortens molecular weights which weakens mechanical strength and reduces processability. The final AM part shows inconsistent properties because of this process which creates difficulties in achieving guaranteed performance. The combination of virgin polymer compounding with additive materials including compatibilizers and stabilizers and reinforcing agents helps restore material properties while fighting degradation and improving melt flow and mechanical characteristics to enhance PCR plastic usage in various AM processes.

16.2.2.2 Short Fiber Reinforced Recycled Composites

Short fiber reinforced recycled composites show great potential to improve mechanical properties of materials during additive manufacturing operations while supporting sustainable practices. Engineers achieve substantial strength and stiffness and dimensional stability improvements in 3D-printed parts through the combination of recycled polymers with short carbon or glass fibers. The effectiveness of these reinforcements depends on three essential factors which include fiber length, orientation, and content. The mechanical properties of printed parts improve when fibers reach longer lengths, but excessive length creates difficulties during specific AM manufacturing processes. The orientation of fibers in printed parts depends on printing parameters such as nozzle speed and infill patterns which results in anisotropic properties because fibers tend to align with the extrusion direction for enhanced properties in that direction. The mechanical performance of materials improves with increasing fiber content until an optimal point is reached after which it results in increased viscosity and poor dispersion and reduced part quality. The composites serve non-critical structural components and jigs and fixtures and consumer goods because their enhanced performance surpasses unreinforced recycled polymers, yet researchers continue to study methods for achieving isotropic properties and high strength consistency.

16.2.2.3 Continuous Fiber Reinforced Recycled Composites

Continuous fiber reinforced recycled composites represent an exciting, albeit emerging, frontier in sustainable additive manufacturing, holding immense potential for high-performance applications. Unlike their short fiber counterparts, continuous fibers (such as carbon or glass) offer superior strength and stiffness, allowing for the creation of AM parts that can rival, or even surpass, the mechanical properties of traditionally manufactured components. However, incorporating continuous recycled fibers presents significant challenges. The primary hurdles include sourcing and processing long, high-quality recycled fibers that maintain sufficient integrity after recovery. Fiber damage during the recycling process, inconsistencies in fiber length and strength, and the difficulty of uniformly impregnating these recycled fibers with a polymer matrix are all critical issues. Furthermore, adapting existing continuous fiber AM processes (like continuous fiber fabrication - CFF) to handle potentially less uniform recycled fibers, or developing new printing techniques specifically for these materials, requires substantial research and development. Despite these difficulties, the potential benefits for high-performance applications are transformative. Imagine lightweight, strong, and durable parts for aerospace, automotive, or industrial machinery, all produced with a significantly reduced environmental footprint by leveraging recycled materials. This could lead to a truly circular economy for advanced composites, drastically cutting down on waste and raw material consumption in industries that demand peak performance.

16.2.3 Thermoset-Based Recyclable Composites

The recycling of thermoset-based composites becomes challenging because their permanent cross-linked molecular structure makes it difficult to recycle. The permanent bonding in thermoset materials blocks traditional reprocessing methods which results in waste disposal through landfills or incineration. Research continues to transform the field by creating vitrimers and other reprocess able thermosets for use in additive manufacturing. The unique properties of vitrimers enable them to function like thermosets while allowing heat-induced reshaping and reprocessing through their dynamic covalent bonds which makes them suitable for stereolithography (SLA) additive manufacturing. The new chemical approach enables controlled de-crosslinking and re-crosslinking operations which creates opportunities for fully recyclable resins. The development of these advancements creates opportunities for closed-loop recycling of thermoset AM parts which establishes a circular economy that recovers high-performance components for reprocessing and reuse to decrease waste and environmental impact of thermoset additive manufacturing.

16.2.4 Metal and Ceramic Recyclable Materials

The sustainable production of metal and ceramic materials in additive manufacturing depends heavily on their recyclability beyond polymers while facing unique obstacles and opportunities. The main focus in metal recycling involves the recovery of metal powders which are used in Selective Laser Melting (SLM) and Electron Beam Melting (EBM) processes. The build chamber's unfused or partially fused powder can be sieved for reuse in a process known as powder recovery. The powder characteristics change after multiple high-temperature exposures so quality control measures including chemical analysis and particle size distribution and flowability tests must be performed before re-sintering. The recycled powder remains suitable for future builds because of this quality control process. The capability to recover expensive metal powders decreases both material waste and economic costs. The recycling of ceramic waste in AM requires researchers to establish methods which convert unfired green bodies and support structures and post-sintered components that fail to meet specifications. The recycling process includes grinding followed by purification and reformulation of new feedstock for binder jetting or material extrusion of ceramics. The development of metal powder recycling remains behind metal powder recycling, but ceramic waste recycling research holds essential value for sustainable additive manufacturing because it reduces environmental impacts from high-temperature processing and raw material extraction.

16.3 Renewable Composite Materials in Additive Manufacturing

This section delves into the exciting realm of renewable composite materials, exploring how bio-based polymers and natural fibers are being harnessed for additive manufacturing. We will discuss their unique properties,

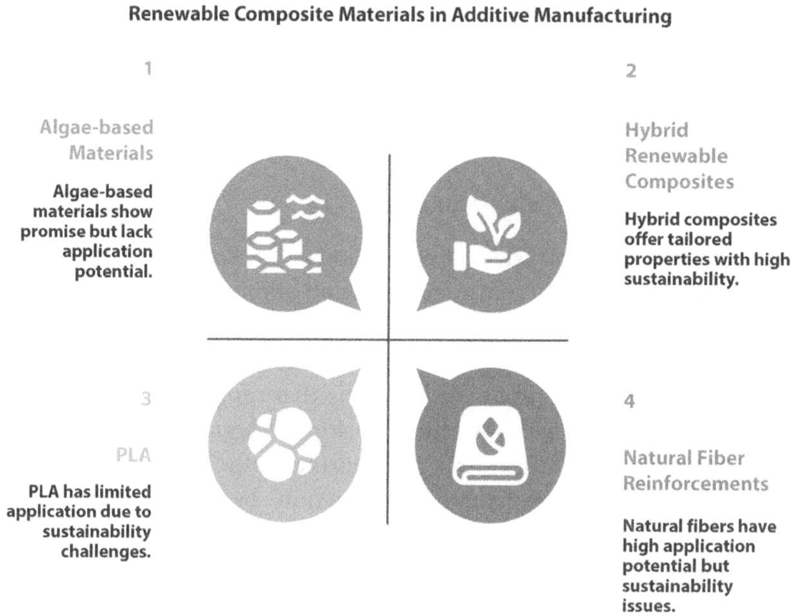

Figure 16.2 Overview of renewable composite materials in AM.

AM processability, and the environmental benefits they offer, alongside the challenges in their widespread adoption. A comprehensive overview of these innovative materials is presented in Figure 16.2.

16.3.1 Bio-Based Polymers as Matrix Materials

16.3.1.1 Polylactic Acid (PLA)

PLA functions as the leading bio-based polymer for additive manufacturing because it originates from renewable materials such as corn starch and sugarcane and agricultural waste while providing an environmentally friendly alternative to petroleum-based plastics. The combination of good stiffness, transparency and easy processing makes PLA suitable for various AM techniques, especially FDM because its low melting temperature enables smooth extrusion and strong layer adhesion. PLA provides biodegradable and compostable properties because it decomposes into natural compounds including CO2 and water during industrial composting processes which creates a sustainable disposal method. PLA has several disadvantages because its heat deflection temperature (HDT) usually reaches only 50-60°C which prevents its use in applications with moderate heat exposure since parts will deform. The brittleness of PLA restricts its application in functional parts that need high impact strength or flexibility but ongoing research on annealing and additives and compounding techniques works to reduce these limitations.

16.3.1.2 Polyhydroxyalkanoates (PHAs)

The biodegradable polyester family known as Polyhydroxyalkanoates (PHAs) emerges from microbial fermentation of renewable carbon sources such as sugars or plant oils to create sustainable plastic alternatives. PHAs show different material properties based on microbial strain selection and feedstock choice which results in materials that range from stiff brittle to flexible elastic. PHAs demonstrate superior biodegradability across various environments including soil and marine water and industrial compost which produces environmentally friendly byproducts that surpass many other bioplastics. PHAs have started to gain traction in additive manufacturing through their use in Fused Deposition Modeling (FDM) and powder bed fusion despite being less popular than PLA. The combination of biocompatibility and biodegradability in PHAs makes them suitable for biomedical applications including tissue engineering scaffolds and resorbable implants and drug delivery systems because AM geometry customization offers significant benefits. The ability to modify mechanical properties of PHAs creates new opportunities for sustainable packaging and consumer goods and niche industrial components which extends the potential of circular manufacturing through AM.

16.3.1.3 Other Bio-Based Polymers

The sustainable additive manufacturing market is expanding its use of bio-based polymers through the adoption of Bio-polyethylene (Bio-PE), Bio-polypropylene (Bio-PP) and different Cellulose-based polymers. The production of Bio-PE and Bio-PP uses renewable sugarcane ethanol feedstocks to create bioplastics which function as direct substitutes for fossil-derived polymers. The chemical structure of these materials matches conventional FDM and pellet extrusion systems so they can be easily integrated with minimal process modifications. The mechanical properties and chemical resistance and processability of these materials match traditional PE and PP while providing substantial environmental benefits. The main obstacle for using these materials in AM is their non-biodegradable nature because they end up in plastic waste accumulation after use but they can be recycled through standard polyolefin recycling processes. The production of cellulose-based polymers from wood pulp and cotton provides excellent stiffness along with biodegradability and high renewability. The development of cellulose-based polymers for AM applications continues to grow especially for FDM and direct ink writing through modifications such as esterification or blending to enhance their thermoplasticity and melt flow. The main difficulties in using cellulose-based polymers for AM stem from their hydrophilic nature which causes moisture absorption problems and processing challenges and their generally lower melt strength than synthetic polymers which requires precise optimization of printing parameters and possibly the addition of reinforcing fillers to achieve suitable mechanical properties in printed parts.

16.3.2 Natural Fiber Reinforcements

16.3.2.1 Types of Natural Fibers

Natural fiber provides a sustainable alternative to synthetic reinforcements for composite materials used in additive manufacturing through plant-based and animal-based fibers. The majority of AM applications use plant fibers because they exist abundantly while being renewable and possessing suitable mechanical characteristics. Most plant cell walls contain cellulose and lignin as their basic structural elements which give stiffness and strength to the cell walls. The most used plant fibers for AM composite production include hemp, because of its high strength-to-weight ratio and fast growth rate and flax, because of its excellent mechanical properties and smooth texture, and jute because it offers affordability and strength, and bamboo because it provides rapid renewability and good stiffness. The resulting bio composites benefit from enhanced tensile and flexural strength and reduced weight along with improved biodegradability because of these fibers. Silk and wool represent limited options for AM applications because their production costs are high, and they come from animals which creates ethical concerns and require complex processing methods. Silk's exceptional strength and biocompatibility have led to research in niche biomedical AM applications. Yet, animal fibers face widespread use challenges because of their thermal stability and processability issues in conventional thermoplastic-based AM composites.

16.3.2.2 Advantages of Natural Fibers in AM Composites

The integration of natural fibers into additive manufacturing composites provides multiple sustainable benefits that enhance environmental sustainability initiatives. The primary advantage of natural fibers stems from their renewable nature because they originate from agricultural crops and forestry byproducts which maintain their supply as a sustainable resource compared to petroleum-based synthetic fibers. The biodegradable nature of these fibers enables composite parts to decompose naturally after their use which helps decrease landfill waste and supports circular economic systems. Natural fibers possess low density which results in AM parts that weigh less. The weight reduction capability of natural fibers makes them ideal for applications needing weight optimization such as automotive and aerospace components which leads to better fuel efficiency. Natural fibers demonstrate excellent specific properties because they provide outstanding strength and stiff performance relative to their weight. The production of high-performance parts becomes possible through this method without the environmental impact that traditional reinforcement materials create. The combination of these features leads to substantial environmental benefits compared to synthetic fibers because they require less production energy and produce fewer greenhouse gases while using fewer non-renewable resources, thus establishing natural fiber AM composites as a genuine green manufacturing solution.

16.3.2.3 Challenges and Solutions for Natural Fiber AM Composites

The adoption of natural fibers in additive manufacturing composites faces particular obstacles despite their multiple benefits. The main obstacle in using natural fibers is their tendency to absorb moisture. Natural fibers possess hydrophilic properties which cause them to easily take in moisture from their surroundings. The absorption of moisture by natural fibers results in unstable dimensions and weakened mechanical properties and creates challenges for AM processing because steam bubbles form inside the melt. The main obstacle in using natural fibers is their weak bond with the matrix material. Natural fibers possess different surface chemical properties than typical polymer matrices which creates weak bonding between them. The weak interfacial bonding between matrix and fibers prevents efficient stress transfer which reduces the mechanical reinforcement potential of composite material. Natural fibers show vulnerability to thermal breakdown when they undergo processing operations. The high temperatures used in FDM AM processes trigger the breakdown of organic components in natural fibers such as hemicellulose and lignin which produces discoloration and charring while decreasing mechanical performance.

16.3.3 Hybrid Renewable Composites

The combination of different bio-based polymers with various natural fibers produces hybrid renewable composites which serve as an advanced sustainable pathway for additive manufacturing. The strategic combination of different bio-based matrices such as PLA with PHAs and the integration of natural fibers including hemp and flax and short and continuous natural fibers represents an advanced approach beyond single-material systems. The main benefit of these hybrid systems emerges from their ability to customize properties for particular AM applications. Engineers achieve precise control of mechanical properties including tensile strength and impact resistance and flexibility and thermal characteristics such as heat deflection temperature by selecting and proportioning biopolymers with natural fibers. The combination of stiff biopolymers with flexible ones and suitable natural fiber reinforcements produces composites that provide both structural strength and desired toughness levels. The customized approach enables the creation of specialized sustainable materials which fulfill various end-use requirements including medical scaffolds with controlled degradation profiles and mechanical support as well as durable lightweight components for sustainable consumer electronics and automotive interiors. Thoughtful hybridization techniques enable sustainable AM applications to expand beyond what single bio-based materials can achieve.

16.3.4 Algae-Based and Fungi-Based Materials in AM (Emerging)

The expanding sustainable additive manufacturing industry now explores algae-based and fungi-based materials as promising novel feedstocks because they derive from nature's most prolific decomposers and photo synthesizers. Microalgae that grow quickly on non-farmable areas and wastewater treatment systems produce polymers such as alginates and PHAs from their biomass. The resulting materials can be processed into bio-inks or filaments for 3D printing applications. Agricultural waste can be used to cultivate mycelium which then binds into lightweight biodegradable composites that can be molded or "grown" into desired shapes. Research teams work to improve extraction and processing techniques which will enhance the printability and mechanical strength and thermal stability of algae-derived polymers for different AM methods. The research on fungi focuses on developing methods to control mycelial growth and optimize substrates and create filaments or printing methods using living mycelium. The emerging field of regenerative manufacturing through waste stream conversion enables the creation of fully biodegradable AM parts which could transform sustainable packaging and temporary structures and biomedical applications.

16.4 Processing and Characterization of Recyclable and Renewable AM Composites

The following section explains the complete process of preparing and processing and evaluating recyclable and renewable composite materials for additive manufacturing. The quality and performance and sustainability of AM parts depend on these methodologies. The composite processes are shown in Figure 16.3.

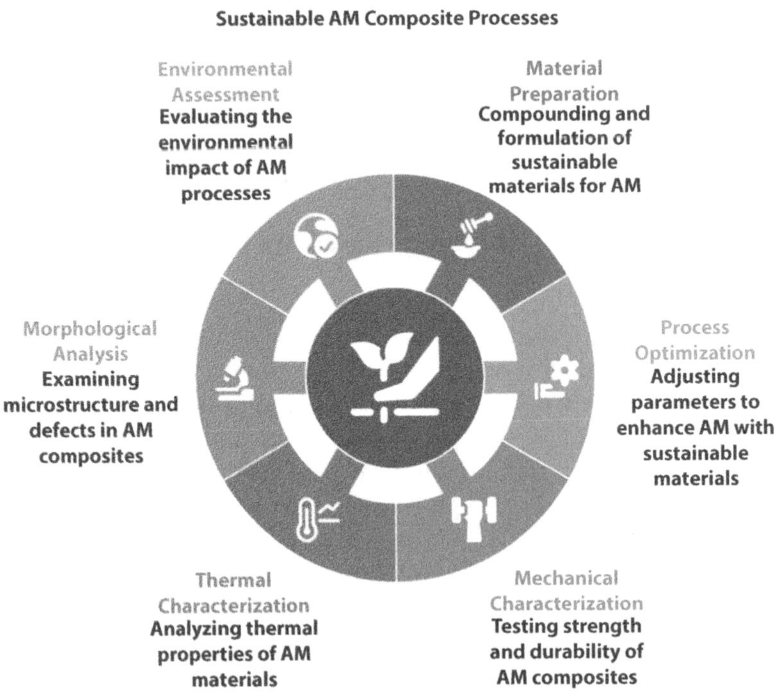

Figure 16.3 Sustainable AM composite processes.

16.4.1 Material Preparation and Formulation

The implementation of recyclable and renewable materials in additive manufacturing requires precise material preparation and formulation to achieve optimal feedstock conditions for specific AM processes. The initial step requires combining recycled or renewable polymers with reinforcing fillers that include natural fibers and short carbon fibers and bio-fillers. The twin-screw extruder serves as the primary equipment for polymer melting and filler-additive mixing (including compatibilizers and thermal stabilizers) to produce a uniform blend with required properties. The compounded material undergoes uniform filament extrusion for FDM printing which demands precise diameter consistency and circularity and void-free material to prevent print failures such as clogging and inconsistent material deposition. The key process for powder bed fusion techniques SLS and SLM involves grinding and sieving followed by spheroidization to produce powders with narrow particle size distribution and optimal flowability and high packing density. The development of vat photopolymerization resins for stereolithography (SLA) and digital light processing (DLP) requires advanced chemical formulation because bio-based oligomers and monomers and photo initiators and fillers need to be carefully chosen to achieve suitable viscosity and cure depth and mechanical properties when exposed to UV light. The fundamental aspects of particle size distribution and homogeneity and flow properties remain crucial for all material forms. The use of inconsistent particle sizes results in poor powder packing density and filament melting irregularities while insufficient filler homogeneity produces anisotropic material properties and printing defects. The print quality and part strength and process reliability of sustainable materials depend on optimized flow properties which enable smooth extrusion in FDM and consistent powder spreading in SLS/SLM and efficient resin flow in SLA/DLP.

16.4.2 Additive Manufacturing Process Optimization

Standard operating procedures need precise adjustments to optimize additive manufacturing (AM) processes for producing high-quality components from recyclable and renewable materials. The thermal and rheological behaviors of sustainable feedstocks require precise calibration of extrusion temperature and print speed and layer height and infill density parameters. The use of recycled polymers and natural fiber composites in AM production creates specific manufacturing challenges because of warping from temperature differences and weak layer adhesion and FDM nozzle clogging from filler agglomeration. The reduction of defects requires multiple strategies including bed temperature adjustments and adhesion promoters for warping prevention and surface treatments and cooling adjustments and extrusion increases and layer adhesion improvement methods for better results. The optimization process delivers reliable and high-quality production results when working with sustainable AM materials.

16.4.3 Mechanical Characterization

The mechanical evaluation of additively manufactured parts made from recyclable and renewable composites serves to determine their operational capabilities and application suitability. Standardized tests must be performed to evaluate fundamental mechanical properties. The tensile strength of a material determines its resistance against breaking forces while flexural strength determines its resistance against bending stresses. The energy absorption capacity before material failure determines toughness through impact strength testing and compressive strength evaluates the resistance against crushing forces. The tests serve as essential tools for determining material behavior under operational stresses. Multiple material and process variables strongly affect the test results. The properties of materials change according to their recycled content because polymer degradation occurs when recycled content increases, thus requiring specific formulation approaches. The addition of fibers through loading increases both stiffness and strength until a specific point after which it produces brittleness or processing challenges. The mechanical properties of AM parts become anisotropic because fiber orientation follows the print path and flow dynamics which results in better strength along the fiber alignment direction; thus, designers need to consider load cases. The required performance specifications for sustainable AM parts can be achieved through comprehensive mechanical characterization which connects environmentally friendly materials to engineering requirements.

16.4.4 Thermal Characterization

The thermal characterization of recyclable and renewable composite materials in additive manufacturing requires knowledge of their thermal properties because these properties determine both manufacturing parameters and end-use performance. The main techniques used for thermal characterization consist of differential scanning calorimetry (DSC), thermogravimetric analysis (TGA) and dynamic mechanical analysis (DMA). DSC serves to determine the melting behavior of crystalline polymers through measurements of melting temperature and enthalpy of fusion while identifying the glass transition temperature (Tg) of amorphous polymers and semi-crystalline polymers' amorphous regions. The glass transition temperature (Tg) represents a critical value because it defines the temperature range during which polymers shift from rigid glassy to rubbery states which affects dimensional stability. TGA analyzes material thermal stability through weight loss measurements at increasing temperatures which reveals both degradation onset points and volatile substance presence. The organic fibers in natural fiber composites need special attention because they degrade when exposed to AM processing temperatures. The viscoelastic properties of materials under different thermal conditions can be precisely determined through DMA measurements which include temperature and frequency and time dependencies. The combination of thermal analysis techniques delivers essential data for optimizing print temperatures which protect materials from damage while achieving proper fusion and enables performance prediction at high temperatures and thermal reliability assessment of sustainable AM composites.

16.4.5 Morphological and Microstructural Analysis

The internal architecture and quality of additively manufactured parts made from recyclable and renewable composites requires morphological and microstructural analysis for complete understanding. Scanning electron microscopy (SEM) functions as a robust analytical instrument which enables researchers to observe microscopic details of surfaces and internal features. The microscopic examination of SEM images reveals essential information about fiber-matrix bonding through observations of the polymer matrix-fiber interface where poor adhesion shows gaps or pull-out marks. The printed part's mechanical integrity becomes compromised by the presence of pores or air pockets which SEM allows researchers to quantify. The examination of mechanical test fracture surfaces with SEM provides essential failure mechanism information about brittle or ductile fracture and fiber-related strength or weakness. The non-destructive three-dimensional imaging of complete internal structures of AM parts is possible through X-ray micro-computed tomography (μ-CT). μ-CT provides researchers with exceptional capabilities to analyze internal structures and defects because it enables the precise measurement of fiber orientation and dispersion and aggregation patterns across the entire part volume without requiring sample destruction. Technology enables precise identification and measurement of voids as well as cracks and un-melted powder particles to deliver a complete evaluation of internal quality. SEM and μ-CT work together to provide complete microstructural analysis of composites which helps researchers link manufacturing parameters to material properties and defects for sustainable AM material development optimization.

16.4.6 Environmental Characterization (Life Cycle Assessment - LCA)

Life Cycle Assessment (LCA) serves as an essential tool for determining the actual environmental effects of recyclable and renewable materials used in additive manufacturing. Through its data-based approach LCA enables the complete assessment of environmental burdens that span from raw material extraction to end-of-life disposal, thus preventing unwanted burden transfers. The LCA methodology verifies whether sustainable AM benefits from recycled or bio-based feedstocks by demonstrating their ability to lower greenhouse gas emissions and energy usage and waste generation compared to traditional methods. The implementation of LCA for AM follows ISO 14040/14044 standards yet encounters major implementation obstacles. The combination of limited data availability for new bio-based and recycled materials and the complicated energy requirements of AM processes creates obstacles for precise inventory data collection. The process of defining functional units and system boundaries for comparison proves to be complex. The implementation of robust LCA remains vital for identifying environmental hotspots and selecting materials and validating environmental claims for sustainable AM to establish verifiable green manufacturing practices.

16.5 Applications for Recyclable and Renewable Composite Materials in AM

This section highlights the diverse and expanding applications of recyclable and renewable composite materials across various industries, showcasing how additive manufacturing is enabling the creation of sustainable products. A visual summary of these key application areas is provided in Figure 16.4.

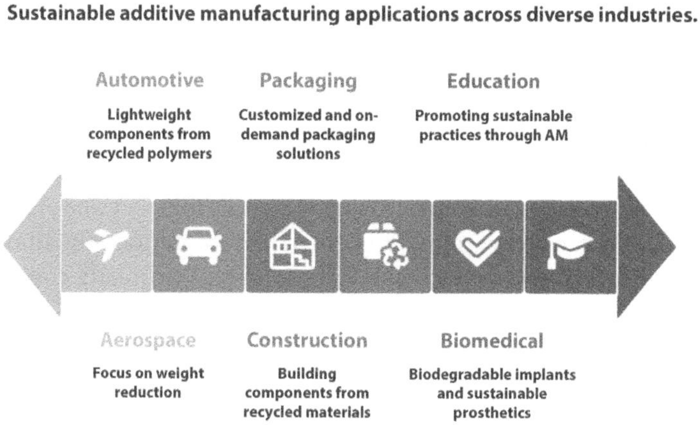

Figure 16.4 Sustainable additive manufacturing applications across diverse industries.

16.5.1 Biomedical and Healthcare

The biomedical and healthcare sector represents an ideal field for implementing recyclable and renewable composite materials through additive manufacturing because it requires biocompatible patient-specific environmentally responsible solutions. The production of biodegradable implants through AM allows the creation of complex tissue engineering scaffolds from PLA/PHA materials which enables cell growth before the implant degrades inside the body without requiring additional surgical procedures. The combination of biodegradable materials such as PLA and PHA with AM technology enables the creation of complex porous structures that mimic natural tissue through the production of intricate scaffolds for tissue engineering. The production of sustainable prosthetics and orthotics undergoes transformation through the use of AM with sustainable materials. Custom-fitted lightweight products made from recycled thermoplastics and natural fiber composites provide an environmentally friendly alternative to traditional methods at reduced costs. The digital manufacturing process of AM enables quick design cycles and local production which reduces transportation-related impacts. The increased need for disposable medical tools during global health crises can be addressed through biodegradable or recyclable AM parts that decrease the substantial medical waste output. The adoption of sustainable materials in biomedical AM solutions addresses both healthcare system requirements and environmental issues that affect medical practices.

16.5.2 Automotive and Transportation

The automotive and transportation sector is increasingly adopting recyclable and renewable composite materials in additive manufacturing to address critical demands for weight reduction, fuel efficiency, and reduced environmental impact. A primary focus is on producing lightweight components from recycled polymers. Using materials like recycled ABS, PP, or even rPET, AM can create complex geometries for interior trim, under-the-hood components, and non-critical structural parts, directly contributing to vehicle mass reduction and consequently, lower emissions. The ability to customize designs and produce parts on-demand also streamlines supply chains. Furthermore, interior components with natural fiber composites are gaining traction, leveraging materials like hemp or flax-reinforced bioplastics for dashboards, door panels, and seating elements. These materials offer not only a reduced carbon footprint but also improved aesthetics and acoustics, contributing to a more sustainable and pleasant passenger experience. Lastly, prototyping and tooling with sustainable materials are already widespread. The rapid iteration capabilities of AM, combined with the use of recycled or bio-based filaments for jigs, fixtures, and rapid prototypes, significantly reduce material waste and lead time in the development cycle. This transition to sustainable materials in AM within automotive and transportation aligns with industry goals of circularity and reduced reliance on virgin fossil resources.

16.5.3 Packaging and Consumer Goods

The packaging and consumer goods sectors are ripe for disruption by additive manufacturing utilizing recyclable and renewable composite materials, offering solutions that cater to personalization, waste reduction, and environmental responsibility. AM enables the creation of customized and on-demand packaging solutions, where unique designs, branding, or specific product protection features can be rapidly produced using biodegradable or recyclable polymers. This approach drastically reduces material waste inherent in mass production of diverse packaging, as items are only printed when needed, often directly addressing the "last mile" of manufacturing. Imagine bespoke packaging for fragile items or personalized gift boxes, all printed with minimal excess material. Beyond packaging, the application extends to a wide array of sustainable consumer products, such as eyewear frames and phone cases. These items can be designed for optimal fit and aesthetics, then manufactured from recycled plastics (like rPET or rABS filaments) or bio-based composites (e.g., PLA reinforced with natural fibers). This not only reduces reliance on virgin plastics but also allows for localized production, potentially cutting down on transportation emissions. The ability to create functional, aesthetically pleasing, and environmentally conscious products on demand makes AM with sustainable materials a powerful tool for circular economic initiatives in the consumer goods market.

16.5.4 Construction and Architecture Applications

The construction and architecture sector investigates additive manufacturing with recyclable and renewable composite materials to achieve waste reduction and construction acceleration and sustainability improvement. The main application area involves manufacturing building components through the combination of recycled aggregates and polymers. Large-scale AM technologies including robotic extrusion produce non-structural or semi-structural elements such as wall panels and insulation blocks and customized formwork from concrete mixes containing recycled plastic granules and polymer-based composites filled with construction and demolition waste. The process eliminates waste disposal in landfills through waste diversion and provides both weight reduction and design adaptability for complex architectural forms. The use of sustainable materials through AM technology has transformed the process of developing sustainable architectural models and building facades. The use of biodegradable PLA or recycled plastic filaments enables architects to create complex building designs at high speed which minimizes the material waste that traditional model making generates. The production of complex lightweight cladding elements through AM technology using bio-based or recycled composites results in improved building aesthetics and thermal performance while decreasing environmental impact in built environments. The combination of AM technology with sustainable materials will transform the process of designing and building and living in future spaces.

16.5.5 Aerospace Applications (Limited But Emerging)

While currently limited but emerging, the aerospace industry is cautiously but increasingly exploring recyclable and renewable composite materials in additive manufacturing, driven by an unwavering focus on reduced weight and material circularity. Unlike other sectors where rapid adoption is more feasible, aerospace demands exceptionally high performance, stringent safety certifications, and long service lives, which makes the integration of novel, sustainable materials a slower, more rigorous process.

However, AM's ability to produce complex, lightweight geometries is invaluable. Consequently, the immediate applications for sustainable AM materials primarily involve non-structural components, such as interior cabin elements (e.g., seat parts, overhead bin components), ducting, and fairings where direct load bearing is minimal. Furthermore, the use of recycled or bio-based polymers for tooling and jigs is gaining traction. These parts, used to aid in the assembly or manufacturing of primary aerospace structures, do not fly with the aircraft, thus alleviating some of the certification burdens. The overarching goal is to achieve significant weight reduction – a perennial objective in aerospace to improve fuel efficiency and lower emissions – while simultaneously advancing material circularity. This involves exploring ways to incorporate high-performance recycled carbon fiber composites or durable bio-based polymers into the aerospace supply chain, moving towards a future where aircraft components can be produced with a minimized environmental footprint without compromising safety or performance.

16.5.6 Education and Research

The advancement of recyclable and renewable composite materials in additive manufacturing depends on the essential partnership between education and research. Academic institutions now focus on teaching sustainable practice s in AM by adding these concepts to engineering and materials science and design curricula. The education of future engineers and designers includes teaching them about environmental material impacts and design for recyclability principles and circular economy advantages in AM workflows. Students learn sustainable manufacturing through practical experiences with recycled filaments and bio-based resins. The development of new materials and processes requires dedicated research initiatives to be successful. The research focuses on basic studies of new biopolymers and natural fibers for AM processing optimization and the creation of sophisticated recycling methods for complex AM waste streams. Researchers investigate sustainable composite materials with multiple components while developing AI/ML methods to speed up material discovery and process optimization and intelligent processing parameters to reduce energy usage. The ongoing educational research connection enables sustainable AM material advancements to be developed and properly shared which leads to an environmentally responsible manufacturing future.

16.6 Challenges and Future Outlook

The development of recyclable and renewable composite materials in additive manufacturing encounters multiple substantial obstacles which combine technical difficulties with economic barriers and regulatory requirements. The technical limitations of recycled polymers exist mainly because their previous processing history creates mechanical performance variations that challenge part quality assurance especially when parts contain high amounts of recycled content. The use of sustainable feedstocks in additive manufacturing faces additional challenges because of different processability requirements which need precise optimization of AM parameters to prevent warping and nozzle clogging. The implementation of multi-material and gradient materials in AM technology presents significant recycling challenges because it becomes difficult to properly separate the intimately combined components. The large-scale production of sustainable AM materials is restricted which limits their market penetration while traditional materials maintain a better performance level in demanding applications that need enhanced mechanical strength and longevity.

Technical complexities encounter major barriers alongside economic difficulties and regulatory challenges in the field. Sustainable AM materials face a cost-effectiveness challenge because specialized recycling procedures for recycling, or bio-based feedstocks increase their prices beyond conventional materials. The current shortage of specialized infrastructure to handle the collection and sorting of recycled feedstock hinders the delivery of high-quality material streams necessary for AM operations. Market acceptance of sustainable AM products faces a major economic hurdle because customers prefer cheaper prices and perceived product performance over environmental benefits. A major standardization problem exists because there are no established testing methods for sustainable AM materials which makes performance evaluation and quality control challenging. The absence of clear guidelines regarding biodegradability and composability creates opportunities for "greenwashing" which makes environmental transparency difficult to achieve. The market will expand through consumer trust-building when sustainable AM products receive strong certification and labeling systems which validate their sustainability claims. The complete realization of recyclable and renewable composites in additive manufacturing depends on solving these diverse obstacles.

16.7 Conclusion

The path to sustainable manufacturing receives essential empowerment from the developments in recyclable and renewable composite materials that occur in additive manufacturing. The combination of AM capabilities with expanding environmentally friendly feedstocks enables transformative industry changes and reduces ecological impact while establishing a circular economy. The combination of sustainable materials with AM technology enables the creation of customized biomedical implants that break down safely and lightweight automotive components that boost fuel efficiency and customized consumer products from recycled waste which together create a production system that uses resources efficiently while minimizing environmental harm. The combination of bio-based polymers and natural fibers and recycled content processed with precision enables AM to evolve beyond prototyping into a fundamental component of sustainable on-demand localized manufacturing which transforms our production and consumption systems.

The achievement of this substantial potential requires an extended collaborative partnership between multiple stakeholders. The advancement of material science research must continue to develop new bio-based and recycled composites which exhibit improved processability alongside consistent properties and equivalent performance to conventional materials. Industry stakeholders including material suppliers and AM machine manufacturers and end-product producers need to dedicate resources to expand sustainable feedstock production while building strong recycling systems for AM waste and integrating these materials into standard manufacturing operations. The establishment of clear regulatory frameworks and standardized testing methods and incentive programs by policymakers will drive sustainable AM practice adoption and circular economy transition while promoting transparent labeling. The achievement of sustainable and resilient future through recyclable and renewable composites in additive manufacturing depends on unified action between innovation and industrial dedication and governmental support.

AI disclosure statement: The author acknowledges the use of AI-based tools for minor language editing or grammatical correction. However, all intellectual contributions, analyses, and conclusions presented in this chapter are entirely the author's own.

References

1. Agarwal, R., Sharma, S., Gupta, V., Singh, J., Khas, K.S., Additive manufacturing, in: *Additive Manufacturing*, 1st ed., pp. 77–97, CRC Press, Boca Raton, 2022.
2. Bourell, D.L., Wohlers, T., Introduction to additive manufacturing, in: *ASM Handbook, Vol. 24: Additive Manufacturing Processes*, pp. 3–10, ASM International, 2020.
3. Kanishka, K. and Acherjee, B., Revolutionizing manufacturing: A comprehensive overview of additive manufacturing processes, materials, developments, and challenges. *J. Manuf. Process.*, 107, 574–619, 2023.
4. Kumar, S., Introduction, in: *Additive Manufacturing Processes*, 1st ed., pp. 1–15, Springer, Cham, 2020.
5. Mehrpouya, M., Vosooghnia, A., Dehghanghadikolaei, A., Fotovvati, B., The benefits of additive manufacturing for sustainable design and production, in: *Sustainable Manufacturing*, 1st ed., pp. 29–59, Elsevier, 2021.
6. Javaid, M., Haleem, A., Singh, R.P., Suman, R., Rab, S., Role of additive manufacturing applications towards environmental sustainability. *Adv. Ind. Eng. Polym. Res.*, 4, 4, 312–322, Oct. 2021.
7. Mohammed, R., *et al.*, Advancements and Challenges in Additive Manufacturing: Future Directions and Implications for Sustainable Engineering. *Adv. Sustain. Sci. Eng. Technol.*, 7, 1, 0250108, Jan. 2025.
8. Alami, A.H., *et al.*, Additive manufacturing in the aerospace and automotive industries: Recent trends and role in achieving sustainable development goals. *Ain Shams Eng. J.*, 14, 11, 102516, Nov. 2023.
9. An Introduction to the World of Additive Manufacturing, in: *Additive Manufacturing for Chemical Sciences and Engineering*, pp. 1–18, Springer Nature Singapore, Singapore, 2022.
10. Ajithkumar, S., Arulmurugan, B., Rajeshkumar, L., Prospects of natural fiber-reinforced polymer composites in engineering and commercial applications, in: *Applications of Composite Materials in Engineering*, 1st ed., pp. 335–364, Elsevier Science Ltd., 2025.
11. Nayak, A., Satpathy, I., Patnaik, B.C., Jain, V., Sustainable manufacturing: A paradigm shift towards sustainable human development, in: *Sustainable Human Development*, 1st ed., pp. 45–68, IGI Global, 2023.
12. Schüpfer, D., Wagner-Wenz, R., Hendrich, K., Weidenkaff, A., Materials come around and go around: Adapting to nature's circularity. *MRS Bull.*, 48, 11, 1184–1189, Nov. 2023.
13. Balaji, A., Liu, X., Plastics in circular economy: A sustainable progression, in: *Circular Economy*, 1st ed., pp. 185–204, Springer, 2021.
14. Surat'man, N.E., Tan, Y.L., Mah, J.J.Q., Qu, C., Wang, C.G., Li, Z., Sustainable materials applications: Current challenges and future perspectives, in: *Circularity of Plastics*, 1st ed., pp. 305–320, Elsevier, 2023.

15. Ramakrishna, S., Ramasubramanian, B., Introduction to materials circular economy, in: *Handbook of Materials Circular Economy*, 1st ed., pp. 1–30, Springer Nature Singapore, 2024.
16. Aranda, F.L., Zúñiga, M., Rivas, B.L., Polymers in Circular Economy: A Comprehensive Approach to Sustainability. An overview. *J. Chil. Chem. Soc.*, 68, 3, 5950–5962, Sept. 2023.
17. Kazakova, E. and Lee, J., Sustainable Manufacturing for a Circular Economy. *Sustainability*, 14, 24, 17010, Dec. 2022.
18. Golinska-Dawson, P., From traditional non-sustainable production to closed loop manufacturing: Challenges for materials management based on PPC and EMIS integration, in: *Corporate Environmental Management Information Systems: Advancements and Trends*, 1st ed., pp. 106–120, IGI Global, 2010.
19. Nataraj, G. and Ramesh Babu, S., Enhancing mechanical properties of PLA-based bio composite filament reinforced with horse gram filler for 3D printing applications. *J. Vinyl Addit. Technol.*, vnl.22182, Dec. 2024.
20. Popescu, C., Dissanayake, H., Mansi, E., Stancu, A., Eco Breakthroughs: Sustainable Materials Transforming the Future of Our Planet. *Sustainability*, 16, 23, 10790, Dec. 2024.
21. Eco Materials, in: *Futuristic Trends in Construction Materials & Civil Engineering*, vol. 3 Book 5, First, pp. 161–170, Iterative International Publishers, Selfypage Developers Pvt Ltd., 2024.
22. 2.31 Material Recycling, in: *Comprehensive Energy Systems*, pp. 1018–1042, Elsevier, 2018.
23. Recycling, in: *Industrial Applications of Renewable Plastics*, pp. 67–114, Elsevier, 2017.
24. Prasad, G., Arunav, H., Dwight, S., Ghosh, M.B., Jayadev, A., Nair, D.I., Advancing Sustainable Practice in Additive Manufacturing: A Comprehensive Review on Material Waste Recyclability. *Sustainability*, 16, 23, 10246, 2024, MDPI AG.
25. Mechanical, Chemical and Thermal Recycling of Bio-Composites: A Review, in: *Lecture Notes in Mechanical Engineering*, pp. 387–395, Springer Singapore, Singapore, 2021.
26. Samuel, H.S., Ekpan, F.-D.M., Ori, M.O., Biodegradable, Recyclable, and Renewable Polymers as Alternatives to Traditional Petroleum-Based Plastics. *Asian J. Environ. Res.*, 1, 3, 152–165, Oct. 2024.
27. Raj, T., Chandrasekhar, K., Naresh Kumar, A., Kim, S.-H., Lignocellulosic biomass as renewable feedstock for biodegradable and recyclable plastics production: A sustainable approach. *Renewable Sustainable Energy Rev.*, 158, 112130, Apr. 2022.
28. Dziike, F., Linganiso, L.Z., Mpongwana, N., Legodi, L.M., Biomass conversion into recyclable strong materials. *South Afr. J. Sci.*, 118, 7/8, July 2022.
29. Recycling of biopolymers, in: *Biodegradable Waste Processing for Sustainable Developments*, 1st ed., pp. 272–298, CRC Press, Boca Raton, 2024.
30. Coltelli, M.-B., Gigante, V., Aliotta, L., Lazzeri, A., Recyclability Perspectives of the Most Diffused Biobased and Biodegradable Plastic Materials. *Macromol.*, 4, 2, 401–419, June 2024.

17

Eco Friendly Bio Composites in Additive Manufacturing for Sustainable Applications

Ganesh Nataraj[1]*, Ramesh Babu S.[2], A. R. Ravikumar[3], Karthikeyan Rajendran[1], Arul Pandian[1] and Ajithkumar Sitharaj[2]

[1]Department of Mechanical Engineering, Sri Shakthi Institute of Engineering and Technology, Coimbatore, India
[2]Department of Mechanical Engineering, KPR Institute of Engineering and Technology, Coimbatore, Tamil Nadu, India
[3]Department of Mechanical Engineering, School of Engineering, Mohan Babu University, Tirupati, India

Abstract

The chapter investigates the connection between eco-friendly bio composites and Additive Manufacturing (AM) as a solution to rising worldwide environmental challenges while demonstrating their sustainable application potential. The text introduces biocomposites as sustainable materials which surpass conventional materials through their renewability and biodegradability features and minimal carbon emissions. The text investigates multiple additive manufacturing techniques which work with these materials including Fused Deposition Modeling (FDM), Direct Ink Writing (DIW), Binder Jetting and Selective Laser Sintering (SLS). The paper explains feedstock preparation techniques through extrusion and resin mixing methods while discussing the problems that arise from fiber-matrix compatibility and moisture sensitivity and processing issues such as clogging and inadequate bonding. This research evaluates the mechanical properties of 3D-printed bio composites through tensile and flexural tests and impact resistance measurements and thermal stability assessments using Thermogravimetric Analysis (TGA) and Differential Scanning Calorimetry (DSC). The chapter demonstrates various sustainable applications throughout packaging and automotive industries as well as biomedical fields and construction and consumer product sectors. The discussion about environmental implications uses Life Cycle Assessment (LCA) to demonstrate

**Corresponding author*: ganmech2002@gmail.com

Gurudas Mandal, Swarup Kumar Ghosh, Snehanshu Pal and Sandip Kunar (eds.) Sustainable Manufacturing of Advanced Materials, (351–376) © 2026 Scrivener Publishing LLC

reduced carbon footprints and circular economy principles while addressing the complexities of methodological approaches. The chapter concludes by recognizing current boundaries while proposing future directions that stress the need for interdisciplinary collaboration and robust legislative backing to advance sustainable materials and design practices in manufacturing systems.

Keywords: Eco-friendly bio composites, additive manufacturing (AM), sustainable applications, life cycle assessment (LCA), natural fibers

17.1 Introduction

Traditional manufacturing creates major environmental problems because it chooses economic efficiency over environmental protection. The entire product lifecycle from raw material extraction to disposal generates significant waste and pollution which leads to environmental degradation [1, 2]. The implementation of sustainable practices faces two main obstacles which include high energy usage and greenhouse gas emissions and complex life cycle assessments and material recycling challenges [3]. The implementation of sustainable manufacturing systems (SMS) faces multiple obstacles because governments do not provide sufficient support for new technologies and industries fail to adopt sustainable principles [4]. The manufacturing sector must make green initiatives and advanced manufacturing methodologies—its top priority to reduce environmental impact while improving sustainability results.

Environmental health together with resource stewardship depend heavily on sustainability practices in materials science and production. The manufacturing process of metals leads to high energy usage and greenhouse gas emissions, thus requiring sustainable production methods [5]. Material scientists need to evaluate the complete material life cycle from extraction to disposal to select options that reduce environmental harm while promoting recycling programs [6]. Research indicates that material scientists lack sufficient understanding about sustainability impacts in their designs which demonstrates the need for better educational programs and sustainability standards in funding systems [7]. The development of sustainable materials stands as a vital solution to address resource limitations while creating a circular economy that promotes recycling and reuse practices [8]. The promotion of sustainability in materials science remains essential because it helps fight climate change while protecting the well-being of future populations [9].

Additive manufacturing (AM) occupies a pivotal position in the mitigation of material waste and the optimization of energy consumption, particularly *via* methodologies such as fused deposition modeling (FDM) among others. In contrast to conventional manufacturing approaches, AM markedly reduces the waste of raw materials, yielding reductions ranging from 35% to 80% in the fabrication of final components [10]. The integration of renewable energy sources and advanced waste reduction techniques in FDM has demonstrated notable decreases in carbon emissions, and recycled materials maintain their mechanical properties while attaining high recovery rates [11]. Additionally, AM processes are methodically designed to improve lifecycle emissions and resource efficiency, which advances a more sustainable manufacturing framework [12]. Despite AM's intrinsic energy requirements, significant energy savings can be achieved through careful methodology and product design selection [13]. All things considered, AM's ability to produce localized, customized goods on demand enhances its sustainability potential by reducing the need for transportation and the resulting material waste.

Eco-friendly bio composites represent composite materials that amalgamate biodegradable polymer matrices with renewable natural fibers, thereby providing a sustainable substitute for conventional petroleum-derived products. These materials are suitable for a wide range of applications, such as the automotive, biomedical, and aerospace industries, because of their renewable nature, biodegradability, and reduced environmental effect [14, 15]. The primary components of bio composites are natural fibers such as sisal, hemp, and flax. They simultaneously make the material lighter and stronger [16]. The advantages of bio composites extend beyond their environmental benefits. By employing agricultural and industrial by-products, they also contribute to a circular economy by reducing pollution and saving energy [17]. As people become more aware of environmental issues, the demand for bio composites is expected to rise. This will lead to more innovation and use in a variety of fields [14, 18].

17.2 Fundamentals of Bio Composites

17.2.1 Definition and Classification

Bio composites serve as sustainable alternatives to synthetic composites because they consist of natural fibers bonded to biodegradable polymer matrices. These materials gain increasing popularity because they possess environmentally beneficial characteristics including biodegradability and renewability and lower carbon emissions [19]. The combination of natural fibers such as sisal, jute, hemp and flax with their availability and affordability and light weight makes them popular reinforcement materials in bio composites [20, 21]. The combination of natural fibers with biodegradable polymers like PLA and PHA has enabled new environmentally friendly additive manufacturing (AM) applications [22, 23].

17.2.1.1 Natural Fiber-Reinforced Composites

When natural fibers are combined with polymer matrices, the result is the production of natural fiber-reinforced composites (NFRCs), which improve the mechanical and thermal properties of the material. The composites show excellent suitability for AM because they are both sustainable and processable. The addition of hydroxyapatite nanocrystals to jute fibers produces enhanced interfacial properties which leads to better mechanical strength and thermal conductivity. The advantages of this material make it suitable for additive manufacturing techniques including direct writing (DW) [21]. Wood fiber-reinforced PLA composites have been successfully employed in FDM to create intricate structures like weaved bowls and honeycombs, demonstrating their long-term promise.

17.2.1.2 Biodegradable Polymer Matrices

Bio composites contain biodegradable polymer matrices which serve as environmentally friendly alternatives to non-recyclable polymers. Among all biodegradable polymers, polylactic acid (PLA) stands as one of the most widely used types. The material serves multiple purposes because it can be recycled while being non-toxic to living organisms and compatible with AM manufacturing processes [22, 23]. Scientists conduct research on polyhydroxyalkanoates (PHA) and polybutylene succinate (PBS) to determine their potential use in bio composite applications. The combination of these matrices with natural fibers produces bio composites that demonstrate strength and heat resistance for particular applications.

17.2.1.3 Types of Natural Fibers

Natural fibers exist in separate categories which distinguish them through their inherent characteristics and their natural geographic locations. The high tensile strength and rigidity of bast fibers including flax hemp and jute makes them suitable for polymer matrix reinforcement applications [19, 21]. Seed fibers including cotton and leaf fibers consisting of sisal and pineapple find common use in bio composite formulations because of their mechanical properties and availability [24]. The use of rice husks and wheat middling's as PLA-based bio composite fillers represents a strategy to decrease material costs while minimizing environmental impact [25].

17.2.2 Advantages Over Conventional Composites

Bio-conscious eco-composites demonstrate superior advantages compared to traditional composites because they enhance ecological sustainability and economic viability and processing efficiency. The main sustainability indicators that show bio-composites outperform conventional synthetic composites are presented in Figure 17.1.

Figure 17.1 Advantages over conventional composites.

17.2.2.1 Renewability

The primary advantage of bio composites stems from their ability to derive resources from renewable materials. Bio composites differ from conventional composites because they use natural fibers extracted from flora (flax, hemp, jute) and bio-based polymers obtained from renewable biomass (PLA from corn starch and bio-PE from sugarcane). The continuous availability of sustainable raw materials exists because bio composites use renewable resources which reduces the depletion of natural resources and decreases geopolitical dependencies on petroleum. The agricultural and forestry practices used for resource cultivation help sequester carbon which enhances their environmental benefits.

17.2.2.2 Biodegradability

Bio composite materials that incorporate biodegradable polymers such as PLA, PHAs, and PCL can decompose naturally after fulfilling their purpose. The degradation process transforms the material into its fundamental elements which include water together with carbon dioxide and organic biomass. The degradation process depends on microbial activity which operates in industrial composting facilities under specific conditions. The environmental problems that traditional synthetic composites create in landfills and natural ecosystems do not apply to biodegradable composites because they break down naturally. Bio composites provide an environmentally friendly waste disposal solution because they naturally decompose which supports circular economy principles.

17.2.2.3 Lower Carbon Footprint

Traditional composites face significant reduction in carbon footprint throughout the complete life cycle of bio composite materials. The growth cycle of natural fibers allows plants to remove atmospheric carbon dioxide from the air while requiring minimal energy for cultivation. Natural fibers and bio-based polymers require less energy during manufacturing while producing lower greenhouse gas emissions than the energy-intensive production of synthetic fibers (such as glass or carbon fibers) and petroleum-derived plastics. The reduced energy consumption and environmental emissions during bio composite material creation through production and usage and disposal result in their essential role for climate change mitigation.

17.2.2.4 Economic Feasibility in Developing Regions

The production of bio-composites shows strong economic potential to boost agricultural and rural regions especially in developing countries. The local abundance of natural fibers enables extensive employment opportunities since their cultivation and initial processing steps that include retting and decortication require many workers. The large number of employment opportunities created by this situation leads to higher income levels among farmers and local residents who experience better rural living conditions and reduced poverty. The lower capital expenses for processing natural fibers instead of synthetic fiber facilities make bio-composite manufacturing more affordable and reachable for these regions which promotes sustainable economic growth.

17.2.2.5 Reduced Abrasiveness

The processing equipment shows lower abrasion when using natural fibers because these fibers possess softness and flexibility; yet, they remain less abrasive than synthetic fibers such as glass or carbon. The reduction of abrasiveness results in decreased wear on expensive manufacturing equipment such as extrusion screws and molds and dies. The longer machinery lifespan results from this occurrence which also decreases maintenance needs and related costs while reducing the need for frequent equipment replacements. The economic advantages from this occurrence pair up with resource efficiency because manufacturing assets can operate for longer periods.

17.2.2.6 Lower Density and High Specific Properties

Natural fibers demonstrate superior softness and flexibility compared to synthetic reinforcing materials including glass and carbon fibers while showing reduced abrasiveness to processing equipment. The reduced abrasiveness of these materials results in decreased wear and tear on costly manufacturing equipment including extrusion screws and molds and dies. The operational lifespan of machinery is extended and the demands for maintenance and the associated financial burdens experience a decrease, thereby reducing the necessity for frequent equipment replacement. This phenomenon not only provides economic benefits but also promotes resource efficiency by prolonging the functional longevity of manufacturing assets.

17.3 Additive Manufacturing: An Enabler of Sustainable Design

17.3.1 Overview of Additive Manufacturing

Additive manufacturing, often known as 3D printing, is a breakthrough manufacturing technology that adds material layer by layer based on digital models. This approach is fundamentally different from traditional subtractive manufacturing processes that start with a solid block. The techniques employed by AM include vat photopolymerization, material jetting, powder bed fusion, directed energy deposition, material extrusion, binder jetting and sheet lamination and each technique has its specific applications and material possibilities that include metals, polymers, ceramics and composites [26–28]. The technology offers great advantages through its ability to create complex geometries and reduce material waste and enable mass customization that traditional manufacturing approaches cannot match [29]. AM has gained widespread adoption in aerospace, automotive, medical and consumer goods sectors because it enables the production of complex components while reducing prototyping and production time [30–32]. The production cost of AM is high, and the material compatibility is limited, and the speed and precision and sustainability of AM needs improvement. Research reveals that natural fiber polymer composites are emerging as environmentally friendly affordable lightweight materials which can substitute synthetic materials across various industries [33]. The combination of natural fillers with PLA-based matrices in green composites enhances mechanical properties that support both sustainability initiatives and waste valorization projects [34]. Root fiber composites (RFCs) are an important bio-composite family because surface modification methods enhance fiber–matrix bonding which leads to better mechanical properties. Bio-based composites continue to develop due to improved processing methods and enhanced production techniques. The combination of AM with operations research has generated new decision-making issues for supply chain management which requires updates to existing operational protocols [35]. The current research investigates the combination of additive manufacturing (AM) with alternative fabrication methods including subtractive techniques to overcome certain inherent limitations of AM [36]. The advancement of additive manufacturing technology holds the potential to transform production operations while delivering substantial opportunities for innovation along with enhanced efficiency across multiple industrial domains [37].

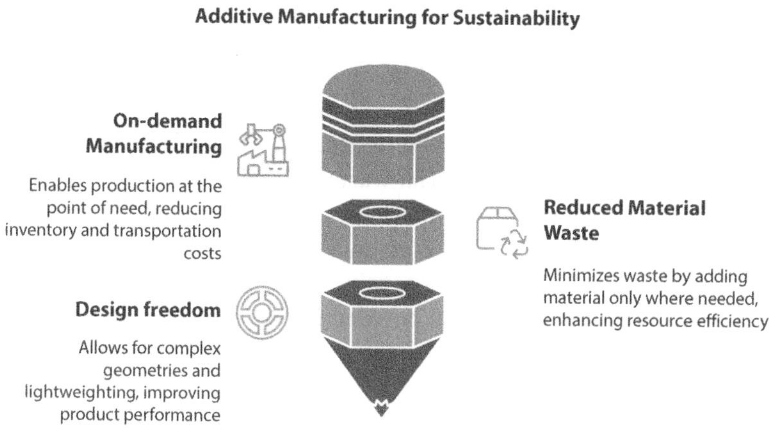

Figure 17.2 Key sustainability benefits of additive manufacturing (AM).

17.3.2 AM Benefits for Sustainable Production

Additive Manufacturing presents numerous inherent advantages that correspond with the tenets of sustainable manufacturing, encompassing effective utilization of resources, energy efficiency, and innovative design methodologies. Figure 17.2 elucidates the fundamental sustainability facilitators of additive manufacturing: localized production on an as-needed basis, diminished material usage, and broadened design capabilities.

17.3.2.1 On-Demand and Localized Manufacturing

Additive Manufacturing (AM) drives a fundamental change in manufacturing operations by moving away from traditional mass production systems toward a decentralized on-demand production model. The new manufacturing paradigm enables goods production exactly where and when customers need them, thus reducing dependence on large inventory storage and complex worldwide supply networks. The practice of localized production which AM promotes reduces transportation costs while simultaneously lowering the carbon emissions that result from cross-country product delivery. The invention enhances regional economic growth and supply chain stability, making them more resilient to unforeseen disruptions. Quick and efficient fabrication of bespoke or replacement components lowers operating downtime and allows for product repair, increasing its lifespan.

17.3.2.2 Reduced Material Waste

The manufacturing process of additive manufacturing differs from traditional subtractive methods like machining or carving because it builds objects by adding material only where needed. The additive nature of this process leads to substantial material waste reduction because it eliminates 70-90% of scrap material and sometimes more depending on the specific material and process used. The efficiency of this process becomes most important when working with expensive or scarce materials. The ability to recycle and repurpose unfused powder during Selective Laser Sintering (SLS) processes reduces waste and creates a substantial contribution to sustainable manufacturing practices.

17.3.2.3 Design Freedom for Complex Geometries

Additive Manufacturing (AM) technologies enable designers to create complex geometries and optimized constructs that traditional manufacturing methods cannot produce at an affordable cost. The capability enables sustainable applications through its ability to:

Lightweighting: The creation of components through optimized internal lattice frameworks or topologies enables lighter products with decreased operational energy usage because they achieve required strength and stiffness with minimal material usage (for example, in aerospace or automotive applications).
Part Consolidation: The process of uniting multiple components into one advanced printed structure eliminates assembly requirements and decreases part numbers and failure points and simplifies supply chain operations.
Functional Integration: The direct integration of functionality into materials and structures produces more sustainable and efficient product designs.

The design flexibility of this approach naturally leads to better resource efficiency and performance enhancement which are core principles of sustainable production.

17.4 Integration of Bio Composites in Additive Manufacturing

17.4.1 Material Compatibility and Challenges

The successful implementation of bio composites in additive manufacturing depends on solving multiple material-related issues. The main obstacles in bio composite production stem from natural fiber properties and their behavior when mixed with polymer matrices during feedstock creation and printing processes.

17.4.1.1 Feedstock Formulation

The creation of suitable bio composite feedstock (e.g., filaments for Fused Deposition Modeling, inks for Direct Ink Writing, powders for Selective Laser Sintering) begins with essential challenges during this initial phase:

17.4.1.1.1 Fiber–Matrix Compatibility

The hydrophilic cellulose structure of natural fibers creates poor compatibility with hydrophobic polymer matrices that originate from conventional plastics. The difference in polarity between the fiber and matrix leads to poor interfacial bonding which results in poor stress transfer and reduced mechanical performance of the printed component. Natural fiber surface modifications through alkalization and silane treatment and grafting become necessary to improve wettability and chemical bonding but these procedures make feedstock preparation more complicated and expensive. The material requires perfect fiber distribution throughout the matrix because poor dispersion leads to fiber agglomeration which creates weaknesses in the material structure.

17.4.1.1.2 Moisture Sensitivity

Natural fibers show high hygroscopic behavior which means they tend to absorb moisture from their environment. The absorbed moisture creates multiple manufacturing issues when performing additive manufacturing:

Hydrolysis: The high temperatures needed for melt processing (e.g., Fused Deposition Modeling) can trigger hydrolytic degradation of natural fibers and specific biodegradable polymers (such as Polylactic Acid) because of their absorbed moisture which reduces their molecular weight and mechanical strength.

Bubble Formation: The fast evaporation of absorbed moisture during extrusion or laser sintering processes creates voids and bubbles in deposited layers which damages both the structural quality and surface finish of fabricated components.

Processing Inconsistencies: The inconsistent moisture levels lead to unpredictable melt flow behavior and irregular extrusion processes and weak layer adhesion which makes it difficult to produce reliable high-quality printed outputs. The essential steps of drying fiber materials and polymer pellets before compounding and storing finished filaments or powders in dry environments increase processing complexity.

The challenges can be overcome through processing strategies and chemical treatments as shown in Figure 17.3.

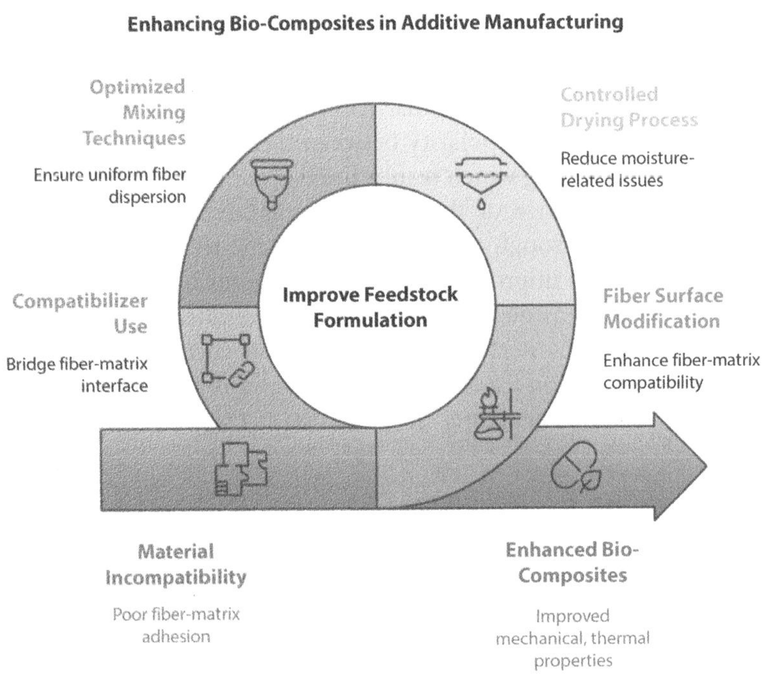

Figure 17.3 Strategies to enhance bio-composites for additive manufacturing.

17.4.1.2 Processing Issues

The Fused Deposition Modeling (FDM) technique among other extrusion-based methods encounters unique manufacturing challenges when using bio composites despite optimized feedstock materials:

17.4.1.2.1 Clogging

Natural fibers containing length, irregularity or substantial concentration tend to block the printer nozzle repeatedly during the printing process. Natural fibers tend to stack up while burning within the heated zone which disrupts material flow. The situation requires the use of larger nozzle diameters that lower print resolution and strict control over fiber length and aspect ratio and optimal nozzle design to prevent blockages.

17.4.1.2.2 Uneven Flow

Natural fibers demonstrate strong influence on the rheological properties of molten polymers. The addition of these fibers tends to increase the melt viscosity while creating non-Newtonian flow behavior in the material. The modified flow characteristics result in irregular material flow patterns which produce variable layer thickness and printing process pulsations. The production of high print quality demands a stable material flow, so it becomes essential to precisely control the extrusion temperature along with pressure and printing velocity. The flow behavior becomes more inconsistent when filament or ink contains varying fiber concentrations or distribution patterns.

17.4.1.2.3 Poor Bonding (Layer Adhesion)

The mechanical strength of three-dimensional printed components depends heavily on achieving strong bond formation between successive layers. Bio composites face interfacial issues because natural fibers rapidly cool down and thermal degradation happens at interfaces during multi-layer depositions while their hydrophilic nature makes fusion between layers difficult. The resulting mechanical properties in the Z-direction become weaker and delamination problems occur. Achieving better inter-layer adhesion requires optimizing the build plate temperature together with the enclosure temperature and cooling fan settings while studying how fibers align through analysis.

17.4.2 Filament and Resin Preparation

The synthesis of appropriate feedstock materials is essential for the effective additive manufacturing of bio-composites, necessitating the modification of traditional polymer processing methodologies to address the distinctive attributes of natural fibers and bio-derived matrices.

17.4.2.1 Single-/Twin-Screw Extrusion for Filament Fabrication

Fused Deposition Modeling (FDM) requires continuous filament as its basic feedstock material. The primary method for making bio composite filaments involves extrusion processes. The extruder receives a mixture of polymer pellets which act as the matrix material and short chopped natural fibers which function as reinforcement. Single-screw extrusion operates at low cost but produces less efficient mixing which leads to poor fiber distribution especially when fibers reach high concentrations. This method works best for materials that have good flow properties and situations where uniform mixing is not the main priority. Bio composite filaments production benefits most from twin-screw extrusion because this method delivers superior mixing performance. The intermeshing screws produce powerful shear forces which enable better distribution of natural fibers throughout the polymer matrix to achieve uniform properties in the final filament. Twin-screw extruders offer more control over temperature profiles and residence periods, reducing thermal deterioration of natural fibers. Extrusion needs precise temperature management, screw velocity control, and die design adjustments to prevent fiber damage and achieve uniform melt and filament diameter. To ensure accurate diameter measurements for 3D printing machine feeding, cooled filament is treated in a water bath before winding onto a reel.

17.4.2.2 Resin Mixing Techniques for SLA/Inkjet Bio Composites

While Fused Deposition Modeling (FDM) employs solid filaments, alternative additive manufacturing technologies such as stereolithography (SLA) and material jetting (inkjet printing) make use of liquid resins or inks. The integration of biocomponents into these systems necessitates the application of specialized mixing methodologies. In the case of SLA, the matrix consists of a photopolymer resin that undergoes curing when exposed to ultraviolet (UV) light. Biocomponents (for instance, cellulose nanocrystals, finely powdered natural fibers, and bio-derived photoinitiators) must be uniformly dispersed throughout this resin matrix.

The mixing process generally entails mechanical stirring, ultrasonication (the application of high-frequency acoustic waves to disaggregate agglomerates and achieve superior particle dispersion), as well as high-shear mixing techniques such as rotor-stator or planetary mixers to impart significant shear forces. For inkjet printing or direct ink writing, stringent control over rheological properties and particle size is imperative to avert nozzle blockages and guarantee proper jettability. Using homogenization methods (such high-pressure homogenizers), biocomponents are added to a viscous ink or hydrogel foundation to create incredibly stable, tiny dispersions. Degassing, which is usually done under vacuum, is essential for avoiding printing flaws and guaranteeing steady flow. Degassing is the process of removing air bubbles from the resin or ink. Significant issues include ensuring that the biocomponents do not interfere with the photopolymerization or jetting processes, preventing sedimentation, and maintaining particle stability.

17.4.2.3 Use of Compatibilizers and Surface Treatments

The preparation of feedstock involves different chemical and physical modification techniques to address both fiber-matrix incompatibility and natural fiber moisture sensitivity. The main technique for enhancing natural fiber-polymer matrix adhesion uses polymeric additives known as compatibilizers which are hydrophilic-hydrophobic interfaces. The compatibilizers act as "bridges" through their functional groups which enable bonding between fibers and matrices. MAPP stands as one of the most commonly used compatibilizers because it belongs to the malleated polyolefin family. Natural fibers receive multiple surface treatments before or during their combination with polymer matrices. The surface modification techniques for natural fibers include alkalization through mercerization with alkaline solutions to remove impurities and increase roughness and silane treatment for chemical bonding between fibers and matrices and acetylation to reduce hydrophilicity and graft copolymerization for surface property modification. The modifications serve to enhance mechanical properties and reduce moisture uptake while improving the manufacturing process of bio-composites for additive manufacturing (AM).

17.5 Mechanical and Thermal Properties

The assessment of effective bio composites produced by additive manufacturing depends on their mechanical and thermal properties because these characteristics decide their suitability for various sustainable applications. Knowledge about these material properties along with their influencing factors stands essential for creating optimized dependable products through design development.

17.5.1 Tensile, Flexural, and Impact Properties of Printed Bio Composites

The standard testing procedures assess three-dimensional printed bio composites to determine their fundamental mechanical characteristics. The tensile properties determine how materials behave when subjected to stretching forces. Tensile strength represents the maximum stress a material can handle before failure or fracture occurs during tensile testing. The material's stiffness and resistance to elastic deformation under tensile forces are measured through its Young's modulus (Elastic Modulus). Young's modulus indicates material rigidity while a higher value indicates greater rigidity. The tensile strength measurement reflects the maximum stress a material can handle before breaking and elongation at break indicates material ductility through its measurement of percentage length increase before material failure. Natural fibers embedded in printed bio composites produce better stiffness and strength compared to pure polymer matrices but might reduce ductility levels. The bending stress response of a material is measured through three-point or four-point bending tests which define its flexural properties. The flexural strength indicates the maximum stress that a material can withstand during bending before it fails or fractures while the flexural modulus measures its bending rigidity. The addition of bio-composite reinforcement results in better flexural strength and modulus values than those of plain polymers which makes them suitable for applications requiring structural stability. Materials undergo energy absorption tests during impact evaluations through Charpy or Izod impact tests to determine their impact properties. The material's toughness becomes evident through this assessment together with its capability to resist fracturing when subjected to impacts. The effectiveness of natural fibers as crack arrestors can lead to enhanced impact strength but poor fiber-matrix bonding results in delamination and reduced toughness. The printing process along with its layer-by-layer operation results in directional dependence of material properties known as anisotropy which emerges from possible fiber alignment.

17.5.2 Influence of Fiber Content, Orientation, and Treatment

Natural fiber-based printed bio composites experience significant mechanical and thermal behavior changes because of multiple elements that affect these properties. Tensile strength along with flexural strength and modulus experience an increase when natural fiber loading reaches its optimal threshold. The mechanical properties decrease after passing the optimal threshold because of poor fiber distribution and fiber-fiber contacts and insufficient matrix material to surround fibers. The use of higher fiber concentrations leads to increased melt viscosity that makes the printing process more challenging. The extrusion process of fused deposition modeling (FDM) leads to short fibers aligning themselves according to the printing direction (raster angle) because fiber orientation remains crucial. The strength and stiffness of these materials increase when tested along the fiber alignment direction, yet decrease when tested against the fiber alignment or in the Z-axis direction. Strategic control of print paths and infill patterns enables the optimization of fiber alignment for different load-bearing applications. Fiber treatment procedures described in Section 17.4.2.3 enhance the bond quality between fibers and matrix materials. Better adhesion between the matrix and reinforcement enables stress transfer which results in significant strength improvements across tensile testing and flexural testing and impact testing. The treatments implemented reduce moisture absorption by fibers which protects the printed component from hydrolytic degradation during processing and improves its long-term dimensional stability.

17.5.3 Thermal Behavior: TGA, DSC Analysis

Bio-composites require thermal dynamic analysis to optimize manufacturing parameters and predict their behavior in different thermal conditions. Thermal analysis techniques include thermogravimetric analysis (TGA) which measures specimen weight changes throughout temperature or time increments. Through this analysis, method researchers determine thermal stability by identifying when material decomposition starts at natural fiber-based bio-composites while also determining moisture content through initial mass loss measurements before major degradation occurs. The amount of char left after complete decomposition helps determine the actual fiber concentration present in the material. DSC, differential scanning calorimetry monitors material thermal exchanges during temperature or time changes. The technique measures three critical temperatures in polymers: Glass transition temperature (Tg) where the material changes from rigid glassy to rubbery state, melting temperature (Tm) which marks the solid-to-liquid transition of crystalline polymers and crystallization temperature (Tc), which indicates the polymer crystallization point during cooling. Natural fibers introduce thermal transition alterations because they either act as nucleation sites or restrict polymer chain movement. Knowledge of these transitions enables the determination of suitable FDM printing parameters including nozzle and bed temperatures for achieving proper melting and flow and solidification of materials.

17.5.4 Comparison with Neat Polymers and Conventional Composites

The evaluation framework for printed bio composites requires comparing them to neat polymers and conventional synthetic composites. The addition of natural fibers to polymers results in superior stiffness and strength measurements for bio composites when compared to unreinforced polymer matrices. The addition of natural fibers results in reduced ductility and break elongation values. The natural fiber addition improves thermal stability of the composite until a particular threshold, but the degradation temperature remains limited by the unstable nature of natural fibers. The density of bio composites exceeds some neat polymers but they weigh less than numerous conventional composites. The specific mechanical properties of natural fiber composites match those of various glass fiber composites yet their absolute strength and stiffness values fall below those of high-performance synthetic fiber composites including carbon fiber. Natural fibers show lower intrinsic strength and higher variability which leads to this performance difference. The impact strength shows inconsistent behavior because natural fibers sometimes improve it through energy dissipation, yet synthetic fibers normally provide better impact resistance. The thermal resistance of natural fiber bio composites remains lower than glass or carbon fiber composites which maintain their strength at elevated processing and service temperatures, thus limiting their use in high-temperature applications. Bio composites surpass conventional composites in environmental performance because they are renewable, biodegradable, and require less energy during manufacturing while producing lower carbon emissions as explained in Section 17.2.2. The environmental benefits of bio composites typically compensate for their performance limitations when applications do not require exceptional mechanical properties.

17.6 Sustainable Applications of Additively Manufactured Bio Composites

The combination of environmentally friendly bio-composites with additive manufacturing technology enables numerous sustainable uses in various sectors because of their unique characteristics and production benefits.

17.6.1 Packaging

Packaging generates large amounts of plastic waste which makes it a vital area for developing sustainable new materials. The production of customized lightweight biodegradable packaging solutions through bio composite additive manufacturing stands as a compelling alternative. The manufacturing process reduces plastic waste problems because it uses renewable plastics and enables biodegradable solutions through composting or natural degradation processes. The materials undergo research for food contact material applications in addition to their standard packaging uses. Bio composites stand as essential materials for food packaging because they can incorporate natural antimicrobial compounds which lengthen product shelf life while being environmentally friendly. Additive manufacturing's design flexibility enables producers to create complicated protective packaging with optimum materials for improved protection and customer appeal.

17.6.2 Automotive and Aerospace

The automotive and aerospace industries require lightweight materials for better fuel efficiency and lower emission levels. The production of various components becomes possible through additively manufactured bio composites as a viable alternative. The automotive sector implements lightweight components for interior trims and dashboard elements and non-structural parts which work together to reduce vehicle weight and produce better fuel efficiency and lower emissions. The aerospace industry evaluates bio composites for usage in interior cabin parts and seatbacks and aerodynamic fairings because they offer significant weight reduction potential. Bio composites become suitable for customized tooling and jigs production through additive manufacturing because this technology enables quick creation of complex geometries. The ability to create molds and production aids quickly through sustainable materials; reduces manufacturing waste, and shortens production lead times for both industries.

17.6.3 Biomedical and Healthcare

The biomedical and healthcare fields will experience substantial growth because bio composite materials possess both biocompatibility and biodegradability features when combined with personalized manufacturing methods of additive manufacturing (AM). The combined approach enables the creation of personalized implants together with bone scaffolds and surgical templates that function as prosthetic devices. The body-compatible implants made of bio composites are engineered to match individual body structures while degrading slowly and safely which eliminates the need for repeated surgeries in particular cases. The controlled breakdown patterns together with adjustable mechanical properties of bio composites make them ideal for regenerative medical procedures. The production of complex drug delivery systems with customized designs and release patterns becomes possible through AM and bio composite technology for pharmaceutical delivery. The fabrication technology enables the creation of customized medical devices together with highly accurate anatomical models for both surgical preparation and educational needs, thus improving medical education and patient care delivery.

17.6.4 Construction and Architecture

Construction represents a major consumer of resources while simultaneously generating significant amounts of waste. The construction industry benefits from sustainable bio composite materials produced through additive manufacturing for building applications and architectural components. The manufacturing process of insulation panels and ornamental features and non-load-bearing architectural elements utilizes natural fiber-reinforced biopolymers. The materials provide excellent thermal insulation along with acoustic benefits which lead to better energy efficiency in buildings. The large-scale structure manufacturing capability of robotic extrusion in additive manufacturing offers potential solutions for affordable housing production. Bio composite materials sourced locally enable rapid and affordable sustainable housing construction in areas with abundant natural resources to solve housing shortages by implementing environmentally friendly methods.

17.6.5 Consumer Products and Design

The consumer products industry benefits substantially from the design flexibility combined with environmentally friendly aspects of bio-composites manufactured through additive processes. The production method allows the creation of multiple sustainable consumer products which include green toys and lightweight furniture and fashionable accessories and strong electronic enclosures. On-demand product design and just-in-time production reduce waste and allow for customer customization. Art and design can use bio-composites in additive manufacturing to make sustainable items with little environmental impact. Artists and designers may push the boundaries of sustainable design by exploring bio-based materials and novel geometries, leading to a new era of ecologically responsible product innovation.

17.7 Environmental Impact and Life Cycle Assessment (LCA)

Evaluating the environmental performance of bio-composites generated using additive manufacturing requires a thorough technique, such as life cycle assessment (LCA). The ISO 14040 and 14044 framework assesses a product's environmental impact from raw material extraction to disposal, providing a comprehensive understanding of environmental loads. The integration of bio-composites with additive manufacturing produces significant reductions in environmental impacts. The bio-composite materials achieve reduced dependence on fossil fuels because they utilize renewable natural fibers and bio-based polymers while sequestering carbon during plant growth. The additive manufacturing method reduces material waste more effectively than traditional subtractive approaches and allows producers to manufacture on demand from localized sources. The new production approach reduces transportation emissions and optimizes supply chain operations. The design capabilities of additive manufacturing technologies enable users to design lightweight optimized structures which reduce operational energy usage and decrease the total carbon impact of the product.

Bio-composites gain a significant environmental benefit from their natural ability to decompose because it strengthens circular economy systems. The majority of conventional synthetic composites remain in landfills; yet, biodegradable bio-composites break down naturally in appropriate environments to restore organic matter to the terrestrial ecosystem and reduce plastic pollution. This approach follows circular economy principles by promoting resource regeneration while decreasing the requirement for fresh resource extraction. The execution of Life Cycle Assessment for additively manufactured bio-composites faces multiple obstacles because of limited complete life cycle inventory information about natural fibers and particular additive manufacturing methods and complex feedstock preparation procedures and various end-of-life scenarios. These emerging technologies exist in an evolutionary phase which leads to continuous changes in data and methodologies that require ongoing research for accurate dependable environmental impact assessments.

17.8 Challenges and Future Perspectives

The extensive integration of additively manufactured bio composites encounters numerous considerable obstacles. The material-related issues including natural fiber-polymer matrix incompatibility and moisture sensitivity lead to material degradation and reduced mechanical properties. Fused Deposition Modeling (FDM) processing faces significant challenges because nozzle obstruction and inconsistent material flow and poor layer adhesion negatively impact print quality and component structural integrity. The execution of precise life cycle assessments faces challenges because of limited data availability and the early development stage of these technologies.

The prospects for environmentally sustainable bio composites in additive manufacturing appear promising despite existing obstacles. Future research will focus on developing advanced functional bio composites and using nanofillers to improve material properties and expanding the use of recycled materials. The implementation of additive manufacturing for sustainable applications will advance through processing innovations that include multi-material printing and real-time monitoring and industry standards development and cost reduction initiatives.

17.9 Conclusion

The combination of environmentally sustainable bio composites with additive manufacturing marks a fundamental change toward creating truly sustainable manufacturing approaches. The combination of bio composites with additive manufacturing technology enables the transformation of various industries by leveraging their renewable and biodegradable nature and low carbon footprint alongside the waste-reducing and localized production and complex design capabilities of additive manufacturing. The powerful combination enables the creation of groundbreaking products which substantially reduce environmental effects throughout their complete life span, thus addressing worldwide problems of resource scarcity and plastic contamination.

The obstacles related to material compatibility and processing complexities and exhaustive life cycle evaluations continue to exist, but interdisciplinary research and development teams actively work to overcome these challenges. The future path for bio composite manufacturing through additive methods will focus on eco-centric innovation which will drive the development of advanced functional materials and improved manufacturing methods and complete standards establishment. The ongoing development of bio composites in additive manufacturing receives support from essential policy measures and rising worldwide interest in sustainable solutions which positions them to become essential for building a greener resource-efficient future for products and industries and the planet.

AI disclosure statement: The author acknowledges the use of AI-based tools for minor language editing or grammatical correction. However, all intellectual contributions, analyses, and conclusions presented in this chapter are entirely the author's own.

References

1. Bambam, A., Gajrani, K., Challenges in achieving sustainability during manufacturing, in: *Sustainable Materials and Manufacturing Technologies, 1st ed.*, pp. 108–124, CRC Press, 2023.
2. Jena, M.C., Mishra, S.K., Moharana, H.S., Environmental Impact of Conventional Manufacturing, Green Manufacturing Initiatives to Mitigate the Impact and Enhancing Sustainability. *Environ. Claims J.*, 36, 4, 485–504, 2024.
3. Environmental Aspects of Manufacturing Processes, in: *Springer Series in Advanced Manufacturing* [Internet], pp. 339–60, Springer London, London, 2011.
4. Patra, S.K., Raj, T., Arora, B.B., An analysis and modelling of selected barriers for a sustainable manufacturing system using ISM technique. *Int. J. Oper. Res.*, 41, 3, 373, 2021.
5. Raabe, D., The Materials Science behind Sustainable Metals and Alloys. *Chem. Rev.*, 123, 5, 2436–608, 2023.
6. Ashby, M., Shercliff, H., Cebon, D., Materials, environment, and sustainability, in: *Materials: Engineering, Science, Processing and Design, 4th ed.*, pp. 581–612, Butterworth-Heinemann, 2019.

7. Vink, K., Sustainable life cycle design aspects: how aware are material scientists? *SN Appl. Sci.*, 2, 8, 2020.
8. Titirici, M., Baird, S.G., Sparks, T.D., Yang, S.M., Brandt-Talbot, A., Hosseinaei, O., *et al.*, The sustainable materials roadmap. *J. Phys. Mater.*, 5, 3, 032001, 2022.
9. Singh, M., Ohji, T., Asthana, R., Green and sustainable manufacturing of advanced materials—progress and prospects, in: *Green and Sustainable Manufacturing of Advanced Material, 1st ed.*, pp. 3–10, Elsevier, 2016.
10. Jung, S., Kara, L.B., Nie, Z., Simpson, T.W., Whitefoot, K.S., Is Additive Manufacturing an Environmentally and Economically Preferred Alternative for Mass Production? *Environ. Sci. Technol.*, 57, 16, 6373–86, 2023.
11. Subramani, R., Ali Rusho, M., Sekhar, KC., Mohammed, S.A., Abdulah, S.A., Hashim, R.D., *et al.*, Utilizing bio-energy and waste reduction techniques in FDM: Toward sustainable production practices. *Appl. Chem. Eng.*, 7, 4, 2024.
12. Egon, A., Bell, C., Shad, R., Sustainability in Additive Manufacturing: Analyzing the Environmental Impact of Additive Manufacturing Processes, MDPI AG, 2024.
13. Maware, C., Muvunzi, R., Machingura, T., Daniyan, I., Examining the Progress in Additive Manufacturing in Supporting Lean, Green and Sustainable Manufacturing: A Systematic Review. *Appl. Sci.*, 14, 14, 6041, 2024.
14. Hoque, M.E., Sharif, A., Jawaid, M., Green biocomposites for biomedical engineering: Design, properties, and applications, in: *Woodhead Publishing Series in Biomaterials, 1st ed.*, pp. 1–474, Woodhead Publishing, 2021.
15. Karak, N., Advances in biocomposites and their applications, in: *Woodhead Publishing Series in Composites Science and Engineering, 1st ed.*, pp. 1–390, Woodhead Publishing, 2024.
16. Pokharel, A., Falua, K.J., Babaei-Ghazvini, A., Acharya, B., Biobased Polymer Composites: A Review. *J. Compos. Sci.*, 6, 9, 255, 2022.
17. Das, O., Babu, K., Shanmugam, V., Sykam, K., Tebyetekerwa, M., Neisiany, R.E., *et al.*, Natural and industrial wastes for sustainable and renewable polymer composites. *Renewable Sustainable Energy Rev.*, 158, 11205, 2022.
18. Kumar, B., Biocomposites, in: *IntechOpen, 1st ed.*, pp. 1–150, IntechOpen, London, 2022.
19. Khilji, I.A., Chilakamarry, C.R., Surendran, A.N., Kate, K., Satyavolu, J., Natural Fiber Composite Filaments for Additive Manufacturing: A Comprehensive Review. *Sustainability*, 15, 23, 16171, 2023.
20. Wu, Y., Yang, Z., Madiyar, F., Jiang, Y., Namilae, S., Hydroxyapatite functionalized natural fiber-reinforced composites: Interfacial modification and additive manufacturing. *Polym. Compos.*, 46, 1, 141–55, 2025.
21. Balla, V.K., Kate, K.H., Satyavolu, J., Singh, P., Tadimeti, J.G.D., Additive manufacturing of natural fiber reinforced polymer composites: Processing and prospects. *Compos. Part B Eng.*, 174, 106956, 2019.

22. Siddique, R., Billings, C., Sherwood, B., Hall, J., Liu, Y., Advancing Environmental Sustainability Through Development of Wood Fiber Reinforced Polymer Composites, in: *ASME 2024 Aerospace Structures, Structural Dynamics, and Materials Conference*, American Society of Mechanical Engineers, Renton, Washington, USA, 2024.
23. Billings, C., Siddique, R., Sherwood, B., Hall, J., Liu, Y., Additive Manufacturing and Characterization of Sustainable Wood Fiber-Reinforced Green Composites. *J. Compos. Sci.*, 7, 12, 489, 2023.
24. Imon, I.H., Sustainable Natural Fiber Based Composites Materials: A Review [Internet]. American Chemical Society (ACS), 2024.
25. Giani, N., Maccaferri, E., Benelli, T., Bovo, M., Torreggiani, D., Campari, E.G., *et al.*, Valorization of Agro-Wastes as Fillers in PLA-Based Biocomposites for Increasing Sustainability in Fused Deposition Modeling Additive Manufacturing. *Materials*, 17, 6, 1421, 2024.
26. Godec, D., Gonzalez-Gutierrez, J., Nordin, A., Pei, E., Ureña Alcázar, J., Introduction to additive manufacturing, in: *A Guide to Additive Manufacturing, 1st ed.*, pp. 1–32, Springer, Cham, 2022.
27. Roschli, A., Post, B., Chesser, P., Love, L., Motion and path planning for additive manufacturing, in: *Woodhead Publishing Series in Additive Manufacturing, 1st ed.*, pp. 1–250, Woodhead Publishing, 2024.
28. Introduction to Additive Manufacturing, in: *Advances in Civil and Industrial Engineering*, pp. 1–24, IGI Global, 2020.
29. Capasso, I., Andreacola, F.R., Brando, G., Additive Manufacturing of Metal Materials for Construction Engineering: An Overview on Technologies and Applications. *Metals*, 14, 9, 1033, 2024.
30. Kanishka, K. and Acherjee, B., Revolutionizing manufacturing: A comprehensive overview of additive manufacturing processes, materials, developments, and challenges. *J. Manuf. Process.*, 107, 574–619, 2023.
31. Chattopadhyay, S., Mahapatra, S.D., Mandal, N.K., Advancements and challenges in additive manufacturing: a comprehensive review. *Eng. Res. Express*, 6, 1, 012505, 2024.
32. Additive manufacturing, in: *Additive Manufacturing* [Internet], 1st ed., pp. 77–97, CRC Press, Boca Raton, 2022.
33. Ajithkumar, S., Arulmurugan, B., Rajeshkumar, L., Prospects of natural fiber-reinforced polymer composites in engineering and commercial applications, in: *Applications of Composite Materials in Engineering, 1st ed.*, pp. 335–364, Elsevier Science Ltd., 2025.
34. Nataraj, G. and Ramesh Babu, S., Enhancing mechanical properties of PLA - based bio composite filament reinforced with horse gram filler for 3D printing applications. *J. Vinyl Addit. Technol.*, vnl.22182, 2024.
35. Framinan, J.M., Perez-Gonzalez, P., Fernandez-Viagas, V., An overview on the use of operations research in additive manufacturing. *Ann. Oper. Res.*, 322, 1, 5–40, 2023.

36. De Yro, P.A.N. and Quiachon, G.M.O., Additive Manufacturing: Selected peer-reviewed extended articles based on abstracts presented at the *2021 ASEAN Conference on Additive Manufacturing*, Trans Tech Publications Ltd., 2021.
37. Lim, C.W.J., Le, K.Q., Lu, Q., Wong, C.H., An Overview of 3-D Printing in Manufacturing, Aerospace, and Automotive Industries. *IEEE Potentials*, 35, 4, 18–22, 2016.

Index

Printed and bound by CPI Group (UK) Ltd, Croydon, CR0 4YY
23/09/2026
14967070-0002